Innovation:
A Cross-Disciplinary
Perspective

Innovation:
A Cross-Disciplinary
Perspective

Edited by
Kjell Grønhaug
and Geir Kaufmann

Norwegian
University Press

HD
53
I56
1988

Norwegian University Press (Universitetsforlaget AS), 0608 Oslo 6
Distributed world-wide excluding Scandinavia by
Oxford University Press, Walton Street, Oxford OX2 6DP

London   New York   Toronto
Delhi   Bombay   Calcutta   Madras   Karachi
Kuala Lumpur   Singapore   Hong Kong   Tokyo
Nairobi   Dar es Salaam   Cape Town
Melbourne   Auckland

and associated companies in
Beirut   Berlin   Ibadan   Mexico City   Nicosia

© Universitetsforlaget AS 1988

British Library Cataloguing in Publication Data
Innovation: a cross-disciplinary perspective.
    1. Innovations
    I. Gronhaug, Kjell   II. Kaufmann, Geir
    303.4'84   HM213

ISBN 82-00-07446-3

Printed in England
by Page Bros (Norwich) Ltd

Preface

For several years we organized an advanced course on innovation at the Norwegian School of Economics and Business Administration. Our purpose was to offer a crossdisciplinary perspective on this important topic. However, when searching for an appropriate main text for the course, we could not find the kind we had in mind. The wanted book was to be wide as well as deep. Wide in the sense of providing a broad and representative coverage of the field, and deep in the way of scholarly incisiveness. We could only find fairly superficial, general surveys or highly specialized in-depth treatments within restricted areas of the general problem space of innovation.

This lacuna in the literature inspired us to embark on the ambitious project of developing a book of readings with contributions from the full spectrum of the most relevant professional fields in the target domain. Furthermore, we decided to put special effort into the task of working out a unifying conceptual model as a map to a clearer and more systematic perspective on the most important research issues and results in this area. We can only leave to the reader to decide if we have succeeded in our efforts.

The book is written for multiple audiences. Among the most important are: (1) Researchers in the field of innovation, (2) graduate students in the many different fields covered by the book, (3) managers and politicians concerned with the innovation issue.

Finally, we want to express our deep gratitude to all the outstanding scholars who have taken part in this comprehensive project and helped shape our thoughts on the important issue of innovation.

Kjell Grønhaug   *Geir Kaufmann*

Norwegian School of   University of
Economics and Business   Bergen
Administration

Acknowledgements

We should like to thank the various publishers of the original articles for their permission to reprint them here:

1. Michael A. Wallach: 'Creativity and Talent' was first published as a documentary report of the Ann Arbor Symposium on the Applications of Psychology to the teaching and learning of music: Session III. Reston, Virginia: Music Educators' National Conference. 1983.

7. Stanley S. Gryskiewicz: 'Trial by Fire in an Industrial Setting: A Practical Evaluation of Three Creative Problem-Solving Techniques'. Parts of this paper were presented at a Division 14 Symposium, 'Creativity in the Corporation', at the Ninety-Second Annual Convention of the American Psychological Association, Toronto, Canada, 1984, and it is based upon the author's doctoral dissertation granted by London University in 1980.

8. L. Richard Hoffman: 'Applying Experimental Research on Group Problem Solving to Organizations' was first published in *The Journal of Applied Behavioral Science*, Vol. 15, No. 3 (1979).

10. Robert A. Burgelman: 'A Process Model of Internal Corporate Venturing in the Diversified Major Firms' was first published in *Administrative Science Quarterly*, Vol. 28 (1983).

14. Arthur L. Stinchcombe: 'On Social Factors in Administrative Innovation' was first published in *Stratification and Organization*, Cambridge University Press and Norwegian University Press 1986.

15. Eric von Hippel: 'Lead Users: A Source of Novel Product Concepts' was first published in *Management Science*, Vol. 32, No. 7 (1986).

16. Modesto A. Maidique and Billie J. Zirger: 'The New Product Learning Cycle' was first published in *Research Policy*, No. 14 (1985).

17. George F. Mechlin and Daniel Berg: 'Evaluating Research—ROI is not Enough' was first published in *Harvard Business Review*, Vol. 73 (1980).

18. Everett M. Rogers: 'The Role of the Research University in the Spin-Off High-Technology Companies' was first published in *Technovation*, Vol. 4 (1986).

Contents

The Authors

Teresa M. Amabile, Professor of Psychology, Brandeis University, Massachusetts.

Bjørn L. Basberg, Research Associate, Department of Economics, University of Trondheim, Norway.

Daniel Berg, Professor of Science and Technology, Carnegy-Mellon University, Pittsburgh, Pennsylvania.

Robert A. Burgelman, Associate Professor of Management, Graduate School of Business, Stanford University, California.

Hans H. Gemünden, Professor of Marketing, University of Odense, Denmark.

Kjell Grønhaug, Professor of Business Administration, Norwegian School of Economics and Business Administration, Bergen, Norway.

Reidar Grønhaug, Professor of Social Anthropology, University of Bergen, Norway.

Stanley S. Gryskiewicz, Director of ICAR (Innovation and Creativity—Applications and Research), Center for Creative Leadership, North Carolina.

Ravenna Mathews Helson, Research Psychologist, IPAR, and Adjunct Professor, Department of Psychology, University of California, Berkeley, California.

Eric von Hippel, Professor of Management of Technology, MIT Sloan School, Massachusetts.

Jonny Holbek, Associate Professor of Management, Agder College, Kristiansand, Norway.

L. Richard Hoffman, Professor of Organizational Behavior, Graduate School of Management, Rutgers University, New Jersey.

Einar Hope, Director of the Center for Applied Research and Professor of Economics, Norwegian School of Economics and Business Administration, Bergen, Norway.

Scott G. Isaksen, Director of the Center for Studies in Creativity, State University College at Buffalo, New York.

Geir Kaufmann, Senior Researcher in Cognitive Psychology, University of Bergen, Norway.

Michael J. Kirton, Director of the Occupational Research Centre, Hatfield Polytechnic, London.

Modesto A. Maidique, President of Florida International University, Florida.

George F. Mechlin, Vice President for Research and Development, Westinghouse Electric Corporation, Pittsburgh, Pennsylvania.

Torger Reve, Professor and Chairman of Organization Science, Norwegian School of Economics and Business Administration, Bergen, Norway.

Everett M. Rogers, Professor of Communication, Annenberg School of Communication, University of Southern California.

Arthur L. Stinchcombe, Professor of Sociology, Political Science and Organizational Behavior, Northwestern University, Illinois.

Michael A. Wallach, Professor of Psychology at Duke University, North Carolina.

Billie J. Zirger, Associate Professor at the Department of Industrial Engineering Management, Stanford University, California.

Introduction

Since the origin of man, innovations have shaped our lives. Innovations have conferred honor, fortune, and prosecution upon inventors and entrepreneurs. The introduction of innovation has created protest, competitive advantage, growth, and success. Innovations attribute to economic wealth. Lack of innovation is often of major concern to industry and government, as emphasized in the following headline: "Innovation: Has America Lost Its Edge?" (*Newsweek*, June 4, 1979), and successful innovations are wanted and pursued by inventors, entrepreneurs, organizations, and governments.

There is no single agreed upon definition of "innovation", but there is consensus on the point that an innovation represents something *new*. Moreover, most writers implicitly agree that to be genuine an innovation has to be *useful* or, more correctly, *perceived* to be useful. This is a prerequisite for the acceptance and impact of an innovation. In most papers the concept is left undefined, their authors presupposing its understanding. For example, only two of the contributions in the present volume explicitly define this term.

The novelty requirement of innovation is reflected in two quite different ways in the literature. One emphasizes the *creation* of something new, the other the *adoption* of something new. The novelty dimension is also related to whether something is *objectively* new—as a new patent—or new in a *subjective* way, as emphasized in the following definition: "An innovation is an idea, practice, or objective that is perceived as new by an individual or other unit of adoption" (Rogers 1983, p. 11). From the contributions in this volume, it will become evident that there are various perspectives on novelty prevailing across disciplines.

The emphasis on "utility" as the basis for the adoption of innovation has led to a distinction between "invention" and "innovation" in some disciplines, indicating that invention represents a new conception whose usefulness, acceptance and/or impact, has not yet occurred.

An innovation is often thought of as a *process*. In some disciplines, emphasis is on the creation of the innovation, i.e. an idea or product, as such, while in others emphasis is on the introduction of the innovation. This is reflected in the following definition: "An innovation is the process by which new products and techniques are introduced into the economic system" (Nelson 1968, p. 339). Definitions are neither right nor wrong, only useful to a greater or lesser extent. It is evident, however, that researchers both within and across disciplines use the term "innovation" differently and emphasize its different aspects.

To be accepted, an innovation has to be introduced and spread. The role of the entrepreneur (who may or may not be the inventor) introducing the innovation is considered crucial in disciplines such as economics, while in marketing and corporate strategy techniques and strategies for introducing the innovation, i.e. emphasis on how to play the entrepreneurial role successfully, are the predominant topics.

Adopting an innovation is often thought of as a rather passive act. An innovation, however, has often to be *changed* or *modified* in the process of adoption and implementation, and has to be "reinvented" in order to be useful. In other words, the adopter, too, has to be creative (cf. Rogers 1984, p. 175).

Any innovation represents a new conception. Concepts are created by man. Thus innovations presuppose *creativity*. Creativity implies the ability to "see" or "imagine" new perspectives.

Creativity involves problem finding as well as problem solving. "Problem finding", however, has not received the same attention that problem solving has, neither in academia nor in practice. Without interesting problems no creative problem solving will take place.

Creativity requires talent and skills. Questions like: "What characterizes creative individuals?"; "Is it possible to identify creative persons?"; and "How do creative persons work?" have been addressed by many researchers. Because creativity is wanted and needed and because creativity is often considered a scarce resource, the question: "Can creativity be taught?" has repeatedly been addressed.

Innovations are created by people, but the individual rarely operates in a vacuum. S/he interacts with others often in the context of *groups*. Groups have been thoroughly studied, in particular by social psychologists. A variety of group-related factors have been found to facilitate and also inhibit creation—and adoption—of innovations. Several group-related factors are dealt with in the contributions to follow.

Innovations are often treated in the context of *organizations*. Creation—and adoption of innovations—is necessary to gain a competitive edge in order to survive and grow. Organizations are collectives assumed to exhibit purposeful behavior towards attaining goals. Frequently assumed (organizational) characteristics are the presence of several individuals, division and specialization of labor, and one or more power outers controlling and directing the organizational efforts. Organizations are often regarded as open systems depending on resources for survival and growth, and thus successful innovation is wanted. Organizations operate technologies, they possess rules and procedures, and they have structure. Various organizational factors facilitate while others deter innovative activities, and may thus be crucial to the organizational innovative capacity. Moreover, organizations must *discover* the "right" problems or market/client needs in order to create useful innovations, and they must *promote* their innovations to be accepted by adopters.

Creation of innovations may be considered as production of *knowledge*, since the prime purpose is to create new and better solutions which will be appreciated and accepted by clients and users. It should be noted, however, that in the process of knowledge production, knowledge will be sought and used as well. Thus type and amount of knowledge possessed and searched by individuals, and how this knowledge is used may be of crucial importance for innovation success.

Individuals, groups, and organizations operate within a larger *society*. In the diffusion of innovation literature, the outside world is recognized as "social systems". Even though environments external to individuals, groups, and organizations are important, influencing creation, adoption, and diffusion of innovations, the research focusing on external aspects in the context of innovations has been relatively limited. The impact of societal and environmental factors on innovative activities is emphasized in several of the contributions in the present volume.

Innovations have been subject to a considerable amount of research in a variety of disciplines, among them psychology, sociology, social anthropology, economics, economic history, engineering disciplines, geography, public policy, marketing and corporate strategy. The list could easily be extended (cf. Rogers 1983). It is more important to recognize that researchers from th᷍ various disciplines to a large extent emphasize *different aspe᷍* innovation. Main *concepts* partly differ across disciplines and—as noted above—one and the same concept may be used differently. No unified "theory of innovation" exists. Moreover, researchers preoccupied with innovation are partly unaware of the research done and conceptualizations used by colleagues from other disciplines. Innovation research still has a long way to go before it becomes truly interdisciplinary, hence the title of the present volume, *Innovation: A Cross-Disciplinary Perspective.*

An Overview of the Book
The volume comprises an introduction and 21 contributions grouped within three main parts. Innovations are studied at different *levels*, i.e., the individual, group, organizational, and societal levels, and various *aspects* of innovations are emphasized, e.g., the creative, adaptive, diffusive, and consequential aspects. By cross-classification of the two dimensions (or factors), we can create a table as the basis for grouping the various topics to be dealt with in the contributions to follow.

Part I deals primarily with innovation related topics at the individual and group levels, but several of the contributions have important implications at the organization level as well. The main focus of the contributions in Part II is on topics listed under the organizational level. Several of the authors in this section also deal with "group level topics". The main focus of the contributions in Part III is on topics related to the interplay between innovations and various societal and external dimensions, but, as readers will soon discover, important implications at the individual and group levels can easily be traced.

Part I contains eight contributions. In focusing on creativity and talent, Michael A. Wallach directs our attention towards intelligence and measurements, factors influencing such measurements, and implications for teaching. He concludes that ". . . we should not be trying to teach creativity, but rather the technical masteries and stylistic sensitivities connected with the fields . . . in which the kinds of achievements we view as reflecting creativity or talent can occur. What needs to be learned is close to the textures of particular fields of work, not removed from them in some ¹ where activity, as such, can be taught". This per- ır-reaching implications for teaching institutions, ıgrams, and course designs.

 ersonality has attracted considerable attention in ₃avenna Helson draws on her extensive research giving an overview of important aspects and

Table 1. Levels, topics and themes.

	Level			
	Individual	Group	Organizational	Societal
Creativity	Creative individuals ● characteristics ● working style ● processes ● behavior ● skills ● measurements Problem finding Problem solving Motivation and creativity Cognitive style Creative techniques Creative environment	Creative individuals ● solutions ● techniques Group characteristics Facilitating/deterring innovation Idea production	● Management and innovation ● Innovative activities ● Innovations within the organization ● Structural properties and innovation ● Creative persons in organizations ● Roles of creative persons ● Organizational structuring	Innovations and Terminations Technological change Sources for problem-finding and innovations Spin-offs Role of entrepreneur Patents Market structure
Focus: Adaption	Problem finding Problem solving	Group characteristics Agents for change	● Resources and adoption of innovations ● Resistance to adoptions	● Institutionalizations of innovation ● Cultural and adoption of users.
Diffusion	Gatekeepers	Agents for change	Promotion of innovations	● Spin-offs ● Structure of serial system
Consequences	of innovations on the individual	Consequences of ● group structures ● commitment on innovative activities	Consequences of organizations and organization members Evaluation of R&D	Consequences of funding cuts ● Research universities

ambiguities of this research. She paints a picture of the general make-up of the creative individual and introduces the reader to a wealth of important findings which should serve as a useful basis for other researchers and as inspiration to managers.

Michael Kirton introduces the reader to the adaption–innovation theory of cognitive style. Adaption and innovation are conceived as contrasts along a continuum of creative problem solving and decision making. Two extreme types of problem-solving behavior are characterized. The adaptor seeks solutions within an existing framework, whereas the innovator is inclined to restructure problems and change paradigms. The measurement instrument used to map cognitive styles, the KAI-inventory, is described, and the theoretical base underlying development and use of the inventory is explained. Kirton also reports on results from several studies conducted in different cultural settings, and concludes, based on his extensive research, that ". . . the adaption-innovation property of cognitive processes is not context specific . . . the theory is essentially value-free in the sense that a test can be culture free". Several applications of the theory and the test instrument are emphasized.

Geir Kaufmann provides a broad review of research on human problem solving with particular emphasis on the creativity aspect. Various strategies, capacities, and symbolic tools involved are described, and conditions that facilitate and inhibit creative problem solving are pointed out. An attempt is also made to expand the general theory of problem solving to cover the creativity aspect more adequately, and the issue of whether creativity engages extra-rational processes is addressed.

In her important contribution Teresa M. Amabile focuses on relationships between individual and organizational creativity. She observes that an individual's creativity in an organization is influenced by the same factors that influence creativity in other contexts; that an individual's creativity within an organization is also influenced by a number of factors that are specific to the organizational context; and that the process of individual creativity is a crucial part of the entire process of innovation within an organization; but also that it is only *one* part. Amabile has much to say about the causes and consequences of creativity and innovation in organizations, and her model of the creative process, emphasizing the interplay between various cognitive processes, task motivation, and skill requirements, has considerable implications for research and management.

Scott G. Isaksen puts creativity and creativity research in a broad perspective, focusing on aspects of creative persons, creative processes and creative products. He employs this multi-faceted conception of creativity as his point of departure for advancing implications of importance to institutions, organizations, and managers.

The effectiveness of three creative problem-solving techniques— brainwriting, brainstorming and excursion—is compared in Stanley S. Gryskiewicz's contributions based on two studies conducted among 280 managers in the U.K. and the U.S. The results are contrasted with previous research and implications for enhancing innovations are highlighted.

In a well-balanced contribution, L. Richard Hoffman treats group problem solving related to an organizational context. He focuses on important aspects of groups and group processes that facilitate and inhibit productive problem solving. He emphasizes omissions in previous research on group problem solving and develops a new theoretical model to structure the field.

Part II contains six papers dealing with important aspects of innovations in our managerial and organizational context.

The first paper in this section is by Johnny Holbek. In his contribution he focuses on the innovation-dilemma, i.e., the two subprocesses, initiation and implementation. Initiation is concerned with how the organization becomes aware of formation of attitudes and development towards making decisions about implementation of the needed activities. Implementation is concerned with the process by which the innovation is realized. Together the two subprocesses form the integrative innovation process. The two phases subsume a variety of different tasks. The tasks require different organizational design and structure to be performed effectively. Holbek reviews perspectives and research on the innovation-dilemma, discusses its relevance, and offers an alternative perspective to the dilemma, by distinguishing between differentiation in space and time, which are combined into a hybrid balancing differentiation in space and time.

Past research has repeatedly demonstrated that organizational structures, rules and procedures may hamper respectively enhance creation of new, more radical innovations. In an important, very thorough contribution based on a study within a large diversified high-technology firm, Robert A. Burgelman describes the internal corporate venturing process. Based on detailed observations he develops a model capturing such processes. His conceptualization maps the process taking place and how it is related to various organizational levels. This contribution is important as its focus on crucial aspects for successful internal venturing is of great relevance for management as well as for future research.

In a very interesting contribution Morris I. Stein directs our attention to the creative person and the various roles this person has to play within the organization. He distinguishes between the scientist, the professional, the administrator, the employee and the social role respectively, and uses role descriptions as the basis for emphasizing the various demands put on the creative person. He also proposes a specifically designed procedure, the *Technical Audit*, as the basis for evaluating and guiding creative work within the organization.

In their paper, Kjell Grønhaug and Torger Reve contrast strategic management on the one hand with innovations and entrepreneurial activities on the other hand as the basis for bringing back the organization's competitive edge. Similarities and dissimilarities emerge when innovation and strategy are contrasted. Both imply new ideas, new conceptions and some formalization of plan (strategy) or a prototype (innovation). To be successful, however, a strategy has to be implemented requiring focus and structuring, which may hamper future innovation. Theoretical and managerial implications are highlighted.

Innovations have to be promoted to be spread and accepted. This is true in particular in the case of complex, industrial innovations. Hans George Gemünden focuses on the "promoter", which he characterizes as the *key person* in development and marketing of innovative industrial products. In his paper the promoter role is described, and it is maintained that "know-how" and "power" may serve as bases for the success of this role. Gemünden's contribution has considerable implications for management.

Arthur L. Stinchcombe directs our attention towards resistance to administrative innovations. His focus is on how innovations may advance versus damage interests within the organization. This very insightful paper points at various types of administrative innovations according to their impacts, and helps us understand the organizational resistance problem.

Part III contains seven papers focusing on how innovative activities are influenced by various external forces.

To create useful innovations, i.e. innovations that will be accepted and used, perceptions of adequate problems are of great importance. Eric von Hippel discusses how "lead users" may be of crucial importance for enhancing useful innovations in fast-moving fields such as the high-technology industries. Moreover, he advises on how to identify lead users, and how to gather, analyze, and use their data.

Modesto A. Maidique and Billie Jo Zirger direct our attention towards important factors influencing the success and failure of new products. Based on information from large-scale research, the authors have identified several factors of the market-manufacturing interface relevant to innovation success, of great importance for management and organization of innovative activities.

Organizational resources are scarce, and evaluations and feedback are needed. Evaluations are also needed for R&D activities. George F. Mechlin and Daniel Berg point out important limitations of commonly used return-on-investment (ROI) measures, and suggest how R&D measurements can be improved.

Spin-offs are well known among researchers and managers. Everett M. Rogers focuses on spin-off of private firms from research universities. In his paper he discusses the key role of research universities in the information society, and emphasizes the growing university–industry relationships, and the benefits and costs of such relationships. This contribution has considerable implications for governmental R&D policies, as well as practical implications for universities and industry.

Some innovative ideas are patented. Bjørn L. Basberg focuses on using patent statistics as measurements of technological change, and on how patents are related to inventions and innovations.

The relationship between market structure and innovation has interested and puzzled economists for a very long time. In his paper Einar Hope gives an overview of this research; he emphasizes important theoretical and methodological problems involved, and offers alternative views.

In the very last contribution in this volume, Reidar Grønhaug introduces the reader to a case study based on his field observations in a community in East Antalya, in the southern part of Turkey. His emphasis is on the processes of creation of innovations and their repercussions. He integrates elements of cultural ideas and rules, population and ecology, groups and networks in social organizations, and identities and status as combined in individual careers and social persons. From there he moves on to delineate the activity fields for adaptions, and the way such fields emerge and are exploited.

REFERENCES

Nelson, R. R. 1968: Innovation. In D. L. Sills (ed.), *International encyclopedia of the social sciences*. New York: The Macmillan Company and the Free Press, Vol. 7, 339–345.

Newsweek 1979: Innovation. Has America Lost Its Edge? *Newsweek*, June 4, 58–65.

Rogers, E. M. 1983: *Diffusion of innovations*. New York: The Free Press.

Part I

Creativity and Innovation

1

Creativity and Talent

MICHAEL A. WALLACH

A great deal of effort on the part of those concerned with understanding and cultivating talent has been aimed at elaborating
abstract theoretical constructs to illuminate the competences of
interest. If misgivings were felt concerning a construct, the usual
response was to switch one's allegiance to a different, but com-
parably abstract, construct. The reason for the misgivings was
usually a growing realization that assessments deriving from a given
construct failed to deliver much information about the real-life
competences characteristic of one child or student but not another,
or characteristic of them in various degrees—what their art work
was like, whether they could play a musical instrument proficiently,
whether they could deliver compelling portrayals in a dramatic
production, how well they could write prose or poetry, whether
they had a sense of what dance as an art form was all about,
whether they could compose a piece of music. Yet it was such real-
life competences as these that were of interest to us. We wanted
to help the young develop their talents so that more individuals
could fulfill more of their potential.

Fulfillment of potential is, after all, one of the goals of trying to
reach a better understanding of talent (aside from our interest in
the knowledge itself). This calls for the defining of criteria by which
students will be selected for scarce educational opportunities and
the providing of instruction. Some forms of human competence
are more rare than others—not as many children learn to play the
flute well as learn to read well, for example; not as many learn to
compose music creditably as learn to play a good game of baseball.
We are not, in fact, clear about the reasons for such differences in
frequency, even though they may indicate more about what
receives greater attention in our current society than something
deeper about the organism. As widespread a form of competence
as literacy falls far short of being as widely distributed as one would

wish, and there are far fewer yet capable of writing cogent prose, playing a musical instrument well, weaving handsome fabrics, composing a sonata or even a song, or painting pictures that give pleasure to the eye. Unfulfilled potential, in short, is a familiar story. Why have psychology and education not been more successful in improving this situation?

THE CONSTRUCT OF INTELLIGENCE

I think there is a straightforward reason why the success has not been greater. And it is a reason with clear implications for music education. In order to describe this, I will discuss some of the recent history of psychological work on the defining and nurturing of student talent.

There is a theoretical construct that we call "intelligence", and we infer a person's status with respect to it from his or her performances on various tasks (such as the ability to repeat the random sequence of digits just stated by an examiner or to name the capital of a certain country). We do not have a specific interest in the respondent's answering of these very questions. Rather, our interest resides in the clues these answers are presumed to offer as to that person's standing in regard to the abstract construct of intelligence, which we think of as sitting in the person's head and sending forth signs captured by the items of intelligence tests. And the reason we care about intelligence is that measuring the amount of it in a person's possession is supposed to inform us in an economical fashion about what that person potentially can accomplish in the world—how well he or she can generate prose, write poems, carry out art work, be an imaginative actor, do research, compose and perform music, and so on. That is why you will find countless instances of classes for the "gifted" and "talented" for which membership eligibility is defined by cut-off scores on IQ tests. What has become clear as evidence has accumulated, however, is that the intelligence construct cannot deliver very well on this promise.

To be sure, there was some relationship between scores on conventional tests of intelligence or intellective ability and meaningful attainments of the sorts mentioned. A child earning an IQ score of 70 would be quite different in terms of real-world competences from one with an IQ of 170. A high school senior scoring at the 95th percentile on the Scholastic Aptitude Test (SAT) would present a very different picture of competences from one scoring at the 20th percentile. But as soon as attempts were made to use these scores in a more fine-grained manner—looking at, say, an IQ of 100 versus one of 110, or an IQ of 125 versus one of 113, or an SAT score at the 95th percentile versus one at the 85th—it became difficult to support the claim that score differences were informative regarding the real-world competences of direct concern to us. Small score differences gave one little basis for predicting who would exhibit such competences and suggested little about how to teach these competences more effectively.

Extensive literature arose documenting that these kinds of score differentials on traditional measures of intellective ability indicated little about which students would manifest talented accomplishments in the writing of poetry or prose, would carry out intriguing scientific experiments in an improvised basement laboratory, or other equally direct manifestations of competences that society has a stake in furthering among its members. The tests predicted grades in conventional courses, but neither the tests nor the grades predicted the talented accomplishments. In various sources, such as, for example, materials by Holland and Richards (1966, 1967), Richards, Holland, and Lutz (1966, 1967), M. A. Wallach and Wing (1969), Wing and Wallach (1971), Munday and Davis (1974), Albert (1975), and M. A. Wallach (1971, 1976a, 1976b), the details of this evidence are spelled out and evaluated.

These studies not only found that such test score differences indicate little about who would manifest talented attainments, but these attainments—once manifested—also were found to persist through time. Their presence in the behavior of some students at a given point gave a good basis for predicting which students would exhibit those competences in the future as well, extending on into the years of adulthood and professional activities. These talented achievements also turned out to be more specific than general. The student who was good in one area tended to keep on excelling in that domain rather than do well in another. For example, earlier manifestations of high competence as a creative writer tended to predict later achievement in creative writing but not in music or in science, earlier musical achievement predicted later achievement in music but not in science or in art, and so on. Such findings could, of course, arise simply because of the exigencies of time—doing one thing well limits the time available for something else. But these results also called attention to the great variety of competences at issue and the many differences among them. The sources cited before again offer relevant documentation. This kind of predictability across time was, of course, useful to know about, since it helps indicate what should be offered educationally to which students and helps students decide what paths to pursue.

In striking contrast to the failure of the customary intellective ability or aptitude tests to demonstrate their usefulness was the emphasis they received as sorting and selecting devices in education. If schooling in the early years is to nourish talent as directly as possible, we must concern ourselves with how students are selected for advancement in the later stages of the educational sequence; the criteria that govern college and university admissions and scholarship awards become the considerations that influence what the earlier stages of schooling will be like. We have already mentioned the use of IQ scores to determine eligibility for classes for the gifted and talented. As this practice suggests, selection and teaching from early in the educational sequence are oriented toward the kind of criteria in effect later—namely, whatever controls access to institutions of higher education possessing greater prestige and hence credentials that are more in demand. It is instructive to consider, therefore, the nature of the criteria being used. Research has provided some clear information on this (see, for example, Wing & Wallach 1971; M. A. Wallach 1976a, 1976b).

Despite the degree to which scores on traditional intellective tests are uninformative about real-life manifestations of competence or talent, such scores receive heavy emphasis in the college admissions process. Moreover, score differences on these tests carry great weight for admissions decisions even within that part of the range where their relationship to manifested talent is lowest—from the middle to the high end of the test score distribution. To find out whether alternative grounds in fact existed for making admissions decisions, a close look was taken at how much use is made in the actual admissions decision process of information systematically offered to the selection committee about candidates' talented accomplishments in such realms as creative writing, musical performance and composition, artistic creation, the conducting of innovative scientific projects, and excellence at directing or acting in dramatic productions. Also available, of course, were traditional test scores.

The actual admissions process turned out to rely very little on the information about real-life achievements. Furthermore, selecting a hypothetical class from the actual pool of candidates by giving weight to high-quality talented attainments of the kinds indicated led to a class quite different in its membership from the class actually selected. In other words, if one did pay attention to candidates' intrinsically meritorious talented attainments when making admissions decisions, the class admitted would contain a substantial proportion of different students from the class that actually gains admission under the customary procedures. A study by Willingham and Breland (1982) published by the College Board itself yielded similar conclusions.

What would happen if such changes in selection criteria for the best opportunities in higher education were to be put into effect? These changes would tend to propel a group of individuals with somewhat different characteristics into leadership roles in the society—individuals with stronger talents in the sorts of real-life domains we have considered. That is to say, more people with the kinds of competences noted would be receiving the advantages conferred by prestigious educational credentials. This means that the activities at which they excel could well gain positions of greater value than now is the case; the effect could be a society in which the functions of the musical performer, musical composer, writer, poet, artist, sculptor, ceramist, weaver, dancer, playwright, and actor were looked upon with greater favor and given more support. The current selection procedures, with their test score emphasis, seem more oriented toward propelling bureaucratic types into positions of influence. After all, by now it seems clear that, within certain limits, scores on the usual entrance tests can even be enhanced by special preparation (Slack and Porter 1980; Messick and Jungeblut 1981). A society with such selection filters simply ends up with more good bureaucrats and technocrats and less artists and musicians defining its values than would otherwise be the case.

There would also be a direct impact on precollege education, for we would have a new set of forces influencing the content of the early stages of formal education. Such a result ought to interest music educators. If college and university admissions were made more dependent on demonstrating talented attainments in fields like music, art, or creative writing, primary and secondary education would become more concerned with teaching that would foster such competences in the child. A field like music education would be viewed as more central to the curriculum and less of a frill, for earlier stages in the educational sequence always will be pushed by parents and school authorities into emphasizing whatever will increase the child's further educational opportunities.

It could well be that bringing forms of achievement such as work in music and art into positions of greater curricular centrality—making them explicit goals of instruction from early on—would give more prominence to many kinds of students who stand relatively neglected now. My argument is that we simply do not know what may be possible along these lines, because little serious attempt has been made to find out—schooling has not really stressed such goals. Music, art, dramatics, and dance have largely been approached as "enrichment" subjects that do not matter the way higher scores on the SAT do. Until they matter in that way, such fields will remain educational stepchildren.

Some new momentum toward their mattering more may be developing, although the sources of resistance to such change are considerable. The good news is that Educational Testing Service, which carries out traditional testing programs like the SAT, has just begun a talent search program explicitly oriented to demonstrated competences in the fields of music, creative writing, theater, visual arts, and dance (Educational Testing Service 1981). Quality evaluations by judges appropriate to the given fields are made on the basis of work samples submitted by student applicants—color slides of paintings, poetry and prose, audiotapes of musical performances or videotapes of dancers and actors. Those whose work is viewed as most deserving receive help that includes the arranging of scholarship support with various colleges and universities and with other relevant programs. Such an approach to selection is, of course, consistent with the evidence we have reviewed. Talent or giftedness does not seem best selected for in terms of some abstract construct like intelligence, but in terms of demonstrated competence at the work of a given artistic field. The same point also holds in the sciences, where the best predictor once again seems to be demonstrated competence based on work samples, as in the case of extracurricular research projects that a high school student may carry out.

The bad news is that this kind of shift is resisted by educational institutions in the name of maintaining and improving "standards". If change along these lines is to become a reality, perhaps the biggest obstacle to consider will be colleges and universities themselves, with their ideological commitment to construing tests like the SAT as somehow reflecting "real" intelligence in the commonsense meaning of the term. They will have to stop evaluating the "quality" of their students by computing the average of their SAT scores. They will also have to stop establishing their status in the institutional pecking order by comparing their students' average SAT scores with the average SAT scores of students from other institutions. Both of these practices, of course, inevitably build pressure for selecting students with the highest test scores (M. A. Wallach 1976a, 1976b).

THE CREATIVITY ALTERNATIVE

Faced with the growth of evidence questioning the real-life utility of traditional tests, many psychologists took the route of pursuing other possible—and equally abstract—constructs that might prove more fruitful than the construct of "intelligence" for making predictions about attainments in fields like art, music, writing, dramatics, or scientific experimentation. The term "creativity" has been widely used in connection with such conceptual alternatives. Although the nature of the constructs used to flesh out the psychological meaning of creativity has varied, one major direction of effort, perhaps the most frequently encountered, has sought to understand the term by reference to an individual's flow of ideas. Thus, the reaction to evidence that the intelligence construct was not delivering on its promise preserved the same logic of inquiry and switched to a different, but similarly abstract, construct, such as "ideational flow"—the disposition to produce ideas fluently. Once again the strategy was to fashion presumptive tests of a person's standing with regard to the construct, and see whether scores on these new tests would relate better to real-world achievements than had scores on intelligence tests. If so, the nature of this new construct's measures was expected to offer clues of use to pedagogy.

It was not hard to find phenomenological warrant for a construct like ideational fluency. For example, there is introspective evidence from undeniably creative practitioners in the arts and sciences that flow of ideas has been important in their creative efforts. Among the observations to this effect collected in a book edited by Ghiselin (1955) were Wolfgang Amadeus Mozart's referring to occasions on which the flow of his ideas was most abundant, John Dryden's referring to his poetry writing as involving the producing of thoughts that first emerged in a confused jumble, and Albert Einstein's referring to his work as calling upon the playful intermingling of diverse ideas or images. One could argue, however, that practitioners who were less talented could still produce an abundant ideational flow, so the case for greater ideational fluency in those with higher quality real-world accomplishments hardly could rest on introspective citations. Could research support it?

Such research began by seeing whether tests viewed as reflecting the ideational flow construct were tapping something different from conventional tests of intellective ability. Here are some examples of ideational fluency measures from work by M. A. Wallach and Kogan (1965) conducted with fifth graders. The child would be asked to name all the uses he or she could imagine for a particular object, such as a shoe or a cork. The child would be invited to suggest as many similarities as possible between two paired items, such as a train and a tractor, or a cat and a mouse. In another task, the child was to give as many instances as possible of a named category, such as things that move on wheels. Still other tasks offered various kinds of abstract visual materials, such as a triangle with three circles around it, with the child to indicate all the things this pattern might be.

M. A. Wallach and Kogan (1965) reasoned that two characteristics of a child's responses that might have the same aura of meaning as was conveyed by the kinds of introspective accounts by creative adults mentioned previously were the number of ideas the child would generate in responding to a task and their relative unusualness. Children were found to vary widely in both characteristics, but the characteristics themselves were strongly related: Children producing more ideas tended to be the same children who came up with ideas of greater uniqueness. What seemed to occur was that children who generated larger numbers of ideas were using up high-frequency ideas first and then reaching the more unusual ones.

Not only were ideational output and uniqueness of ideas correlated, but the individual who had more ideas, and more that were unusual, about, say, uses for a newspaper or a shoe, also was found to have a larger number of ideas, and more that were unique, when it came to saying what a given abstract design might be or indicating the possible similarities between a carrot and a potato. And, of most importance, these differences from child to child in flow of ideas were unrelated to the differences in their scores on conventional intelligence tests. Measures of output and uniqueness of ideas might be high, whereas the measure of intellective ability was low, or might be low even though tested intelligence was high, at least across the upper half or so of the intelligence scale. Numerous replications of this lack of relationship between ideational fluency and intelligence have been reported: For samples of preschoolers, grade schoolers, high schoolers, and college students, measures of ideational flow, while related among themselves, had little relationship to intelligence among respondents varying from around the middle to the upper end of the intelligence score range (M. A. Wallach 1970, 1971).

If measures of ideational flow thus had little in common with the usual measures of intelligence, perhaps ideational flow would provide a better basis than had intelligence tests for predicting real-life manifestations of talent. The next step, therefore, was to explore this possibility. Research by M. A. Wallach and Wing (1969), for example, found that intellective aptitude test scores failed to indicate the presence of talented accomplishments of various kinds during the high school years, whereas measures of ideational flow gave some indication about this. Across the range studied, higher intellective ability scores did not reveal which students, for instance, were more likely to have had art work exhibited or published, to have won an award in an art contest, or to have produced creative writing of sufficiently high quality for it to be published. On the other hand, higher ideational flow scores did so to a certain extent. Similar results for high school seniors in Israel—ideational fluency but not intelligence predicting talented real-life attainments—were reported by Milgram and Milgram (1976). Singer and Whiton (1971), in work with kindergarten-age children, analogously found level of ideational flow but not intelligence predicting the making of expressive drawings. Similar results were found by Wallbrown and Huelsman (1975) in the case of children's work with clay, although not their crayon drawings. But negative evidence has also been reported (see Kogan and Pankove 1974; Jordan 1975). And at best, the positive evidence was of limited strength.

Thus, a modicum of research support could be claimed for the introspective accounts by creative adults. Something about a more ready flow of ideas may be implicated to a degree in some kinds of talented achievements. Perhaps there is a more energetic cognitive motor generating information to be evaluated, as M. A. Wallach and Wing (1969) have hypothesized; or greater attention to peripheral cues, as M. A. Wallach (1970) has suggested; or a greater bent to toy with and savor the hypothetical, as M. A. Wallach (1967) has proposed; or use of a larger repertoire of templates offering possible ways of relating ideas, as elaborated by Crovitz (1970). Whatever it is, the outcome can be a marginally enhanced likelihood of showing one or another of various real-life attainments. But the gain in predictability is small.

WHAT FOCUS FOR TEACHING?

We have looked, then, at two trends of evidence about talented achievements. One is that scores on conventional intellective tests give much less basis for predicting such achievements than had typically been assumed. The other is that scores on tests of ideational fluency, while unrelated to intellective test scores, are slightly related to talented achievements. Keep in mind that what we care about is fostering the growth of these achievements themselves—of competence at music, art, dance, writing, or whatever. How have psychologists concerned with creativity typically tried to help with the task of bringing about such attainments?

A number of researchers, given that the construct of ideational productivity relates in some degree with real-world talented attainments, place their bets on that construct and try to raise the individual's scores on tests that presumably tap it. Since ideational fluency relates to talented achievements, we are justified in trying to teach people to become more ideationally fluent. The error here, however, is to assume that this relationship can be treated as if it were a linkage of perfect one-to-one correspondence, when actually relationship at best is partial, limited, and marginal. Talented achievements are the result of a variety of factors, of which ideational fluency is but one. And responses to ideational fluency tests can be made for various reasons, of which something like cognitive motor energy level or the other possibilities alluded to may not be the most likely candidates.

What does it mean to give, in response to a tester's request, more uses for named objects, or more similarities between pairs of objects, or more instances of a given category, or more interpretations for an abstract visual configuration? To be sure, it may mean a cognitive motor of greater energy or an inclination to toy with the hypothetical. But it alternatively may mean something much more commonplace, like a greater desire to please the examiner and give what is wanted—or simple conformity to the tester's at least implicit suggestion about what to do. Such suggestibility or conformity hardly qualifies as a distinguishing characteristic of creativity. Obtaining a higher ideational fluency test score also may come from greater obsessiveness or hairsplitting—the compulsive listing of larger numbers of relatively similar responses. Once again, this hardly seems to be a distinguishing attribute of creativity (M. A. Wallach 1971, 1976b). Does it make sense, then, to try to increase real-world talented attainments by enhancing a person's score on ideational fluency tests? Probably not, because scores on such tests can go up for any of various reasons besides their tenuous link to certain real-world achievements.

Yet many psychologists and educators interested in creativity see the pedagogical issue as one of teaching such characteristics as ideational fluency. In short, their aim is to raise scores on creativity tests. For example, Reese and Parnes (1970) gave students tests for assessing ideational flow before and after a course involving work with materials they had been told would teach them how to solve problems creatively. Compared to control subjects not given the course, ideational fluency tests scores went up a modest amount. The most parsimonious interpretation would seem to be not that the course had increased these students' competence at real-life pursuits, but that the rise in test scores had other causes. Thus, the students knew that taking the course was supposed to make them more creative and that they were expected to do better on the ideational fluency tests after the course than before. Since the tests asked for as many responses as possible, a desire to conform to what the teacher seemed to want is sufficient to explain the results. This is a far cry from increasing the student's competence at a real-life form of talented achievement. Yet this approach is characteristic of study after study aimed at enhancing creativity. By now it is becoming clear that training that raises scores on creativity tests may tell us about responsiveness to the implicit demands of authority figures or about improved understanding of what kinds of test responses are valued by the teacher, but says little about how to teach someone to play a musical instrument, compose a fugue, do laboratory research, or paint a picture (Mansfield, Busse and Krepelka 1978).

The essence of the problem with the foregoing type of approach is the huge separation between the pedagogical concerns and the real goal one is after. One is trying to render a person more competent at work in art or music or science with training that centers on doing better at something far away from all that, a presumptive test of creativity. But scores on such tests can go up, if they do, for more than one reason. It is a poor bet to think they go up for a creativity-related reason when there are other reasons, located closer to the test-taking situation, that can account for the results. One can be trapped, in short, by a particular abstract construct, such as ideational fluency, into thinking that scores in presumptive measures of it are adequate proxies for the talented attainments themselves.

The implication seems clear. We should not be trying to teach creativity, but rather the technical masteries and stylistic sensitivities connected with the fields—from sculpture to composing, from dance to harpsichord playing—in which the kinds of achievements we view as reflecting creativity or talent can occur. What needs to be learned is close to the textures of particular fields of work, not removed from them in some abstract realm where creativity, as such, can be taught. Disciplines like the learning of a musical instrument, the craft of poetry writing, or the command over one's body involved in dance all primarily seem to involve matters that are field specific. There are huge conquests of technique that must take place in any of these domains, together with the cultivation of knowledge about styles and forms that lets a person come to understand and hear the difference between a more and a less sensitive rendering of, say, a Mozart piano sonata. From all this comes creativity. But what needs teaching are proficiencies and sensitivities specific to a particular discipline and requiring considerable immersion in it.

The major contribution from psychology to this endeavor is not methods for encouraging creativity; the creativity seems to come from having a field's techniques and styles sufficiently under one's belt that they function as tools and means rather than obstacles and hurdles. Instead, what psychology validly contributes here is emphasis on the importance of teaching in ways that do not presuppose prior knowledge but in fact help the student acquire whatever needs to be learned. This is a concept that certainly is familiar to music educators. The teaching, in other words, should not be cryptic—intending to teach from scratch but in fact taking prerequisites for granted. Otherwise, we are restricting the number of students able to learn and doing so in ways that inevitably favor home background factors, so that, for example, the form of the teaching communicates excellently to a child who comes from a family of musicians, but not so well to a child who starts out minimally acquainted with music. For illustrations of the application of this kind of point to instruction in beginning reading, see M. A. Wallach and L. Wallach (1976, 1979), and L. Wallach and M. A. Wallach (1982). What it implies for helping children become proficient at musical performance and composition seems especially pertinent for a society like ours in which there is so little active making of music carried out by the family in the home. Compare that with the "music literacy" level of many European families a century ago!

An example of the importance of avoiding presuppositions in musical instruction is Sudnow's (1979) account of his learning of jazz piano improvisation. The graduated nature of the proficiencies to be acquired, the patient attempts at specifying different parts of the task that could be approached and mastered, strikingly contrast with the apparent effortlessness of the final expert performance. The problem from the student's point of view of pinning down an adequate explication of the steps needed for the learning is compellingly described by Sudnow as he shows time after time the crucial role of trying to get his teacher to unpack the complexities involved. Sudnow as student was a bulldog who would prod, push, and keep after his teacher until he could better see what was taken for granted in the teacher's presentation. This is hardly what happens in most instructional situations, where the student has no idea what to ask for—and anyway is too shy to try. The unpacking of complexities usually depends fully on the teacher.

The approach to instruction that is called for, then, is one that closely hovers around whatever form of competence is to be assessed and nourished, rather than rapidly ascending to abstract constructs like intelligence or ideational fluency. The place to start seems to be with a task analysis of the particular competence at issue—an attempt to specify what students really need to learn if they are to exhibit that form of competence.

In some ways this adds up to a request to music educators not to let psychologists lead them astray. After all, strong traditions exist concerning how to teach a given musical instrument and how to acquaint a student with the forms of musical composition, and these traditions are intended to supply, rather than presuppose, the needed prerequisites. But these traditions are not unequivocal or without controversy; and, as we saw with Sudnow's jazz teacher, much that should be taught may still be taken for granted. Psychological abstractions seem to give misleading directives both when it comes to selection of the gifted and talented and when it comes to instruction. Psychological knowledge has been emerging about these matters, but, to paraphrase the philosopher Susanne K. Langer, it is psychology in a new key. This new key emphasizes that what comprises talent and creativity is more specific than general, more tied to particular forms of demonstrated competences than conceptually removed from them. And it emphasizes the need to discern and assure mastery of prerequisites if effective instruction is to take place in any given field.

Selecting the talented calls for considering manifested attainments in the real world, not scores on tests of intelligence, or, for that matter, on tests of creativity. It calls for information about particular accomplishments, whether of musical performance or composition, acting, dance, writing, or anything else. And instruction, in turn, calls for a close look at the structure of any full-blown form of competence, with an eye toward discovering its natural components—the lesser parts that must be mastered first if the student is to make progress at achieving that particular type of competence. Carrying out such an attempted decomposition of the mature competence into its parts and determining what parts are prerequisites to what others en route to mastery of the whole is the psychological enterprise that a good teacher undertakes. For a hallmark of appropriate instruction is supplying what needs to be learned, rather than taking it for granted. In music education, as in other domains, we ought to make this process as explicit and clear as we can, in order to help improve the quality of teaching, and increase the number of those who successfully learn. And we ought to select students who are to receive benefits for the talented, gifted, or creative, not on the basis of tests but of sampling actual accomplishments in relevant fields of work.

REFERENCES

Albert, R. S. 1975: Toward a behavioral definition of genius. *American Psychologist, 30*, 140–51.
Crovitz, H. F. 1970: *Galton's walk: Methods for the analysis of thinking, intelligence, and creativity.* New York: Harper and Row.
Educational Testing Service 1981: *ETS 1981 annual report: New ways of assessment in a changing time.* Princeton, N.J.: Educational Testing Service.
Ghiselin, B. (Ed.) 1955: *The creative process.* New York: Mentor.
Holland, J. L. & Richards, J. M., Jr. 1966: Academic and non-academic accomplishment in a representative sample taken from a population of 612,000. *ACT Research Report*, No. 12. Iowa City: American College Testing Program.
Holland, J. L. & Richards, J. M., Jr. 1967: The many faces of talent: A reply to Werts. *Journal of Educational Psychology, 58*, 205–9.
Jordan, L. A. 1975: Use of canonical analysis in Cropley's "A five-year longitudinal study of the validity of creativity tests." *Developmental Psychology, 11*, 1–3.
Kogan, N. & Pankove, E. 1974: Long-term predictive validity of divergent thinking tests: Some negative evidence. *Journal of Educational Psychology, 66*, 802–10.
Mansfield, R. S., Busse, T. V. & Krepelka, E. J. 1978: The effectiveness of creativity training. *Review of Educational Research, 48*, 517–36.
Messick, S. & Jungeblut, A. 1981: Time and method in coaching for the SAT. *Psychological Bulletin, 89*, 191–216.
Milgram, R. M. & Milgram, N. A. 1976: Creative thinking and creative performance in Israeli students. *Journal of Educational Psychology, 68*, 255–9.
Munday, L. A. & David, J. C. 1974: Varieties of accomplishment after college: Perspectives on the meaning of academic talent. *ACT Research Report*, No. 62. Iowa City: American College Testing Program.
Reese, H. W. & Parnes, S. J. 1970: Programming creative behavior. *Child Development, 41*, 413–23.
Richards, J. M., Jr., Holland, J. L. & Lutz, S. W. 1966: The assessment of student accomplishment in college. *ACT Research Report*, No. 11. Iowa City: American College Testing Program.
Richards, J. M., Jr., Holland, J. L. & Lutz, S. W. 1967: Prediction of student accomplishment in college. *Journal of Educational Psychology, 58*, 343–55.
Singer, D. L. & Whiton, M. B. 1971: Ideational creativity and expressive aspects of human figure drawing in kindergarten-age children. *Developmental Psychology, 4*, 366–9.
Slack, W. V. & Porter, D. 1980: The Scholastic Aptitude Test: A critical appraisal. *Harvard Educational Review, 50*, 154–75.

Sudnow, D. 1979: *Ways of the hand: The organization of improvised conduct.* New York: Bantam Books.

Wallach, L. & Wallach, M. A. 1982: Phonemic analysis training in the teaching of reading. In W. M. Cruickshank & J. W. Learner (Eds.), *Coming of age: Volume 3, The best of ACLD.* Syracuse, N.Y.: Syracuse University Press.

Wallach, M. A. 1967: Creativity and the expression of possibilities. In J. Kagan (Ed.), *Creativity and learning.* Boston: Houghton Mifflin.

Wallach, M. A. 1970: Creativity. In P. H. Mussen (Ed.), *Carmichael's manual of child psychology* (3rd ed., Vol. 1). New York: Wiley.

Wallach, M. A. 1971: *The intelligence/creativity distinction.* Morristown, N.J.: General Learning Press.

Wallach, M. A. 1976a: Psychology of talent and graduate education. In S. Messick & Associates, *Individuality in learning: Implications of cognitive styles and creativity for human development.* San Francisco: Jossey Bass.

Wallach, M. A. 1976b: Tests tell us little about talent. *American Scientist, 64,* 57–63.

Wallach, M. A. & Kogan, N. 1965: *Modes of thinking in young children: A study of the creativity-intelligence distinction.* New York: Holt, Rinehart & Winston.

Wallach, M. A. & Wallach, L. 1976: *Teaching all children to read.* Chicago: University of Chicago Press.

Wallach, M. A. & Wallach, L. 1979: Helping disadvantaged children learn to read by teaching them phoneme identification skills. In L. B. Resnick & P. A. Weaver (Eds.), *Theory and practice of early reading* (Vol. 3). Hillsdale, N.J.: Erlbaum.

Wallach, M. A. & Wing, C. W., Jr. 1969: *The talented student: A validation of the creativity-intelligence distinction.* New York: Holt, Rinehart & Winston.

Wallbrown, F. H. & Huelsman, C. B., Jr. 1975: The validity of the Wallach-Kogan creativity operations for inner-city children in two areas of visual art. *Journal of Personality, 43,* 109–26.

Willingham, W. W. & Breland, H. M. 1982: *Personal qualities and college admissions.* New York: College Entrance Examination Board.

Wing, C. W., Jr. & Wallach, M. A. 1971: *College admissions and the psychology of talent.* New York: Holt, Rinehart & Winston.

2

The Creative Personality

RAVENNA HELSON

Personality psychology is concerned with the person as a unit and with individual differences that are relatively complex and enduring. In the study of creativity, personality psychologists have been interested in the identification of creative persons and in the conceptualization, description, and measurement of cognitive-motivational structures characteristic of these individuals.

Recurring issues in the study of the creative person include whether creative individuals are unusually frail and prone to mental illness or are unusually healthy, resourceful, and persevering; whether there are personality characteristics that describe creative individuals across fields, and how creative artists, scientists, and leaders differ from each other; whether conscious or unconscious—or perhaps the cognitive or motivational—aspects of creativity should be emphasized, and how the two are related; whether creativity is evident from childhood or adolescence, and how creative potential is affected by parental models, education, and opportunities; how creative personality traits are related to eminence, productivity, and intelligence; whether creativity in the genius is on a continuum with that of the moderately creative individual, or indeed whether everyone has creative potential. All of these issues are related to questions of how to conceptualize, identify, and assess creativity and creative potential.

The concept of creative person calls up different images in the mind of the poet (himself), the businessman (someone who, in the right work environment, would produce a profitable invention), and the educator (cultivating the potential of a child). Although clarity may come from adopting one perspective, it is unwise to narrow one's scope too soon. Psychologists with different interests have approached the subject in diverse ways, and this chapter begins with a brief discussion of the history and concerns of three main approaches: psychodynamic formulations, early archival studies of the eminent, and personality psychology between about 1950 and 1970. Then the chapter moves on to recent trends and active areas of investigation. It emphasizes studies of whole persons to the neglect of experimental studies of trait-elements and issues related to cognitive mechanisms involved in creativity.

PSYCHODYNAMIC FORMULATIONS

The Unconscious as the Source of Originality

The creative product has an aspect of surprisingness. The creative insight often seems to come out of nowhere. Creative artists, at least in some periods of history, have experiences of inspiration or a compulsion to perfection that are like a "divine madness". Their works may use the imagery of dreams and arouse us in mysterious ways. Since the late nineteenth century, the conceptualization of the not seen, the irrational, and emotional in our personalities has come increasingly from psychoanalysis. Perhaps for this reason, a recurring idea in the study of creativity has been that the primary source of originality is the unconscious, and that the creative person has more access to this source than other individuals. Many influential formulations of the creative personality have come from psychotherapists, who deal daily with thoughts and emotions outside of ordinary awareness and tend to have creative personality traits themselves (Gough 1976). Because the psychoanalytic literature is so extensive and theoretically complex, I can only suggest a few of the issues and contributions. Abstracts, excerpts, and additional references are provided by Gedo (1983), Rothenberg and Hausman (1976), and Stein and Heinze (1960). Unfortunately, many psychoanalytic studies suffer from lack of scientific rigor and from excessive or outworn jargon.

Views of the Unconscious as Positive or Negative

A major division in the psychoanalytic literature on creativity is between those who conceptualize the unconscious as having positive features and those who see its features as negative. According to Jung (1930), the creative process consists in the unconscious activation of an archetypal image and in the elaboration and shaping of this image into the finished work. "The unsatisfied yearning of the artists," he wrote, "reaches back to the primordial image in the unconscious which is best fitted to compensate the inadequacy and one-sidedness of the present . . . Whoever speaks in primordial images . . . lifts the idea he is seeking to express out of the occasional and transient into the realm of the ever-enduring" (p. 82). Thus, the unconscious is seen as having enormous power and a valuable compensatory relation to collective values.

In Freudian psychology, the unconscious is usually seen as negative. How, then, does something creative result from its influence? Freud (1908) described the writer as a pathetic trickster who could not satisfy his desires for women and fame but was somehow able to sublimate these id impulses in the form of disguised daydreams, using esthetic form as "forepleasure" to entice the reader to share his fantasy. In another essay (1905), Freud had made the suggestion that in the production of wit, which in its surprisingness and aptness resembles the creative, a preconscious thought is "entrusted for a moment to the unconscious for elaboration". In this way it gains access to associations ordinarily repressed. The idea was taken up by Ernst Kris, who was both an art historian and one of the psychoanalysts who developed psychoanalytic ego psychology. Kris (1952) offered a rather extensive and systematic theory of creativity in which he gave more power to the ego than his predecessors in psychoanalysis. Especially influential was his idea that the creative person, unlike the psychotic, has "flexibility of repression" and is able to "regress in the service of the ego". Kris distinguished inspirational and elaborational phases of the creative work, discussed processes underlying each phase, and described artists—romantic and classical—who emphasized one set of processes over the other. "Primary process" cognition (fluid, analogical) was said to be conspicuous in the inspirational phase of creative work and "secondary processes" cognition (logical, reality-oriented) in the elaborational phase. Kris suggested that the scientist's experience of "chance" was comparable to the artist's experience of inspiration.

Kubie (1958) attributed creativity to the preconscious, arguing that the unconscious was rigid and always a negative influence. In an essay published in the *American Scientist* (1953), he discussed ways in which unconscious needs can affect the young scientist's choice of career, block his productivity, distort his interpretation of results, and make him dissatisfied in his work.

Ehrenzweig (1953), an art teacher interested in gestalt psychology and psychoanalysis, argued that some aspects of "primary process" cognition, such as the ability to scan large amounts of information rapidly and without regard for the ordering principles of logic, was invaluable in creative work and should be considered not as regressive (as Kris said) but as skills that could be developed and drawn upon as needed.

We will see in a later section that academic personality psychologists have been much interested in these ideas about the relation between the ego and the unconscious.

Motivation for Creative Work

Another classic topic in psychodynamic theories has been the nature of motivation for creative work. Sublimation of libidinal desires received early emphasis, but other motivations that have been discussed include relief of guilt through sharing participation in fantasy with the reader or audience, reparation—making the creative product as a gift to atone for hostile impulses or the redoing in the creative work of traumatic materials from the past, the attempt to become independent of the mother by taking into oneself her functions of emotional support and ability to create children, need to compensate for inferiorities, and narcissistic needs for self-esteem and attention. Self-actualization as a motive was emphasized by Rank (1945) and humanistic psychotherapists (Rogers 1959; May 1975) who discussed qualities in the creative person such as openness, courage, and the ability to make commitments.

Development of the Creative Individual

Greenacre (1957) studied the early development of the creative person, drawing upon both her own patients and biographies and works of figures of the past. One of her ideas was that exceptionally creative children formed "collective alternative relationships", such as an identification with a great figure of history, that enabled them to be less limited by relationships with their family members and less bound by the usual fixations and progressions in libidinal development.

A developmental theory that has become increasingly influential is that of Rank (1945). He conceptualized creative, conflicted, and adapted types of adjustment, centering upon the problem of separation from parents. According to Rank, the adapted person does not rebel very much; quite early he is able to identify his will with that of his parents. He is spared the pain of the guilt that comes from separation, but his personality development is limited. The conflicted type feels divided in personality, being aware of both his own will and that of parents or wider society. There is opportunity here for a more complex development of personality, but unless the individual is able to overcome his conflicts, he suffers from irresolution, guilt, and inferiority. For the creative type, the sense of separateness is great enough to reach the third level of development, characterized by a transcendent goal-seeking force that has been generated by the conflict between the individual's will and that of parents or other social agents. Rank's views were admirably exposited by MacKinnon (1965), who also provided supportive findings from his study of architects at three levels of creativity. See also the work of Helson and Crutchfield (1970) and Helson (1977, 1985).

Psychobiography

Psychoanalysts contributed a large number of case histories or psychobiographies of creative individuals. The prototype was Freud's study of Leonardo da Vinci (1910). Early studies were concerned more with illustration of psychoanalytic concepts and with the sources of an author's themes in his personal experience than with creativity (Gedo 1983). Among major contributors are Erikson (1958, 1969), who studied the development of creative personality in leaders such as Martin Luther and Gandhi in the cultural context and through the life span; and the Georges (1956), who attempted to bring academic method to the use of psycho-analytic concepts in psychobiography.

ARCHIVAL STUDIES OF THE EMINENT

Heredity

Galton (1869) was interested in proving the hereditary nature of "genius". Reasoning that eminence was an index of natural ability because the truly able individual could not be repressed by social obstacles, he made statistical use of biographical sources for demonstrating the extent to which eminence ran in families. Like most of the archival studies he inspired, Galton's work was studded with provocative findings. He reported, for example, that the most illustrious men had more eminent relatives than the less illustrious, that eminent relatives of poets were largely confined to the immedi-ate family—"poets are clearly not founders of families"—that there were fewer transmissions of ability along the female than the male line, and that the mortality curves for eminent men were bimodal—the eminent tend to be either "weak" or long-lived.

Creativity and Pathology

Galton's method aroused much interest in Europe as well as in England and America. Lombroso (1895) used extensive if biased biographical material to support his thesis that genius was related to insanity. However, Ellis (1904) did a careful study of 1,030 British "men of genius", showing that only 4 percent could be classified as clearly insane or psychologically pathological, a rate certainly no higher than that for the general population. White (1930) analyzed Ellis's data and those of Cox (see below) and added the finding that there was more pathology among esthetic types than among scientific and practical types. This conclusion was supported in an archival study of nineteenth-century writers and scientists by Raskin (1936).

Common Traits in the Eminent

In the early twentieth century, James McKeen Cattell did numerous statistical investigations of the eminent. Like Ellis, he measured degree of eminence more carefully than Galton, by the amount of space allotted to the individual in biographical dictionaries. For American scientists, he worked out methods for assessing contemporary eminence through ratings by peers (1906).

Cox (1926), using Cattell's list of 1,000 eminent people, obtained estimates from experts of the IQs of 282 "geniuses" of the past, based on reports of their achievements in childhood. These estimated IQs were quite high. She showed also that the most outstanding of the eminent came from advantaged home backgrounds and were characterized in childhood by personality traits that included not only intellectual power and energy but also originality, confidence, persistence, and ambition. As we will see in later sections, Cox's findings have been confirmed by others.

Age and Creativity

The relation between age and productivity was yet another question addressed in archival research. Lehman (1953) found curves of creativity to rise rapidly in early maturity and then decline slowly. He reported that the age of productivity of best work varied from one field to another, and that in a number of fields recent workers demonstrated their creativity at younger ages than those in earlier eras. He believed that older people are handicapped in fields that require new learning and unlearning of the old, but at an advantage when the accumulation of experience is a requirement. Lehman's work has been challenged, supported, and extended by others (Dennis 1958, Simonton 1984).

With a few exceptions, such as Lehman's work and the Goertzels' interesting and useful abstracts of biographies (1962), there was little archival research relevant to the creative personality between the 1930s and the 1970s. Archival studies had the disadvantage that information was more complete about some individuals than others, biases were hard to evaluate, and many measures that had become of interest in the newly consolidated field of personality psychology—such as indices of motivational dynamics—were impossible to obtain. Personality psychologists were studying living subjects.

THE CREATIVE PERSONALITY IN PERSONALITY PSYCHOLOGY

Galton is considered the father of the field of "differential psychology", and James McKeen Cattell was another outstanding figure in this field. As time went on, workers in differential psychology (or individual differences) became increasingly absorbed in the measurement of intelligence and aptitudes. In the U.S. there was a general lack of concern for the whole person and a lack of theoretical perspective other than behaviorism. Dissident psychologists were stimulated by gestalt psychology, psychoanalysis, and several other theories that had been developing in Europe. Psychologists in the U.S. began to undertake programs of research in the newly consolidating field of personality psychology, and in the late 1930s several outstanding books appeared (Allport 1937; Murray 1938). In the 1940s, two of the best known projective measures of imagination were published, the Rorschach (1942) and the Thematic Apperception Test (Murray 1943). Two important self-report inventories, both shown later to assess themes relevant to creativity, also appeared at about this time: Strong's inventory of interests (1938) and Hathaway's and McKinley's (1943) inventory of psychiatric dimensions.

During World War II, personality psychologists were called upon to apply their knowledge to problems of selection and diagnosis. They felt the lack of measures of characteristics such as resourcefulness and inventiveness. After the war, Guilford was one of those who went to work to develop such measures: in his presidential address to the American Psychological Association, he distinguished convergent thinking, such as intelligence tests assessed in the individual's ability to give the "right" answer to questions, and divergent thinking, in which the goal is to think of many different and original ideas or solutions. He gave examples of some approaches to the measurement of divergent thinking. Guilford's address (1950) is often used to mark the beginning of the modern era in research on creativity.

Several very different factors supported the research that was to continue over the next 20 years. One was the fact that personality psychology now had theories, methods, and tools, along with skills and zeal to develop more as needed. Guilford's suggestions and their obvious practical import aroused the psychologist's urge to measure. Another important factor was the uncomfortable conformity in anxiety-ridden American society in the period after World War II. To study creativity was to strengthen the expression of individuality and humanist values against these conformist pressures, manifested in political harassment of "Reds" ("McCarthyism"), rigidities in the educational system, narrow, reductionistic learning theories within psychology and psychoanalysis, and functionalist theories in sociology. The anxieties of this "Cold War" era were exacerbated by the Russians' launching of Sputnik in 1957, with increased concern for the development of scientific talent. The economic affluence of the period supported funding from both government and private sources for projects of a scope that would not be possible today.

Measures of Creative Traits and Personality Patterns

Guilford's emphasis remained on cognitive aspects of creativity. Torrance (1962) and later Wallach and Kogan (1965) adapted Guilford's measures for use with children, and Mednick (1962) devised an ingenious measure of associational fluency, the Remote Associates Test, presumed to be relevant to creative processes. Other psychologists developed measures of aspects of the creative personality that were as much affective as cognitive. Thus Barron and Welsh (1952) took advantage of Welsh's finding that preferences for line drawings proved to have a dimension of symmetry and simplicity vs. asymmetry and complexity. They found that artists chose complex, asymmetrical drawings much more than non-artists, and that in samples of normal individuals preference for the complex was positively correlated with verbal fluencey, impulsiveness, originality, independence of judgment, and breadth of interest; and negatively correlated with conservatism, overcontrol, and ethnocentrism. The Barron-Welsh Art Scale has differentiated more from less creative individuals in many studies, though the preference for complexity and asymmetry also varies with field.

The story of the development of this and several other measures of traits associated with creativity has been well told by Barron (1963, 1965).

Barron did not claim that the Art Scale was a measure of creativity itself, but that it assessed an aspect of the creative personality. He claimed that the preference for complexity and disorder represents a basic perceptual choice of what to attend to. The creative person is interested in the unstable and unorganized and in the integration of this material into a high-order synthesis (Barron 1963).

Other works in measuring traits related to creativity include Crutchfield's situational procedure to assess independence of judgment (1955), Hall's mosaic design procedure to assess esthetic abilities and styles (1972), and the development of a variety of scales from personality inventories (Barrow 1963; Gough 1957). Among the scales most widely used today are those scored from the Adjective Check List (Gough 1979). In the area of projective tests, Holt (1956, 1968) developed scoring procedures to measure primary process and adaptive regression on the Rorschach; and Maddi (1965) scored types of novelty motivation on the TAT.

Gough and Woodworth (1960) developed a Q-Sort consisting of items that a research scientist could rate to describe his own research style. On the basis of a factor analysis of the self-descriptions obtained from 40 research scientists, Gough and Woodworth described the styles of initiator, artificer, esthetician, and several others. This was one of the first articles that dealt with specialization of talent within a field.

Helson took the idea of Gough's Q-Sort, but devised a more phenomenological set of items with which mathematicians (Helson and Crutchfield 1970) and writers (Helson 1978b) could describe their ways of working and their emotional experiences while working. All these and other measures were put to work in studies of the creative personality.

Creativity and the Unconscious

Whether creative individuals have unusual access to the unconscious has been investigated in a variety of ways. Studies of this complex issue may be used to illustrate questions about how the creative individual is to be identified, what access to the unconscious means, and how it is to be assessed.

Most researchers have used in their studies students who were identified as creative on the basis of their field of interest, such as art, or their scores on tests presumed to measure originality. Access to the unconscious or to "primary process" has usually been measured by libidinal, irrational, unusual, or symbolic content of responses to projective or free response measures, such as the Rorschach, TAT, or word association tests. Some researchers have distinguished amount of primary process from control of primary process—a large amount of controlled primary process being assumed to show what Kris called "regression in the service of the ego". Control has been assessed by measuring crudity vs. differentiation of responses or by measuring differences in the nature of response when subjects are asked to change attitudes. Some researchers have conceptualized access to the unconscious as an ability, others as an aspect of a motivational pattern, still others as a factor of body chemistry or brain structure. Some have accepted Kris's hypothesis with or without interest in his model of conscious and unconscious, which was based on energy theory. Today most investigators feel that the idea of primary and secondary process is clear enough, but prefer other language and revisions in conceptualization (Bush 1969; Martindale 1981; Suler 1980).

Let's consider a few studies. Wild (1965) compared art students with teachers and schizophrenics on object-sorting and word-association tasks. She found that the art students enjoyed the tasks much more than the other groups, gave more unusual and clever responses, and showed bigger shifts in responses given under instructions to take the role of a "regulated" and then an "unregulated" person.

Pine and Holt (1960) found, as they had hypothesized, that male undergraduates classified as creative on the basis of a number of behavioral tests showed no greater *amount* of primary process on the Rorschach than other men, but better ability to control it. However, among the women in the sample, creativity was related to amount of primary process. These results were not replicated in a sample of unemployed actors (Pine 1962). Pine and Holt concluded that the expression and control of primary processes was relevant to an understanding of creativity, but that the specific nature of the relationship depended on variables such as age, sex, type of creativity, and level of creativity. They also suggested that primary process was not directly elicited in response to the tests, but that modes of expression and control of primary process had become generalized as broad cognitive styles.

Rothenberg (1971) attempted to explain how "regression" is conducted in the service of the ego, a question Kris had not handled very well. He described "Janusian thinking", a process of using the defense of "negation" to call up opposite or contradictory ideas simultaneously. In this way, an ego process allows primary-process-like associations to appear in consciousness. He illustrated his concept in a discussion of "The Iceman Cometh", in which he says that O'Neill worked simultaneously with the ideas of sex and death.

Helson (1970) analyzed "real-life" creative products. Works of imaginative literature for children that had been rated by judges as highly creative were found to contain more gripping emotional and irreal content than works rated as less creative. Data from authors of these books showed that those who made fantasy worlds vivid and charged with significance were themselves fascinated by these worlds (Helson 1978b). In this study access to the unconscious was not merely an ability or even a cognitive style, but part of a process of self-discovery and self-actualization. However, we do not know how much authors of other genres resemble these.

Domino (1976) found that high school boys high in creativity reported dreams with more primary process than boys low in creativity. He said that the results might reflect a richer fantasy life on the part of the high creative group or perhaps their ability to tolerate and express less logical material.

In the last decade, activation of the right and left hemispheres has become a partial analogue of primary and secondary process. Katz (1983) and Martindale, Hines, Mitchell, and Covello (1984) show that under certain circumstances or by some inferences individuals with high scores on tests of originality make more use of the right cerebral hemisphere.

These and other interesting studies produce diverse kinds of evidence that individuals who are by some criterion creative show more access to the unconscious, conceived in some way or another, than other individuals. These studies usually raise as many questions as they answer. Failures to obtain or replicate results and limitations of conceptualizations and measures should be kept in mind. One awkward fact is that eminent creative people do not distinguish themselves on artificial tests of originality. The tests do not seem to engage their serious interest or tap their particular talents. Let us turn to studies of such groups.

STUDIES OF EMINENT CREATIVE PEOPLE

Roe and Rorschach

Roe's study of the relation between alcoholism and creativity in visual artists (1946) was the first investigation of creative individuals to employ the recently developed techniques of the personality psychologist. Roe went on to studies of eminent natural scientists and social scientists (1952). What she emphasized most in her conclusions was the degree of involvement, hard work, and intellectual independence of her subjects. She found many of them "feminine", a characteristic she concluded to be common in men attracted to intellectual work. The Rorschach and TAT seemed less useful in demonstrating the originality of the men than in differentiating scientists in different fields of endeavor.

Cattell and the 16PF

Other early studies, built around the 16PF Questionnaire, were conducted by R. B. Cattell and Drevdahl (1955). In a summary article, Cattell and Butcher (1968) said that across fields, creative men showed a "dominant independence", seriousness, low conformity, high self-sufficiency, a willingness to experiment rather than to take a conservative attitude, and a disregard for sentimentality. Cattell and Butcher said that these personality characteristics are not particularly likeable and would not be advantageous in many settings, but that they facilitated the development of new ways of thinking. As compared with scientists, artists and literary men were less practical and more emotionally sensitive and tense.

The IPAR Studies

In the late 1950s there began the "living-in" assessments of creative architects, mathematicians, research scientists, and writers, who were invited in groups of about ten to spend a weekend (sometimes a day) at the Institute of Personality Assessment and Research (IPAR). A variety of perceptual and cognitive tests. projective tests, inventories, and interviews were included in the assessment, and a team of psychologists observed the subjects' social behavior during meals, cocktail hours, and group activities. These observations were recorded by means of ratings, adjective check lists (Gough 1960), and a clinical Q-Sort (Block 1961).

   From the IPAR researches a rich picture of the creative personality emerged, extending far beyond that of Roe or Cattell (MacKinnon 1975). Much care went into the criteria for selection of subjects. Since each study directed by each investigator (Barron, Crutchfield, Gough, Helson, and MacKinnon) was done somewhat differently, it is not easy to describe results across samples. Still, even in an early report MacKinnon (1962) was able to state—for example—that all of the creative groups had peak scores on the theoretical and esthetic scales of the Allport-Vernon-Lindzey Study of Values (1951), that they were virtually all intuitive (as opposed to sensing) and a high proportion introverted and perceptive (as opposed to extraverted and judging) on the Myers-Briggs Type Indicator (Myers 1962); and that they had a characteristic pattern of interests on the Strong Vocational Interest Blank, which he described as indicative of concern with meanings and their implications more than small details or facts for their own sake, cognitive flexibility, intellectual curiosity, interest and accuracy in communication, and relative disinterest in "policing their own impulses or those of others". He said that creative men were unusually open and accepting of aspects of personality that other men suppressed as "feminine".

Barron (1965) showed that staff descriptions of several of the creative groups (writers, architects, and mathematicians) had much in common: ascriptions of intellectual values, inquiringness, intelligence, high aspirations, need for autonomy, and esthetic sensitivity were consistently among the most distinguishing characteristics of each group. Having discussed the scores of several groups on the California Psychological Inventory (CPI) (Gough 1957) and the Minnesota Multiphasic Personality Inventory (MMPI), Barron addressed the question of pathology. He concluded that creative people are more troubled psychologically than people in general, but have more resources for dealing with their troubles. "They are clearly effective people who handle themselves with pride and distinction, but the face they turn to the world is sometimes one of pain, often of protest, sometimes of distance and withdrawal; and certainly they are emotional" (p. 64).

Comparisons Across Fields and Age Groups

The studies of the creative eminent, utilizing the new measures of personality, gave at least partial answers to long-standing questions about the creative personality. It was difficult, however, to put the findings together in any precise way. Eiduson and Beckman (1973) said that summaries of personality characteristics common to individuals across fields must be interpreted with caution, because studies differed substantially in definition of criterion groups, selection of control groups, and choice of personality measures. They pointed also to the problem of the confounding of eminence and creativity.

Both problems were addressed in a study by Parloff, Datta, Kleman, and Handlon (1968). It compared four male IPAR samples (architects, mathematicians, research scientists, and writers) and a highly selected sample of male adolescent aspirants in a nationwide search for scientific talent. Analyses were limited to four main factors from the CPI, which were labeled in this study as disciplined effectiveness, assertive self-assurance, adaptive autonomy, and humanitarian conscience.

Along with evidence of considerable variation from one sample to another, Parloff et al. found that creative adults were distinguished from their less creative peers on a shared set of personality dimensions, and that findings were "remarkably consistent with the general picture of the creative personality [as being] relatively uninhibited and unconventional; appearing to have a greater need for independence, self-direction, and autonomy; and being more assertive, forceful, and seemingly assured". They add, however, that prediction and identification could be enhanced by the study of different personality types associated with creative performance, because many creative individuals did not fit the pattern based on central tendency. (This is a topic to which we will return.) In the adolescent sample, creativity was associated with high rather than low scores on disciplined effectiveness, but in other respects adolescent and adult samples showed the same pattern of relationships. Parloff et al. concluded that autonomy and assertiveness were characteristics predisposing to creative achievement, rather than consequences of eminence. Other personality characteristics that facilitated creativity, such as disciplined effectiveness in adolescents, might vary from one age period to another.

LONGITUDINAL STUDIES OF GIFTED ACHIEVERS

The findings of Parloff et al. were consistent with longitudinal studies of achievement in gifted young people. Terman (1925) had begun his well-known study of gifted children in 1922, and Oden (1968) compared the males of this sample who had been classified as most and least successful on the basis of accomplishments when most of them were in their late twenties. The men who became successful were characterized in adolescence and even more in early adulthood by psychological health, social adjustment, curiosity, and originality. They also came from more advantaged backgrounds.

In the 1960s, longitudinal studies of gifted achievers were published by Holland and his associates at the National Merit Scholarship Corporation. These studies again showed that the gifted who reported creative achievement in college were self-confident, self-controlled, and self-directed as adolescents (Holland & Astin 1962). They were also advantaged in social background. See the excellent review of studies of gifted adolescents by Hogan (1980), himself a contributor to the studies of gifted youth at Johns Hopkins University (Stanley, Keating and Fox 1974). See also recent books edited by Albert (1983) and Bloom (1985).

The incidence of creative achievement was infrequent in both the Terman and Holland samples. High IQ is not an indicator of creativity.

CONCEPTUAL AND METHODOLOGICAL ISSUES

In the 1960s there began to be increasingly animated discussion of how creativity was to be conceptualized and how the creative person was to be identified. Three positions may be distinguished. Some researchers thought of creativity as a cognitive ability and proposed to measure it through instruments similar to intelligence tests, which were considered highly successful at the time. Others thought of creativity as an achievement—the production of something novel and adaptive by real-life criteria. They proposed to identify creative individuals through ratings of creative products or through evidence of recognition for creative accomplishment. Others thought of creativity as a personality pattern, and they proposed to identify creative individuals through measures of personality variables and the demonstration of construct validity through a nomological net of relationships.

Creativity and Intelligence

In 1962 Getzels and Jackson had published a provocative book, *Creativity and Intelligence*, in which they contrasted adolescents who were high on conventional intelligence and low on measures of divergent thinking with boys who were relatively low (but still fairly high) on the former and high on divergent thinking. Illustrations from the TAT stories told by members of the two groups made an effective demonstration of the inventiveness and zest associated with what the authors called "creativity". The authors showed also that "creative" youths tended to be less popular with teachers and peers than "intelligent" youths. Similar findings were reported by Torrance (1962) in studies of creative talent in boys and girls in elementary school. For some investigators, the important issue was the criticism of intelligence tests and the anti-creative prejudices of teachers and peers, which they thought could be alleviated if attention was directed to the problem.

However, other psychologists were intrigued by the question of whether creativity was a cognitive ability that could really be distinguished from intelligence. Wallach and Kogan (1965) decided that it was, but they said that divergent thinking tests must be administered differently from intelligence tests—in a more relaxed, playful manner—if the abilities were to be assessed as adequately as intelligence. Tryk (1968) called attention to large individual differences in motivation to become involved in creative activity, and suggested that lack of attention to motivational variables was a major deficiency in the assessment of creativity through creativity tests.

Many investigators had seized uncritically on the use of divergent thinking measures as a criterion of creativity for all sorts of captive classroom samples. In an article entitled "Creativity in the Person Who Will Never Produce Anything Original or Useful", Nicholls (1972) reviewed the research of the 1960s and wrote a blistering critique of the use of measures of divergent thinking as criteria of creativity. He thought that the creative personality should be conceptualized as consisting of a combination of traits that had been shown to characterize eminent creative individuals across fields. He suggested that this might include not only divergent thinking but also intelligence, intrinsic task involvement, and preference for complexity. However, since he considered insufficient the evidence so far accumulated that these traits cohered in persons of different ages and fields of endeavor, he joined those who recommended that the study of creativity be linked to real-life achievement criteria, especially the evaluation of creative products.

Creativity as Achievement

One way of measuring creative achievement was to obtain ratings from experts who could evaluate the contributions of the group to be studied. This, of course, was the method that had been used in the IPAR studies of eminent creative individuals. Another method was to use as a criterion some form of already achieved recognition of creative performance, such as listings in biographical dictionaries or membership in elite societies. Holland and Nichols (1964) used a checklist with which high school students could report their past creative activities and achievements. The items included achievements that are relatively uncommon, such as having won a prize or award for a scientific paper or project, having started a business enterprise, or having published poems, stories, or articles in a public magazine or anthology. This checklist predicted creative achievement in college better than aptitude tests, grades, or self-conceptions. Others (Wallach and Wing 1969) developed checklists modeled on Holland's, and the idea has won many adherents.

Nevertheless, the concept of the creative person as outstanding achiever was not entirely satisfactory. Were characteristics of creative achievers attributable to their creativity or to their success skills or achievement motivation? Were these characteristics appropriate as a criterion for studies of creativity in children, or in adults less gifted or ambitious?

Another way of using real-life behavior as a criterion was to focus on the creative products. Products could be those of high-achievers, but they could also be research reports by laboratory workers or drawings of children. Of course, there were difficulties here, too. When quantity is used as a measure, there is the implicit assumption that productivity is an essential aspect of creativity. This is an interesting and controversial issue. On the other hand, methods used to assess quality or significance of products, though sometimes quite successful, have often shown unreliability, lack of discriminant validity, or discouragingly low correlations with other criteria of creativity (Hocevar 1981; Taylor, Smith and Ghiselin 1963; Tryk 1968). The problem may be in representativeness of the product or products chosen, the range of quality in the products to be rated, the qualifications and biases of raters, fallibility of other criteria, and sometimes the inherent difficulty in rating something that is highly creative. Jackson and Messick (1965) distinguished four properties of the creative product: unusualness, appropriateness, transformation, and condensation. Moderately creative products are unusual and appropriate, they said, but the most creative products are also *transformative*, meaning that they make one think about something in a new way, and *condensed*, meaning that they have a multiplicity of compactly presented implications that reward repeated "savoring". Agreement about transformation and condensation is difficult to obtain, they suggest, and good judgments require knowledge, time, and effort.

Creativity as a Personality Pattern

Some theorists conceptualized creativity in terms of self-actualizing attitudes and motivations that were not necessarily associated with either products or outstanding achievement (Maslow 1959; Rogers 1959). Other theorists wanted to anchor their concept of the creative person in the outstanding creative achiever but to conceptualize and identify this person in terms of personality traits. Nicholls had considered this approach but rejected it until the field was further advanced. Others pursued it (Golann 1963; Maddi 1965; Welsh 1975; and several of the IPAR researchers).

Welsh (1975) argued that though intelligence and originality were both important in creativity, they were more usefully conceived as personality constructs than as cognitive abilities. In a highly gifted sample, he reasoned, scores on the Terman Concept Mastery Test (a high-level measure of cultivated intelligence) would have personality correlates indicative not of intelligence, but of what he called intellectence—*interest* in symbolic expression and principles of comprehension as opposed to the demonstrable, practical, and concrete. Similarly, personality correlates of scores on the Revised Art Scale would assess origence—*preference* for an open world that can be structured in an individualistic way rather than an explicit world defined by the application of impersonal rules. Having access to a sample of 1,100 talented adolescents, Welsh developed scales to measure four types of personality characterized by each of the combinations of high or low scores on his two criterion variables. (These scales may now be scored from the Adjective Check List [Gough and Heilbrun 1980]). Using data available from the gifted adolescents and other samples, Welsh gave a convincing picture of differences in vocational choice and in personal and intellectual style associated with the four types.

Need for an Overview

Each point of view I have described had merits, and research frequently adopted a mixed or global approach. For example, when teachers or supervisors were asked to rate the creativity of students or employees, they usually looked for signs of creativity as a cognitive ability, for creativity in products, and for interests and attitudes that would indicate a creative personality. (Like ratings of products, ratings of persons worked out well in some cases and badly in others.) Many researchers went their ways studying different phenomena all called by the same name but not very highly correlated. Several reviewers of the assessment of creativity emphasized the need for a more general theoretical framework that could prune and integrate, or at least assign a place to, the many existing theories and lines of research. Tryk made this comment in 1968; Hocevar in 1981.

CHANGING CONCEPTUALIZATIONS AND EMPHASES

The study of creativity in personality psychology continued to change within the context of its own growth, that of the rest of psychology, and changes in American society.

By the late 1960s, several social movements had redirected attention and priorities. There was a massive effort to relieve discrimination against the disadvantaged. Behavior therapy and experimental social psychology were thriving fields. Personality testing was criticized as an invasion of privacy and as a tool for "blaming the victim"; the findings of personality research were criticized as weak and doubtful; and even the concept of personality was said by some to be an illusion. Critics tended to ignore the strong findings of research on the personality of outstanding creative people, but this research was isolated by its very elitism. By 1980, Feldman could articulate his sense that research on the creative personality had made insufficient allowance for change in individuals and for the interaction between the individual and the environment. His own work on child prodigies convinced him that the prodigy is a result of unlikely and unusual special abilities that receive unusual encouragement in their encounter with unlikely and unusual bodies of knowledge.

Though one can easily point to early articles concerned with the influence of the environment or with changes in the creative person with age, I agree that most research of the period between 1950 and 1970 had a strong individualistic slant. I have tried to show that this individualism was, in part, a protest against a conformist society. But emphases in the field changed as the world changed and as several areas of inquiry received new attention.

Creativity in Women and Sex Differences in Creativity

There were early archival studies of eminence in women (Castle 1913), and in the late 1950s creativity in women was one of the topics in the research begun at IPAR (Barron 1965). Torrance (1962) called attention to sex differences in divergent thinking that increased from the first to the third grade and seemed to be rooted in peer prejudices. However, most investigators gave the subject scant attention. A few alluded, sometimes with jocularity, to the "curious" or "fascinating" question of why there were so few creative women. It was said that normal women lacked traits such as originality, the ability to think abstractly, or the motivation to concern themselves beyond personal relationships. Such remarks reveal the continuation of Galton's assumption that the "truly creative" individual will overcome all obstacles, a lack of awareness of phallocentric biases, and a lack of sophistication about the influence of social organization (Helson 1978a). As the Women's Movement progressed, gender differences in creativity came to be seen less as a "curious question" and more as a key illustration of the effects of the difference in power between men and women and the complex factors related to this difference.

In early work, Helson tried to counter the impression that women with creative personality characteristics were rare. She showed that they were conspicuous in the senior class of a women's college (1967a), and could even be found in the unlikely field of mathematics (1971). The study of women psychologists by Bachtold and Werner (1970) made a similar point.

Helson demonstrated differences in personality and style between creative men and women (1967b). She showed that differences were great when the environmental conditions under which men and women worked were great, and that these were often striking indeed (1978a, 1983). She showed that the "matriarchal style", characterized by low ego-assertivenes and a "pregnant" unconscious, could be identified in both creative men and women, though it was more common in women. It was associated in creative writers not with any deficiency but with certain features of form and content, such as more emphasis on setting and character than on plot, and a concern with tender emotion rather than heroic struggle or comic ambivalence.

Studies by non-psychologists have broadened our perspectives about the cultural conditions under which men and women have pursued creative endeavors and about the influence of sex roles on the creative process (Gilbert and Gubar 1979; Keller 1985; Nochlin 1971). Keller, a mathematical biophysicist knowledgeable about recent developments in psychoanalysis, argues that the customs and mores of modern science are based on a particular solution to male problems in separating from the mother. She makes her case through studies of the history of science and analysis of the effects of sex differences in early object relations.

An intense desire to know, she says, is so consistently libidinized that the pursuit of knowledge becomes shaped by sexual metaphors. In Plato, she shows, the model was homoerotic. In seventeenth-century England there was a battle between two factions of early scientists, the alchemists and the future founders of the Royal Society. At this time there was also a battle between the sexes, with witch-hunting a part of it. The Alchemists envisioned their work in terms of the unification of Sol and Una, often symbolized as a sexual union between these masculine and feminine principles. The future members of the Royal Society found the

Alchemists obscene. They no longer, like Bacon (who wrote a bit earlier) conceived Nature as a bride and the scientist as a bridegroom. Matter (mater) was desexualized, but it remained the task of the scientist to subject Nature to his control. The consolidation of modern science was accompanied by a drop in the status and rights of women. Keller goes on to discuss problems in science, such as those in quantum theory, that she sees as related to the accentuation of separation between subject and object; and the dilemmas of women scientists for whom assumptions about the scientist and his role constitute a distortion or inauthenticity. She shows how Barbara McClintock achieved her insights, not through distancing herself from what she studied, but through her empathy and respect for it.

In the 1950s and 1960s, a number of psychoanalysts and researchers had suggested that creative men were more feminine than ordinary men, or more open to their femininity. In the 1970s the interest in "androgyny" contributed to studies of the relation between creativity and measures of masculinity and femininity, often newly assessed as separate dimensions. To the extent that creative achievement requires some traits characteristic of men and others characteristic of women, one would expect some truth in the idea that creative men and women tend to be androgynous. Conclusions as to the validity of this hypothesis, however, depend on how masculinity, femininity, and androgyny are measured, how creativity is measured, what masculine and feminine traits contribute to creative performance in the particular field or circumstances studied, and nature of the sample, and how creative and comparison groups are selected (Harrington and Andersen 1981; Helson 1967b, 1978a; Kanner 1976; Welsh 1975). Interest in this topic has declined, I believe with decreasing tendency to characterize traits as masculine or feminine and more open acceptance of homosexuality and bisexuality.

Increased awareness of the influence of gender roles has taken most of the "mystery" out of the question of why there are many more creative men than women. There are interesting theoretical questions and practical problems, however, some of which will be mentioned in subsequent sections.

Creativity and the Study of Lives

We have seen that several recent studies of young people (Hogan 1980) confirm findings of Terman and Oden, Cox, and others that talented achievers are characterized by intelligence combined with environmental advantage, originality, ambition, energy, and per-severance. The durability of the talented achiever has made him an excellent subject for the personality psychologist.

And yet, there are other aspects of the fate of talent. Galton found "weak" as well as "strong" eminent people. Jaques (1965) called attention to late beginnings and early silences in creative individuals and to changes in direction or in form of work in middle age. He related these changes to critical phases in adult development, early difficulties being connected with problems in the area of intimacy, midlife problems with wounded narcissism and destructiveness associated with the anticipation of missed opportunities, decline, and death.

Jaques used concepts from object relations theory, an increas-ingly influential development within psychoanalysis, starting per-haps with Rank, that takes relationships rather than drives as central concerns. The yearnings for wholeness and numinous sig-nificance important in the psychology of many creative people may be related to roots in archaic object relations. For example, Homans (1979) and Stolorow and Atwood (1979) explained Jung's theoretical system in terms of his own early object relations.

Since the mid 1960s psychologists have been increasingly inter-ested in the life-span perspective, a differentiated conceptualization of the adult life course, psychohistory and psychobiography, and methodologies for the study of lives. I can only illustrate some of the ways in which these interests have been manifested in the study of creativity. See Barron and Harrington (1981) for additional discussion and references.

Longitudinal Studies
A new level of methodological care and computer technology is about to bring a wealth of information about the development of originality and creativity in children and adolescents. An example is the excellently executed demonstration by Harrington, Block, and Block (1983) that measures of divergent thinking (DT) obtained from 75 preschool children showed a correlation of 0.45 with teacher evaluations of creativity at age 11. DT quality in early childhood accounted for 14 percent of the variance in preadolescent creativity *beyond* that accounted for by sex, intelligence, and DT fluency. Noteworthy is the care used in the definition and administration of the DT procedure.

Prospective studies of creativity that extend into adulthood are rare, for reasons helpfully discussed by Kogan (1973). The coher-ence of the creative personality is one issue to be investigated, but change is equally interesting. We are just beginning to study pressure points in the creative career and how the individual's values, social skills, and vulnerabilities contribute to the resolution or worsening of crises and hard times. Getzels and Csikszentmihalyi (1976) studied young male artists in art school and several years afterward. They emphasize the discontinuity between the characteristics important for creative performance in school and characteristics needed for success or even survival as artists in the world outside.

Helson (1985, in press) restudied at ages 27 and 43 a sample of women who had been nominated by faculty members for their creative potential in college in 1958 and 1960. She contrasts creative nominees who became successful careerists with other nominees. In college, the former group were more forceful and ambitious. They scored higher on inventory measures of creative traits, but not on DT tests. The careerists had identified negatively with dependent mothers, and at age 27 they were relatively unsure of themselves as they engaged with issues of femininity and gender role. But by age 43 they again showed the familiar traits of the talented achiever. Of the creative nominees who did not have successful careers at age 43, many had problems in making and sustaining commitments. But almost half showed a high level of ego development in combination with unusual interest in relationships, inner development, and spiritual life.

Rieger (1983) followed up 83 women who were classified as high or low in creativity on the basis of the Torrance Tests of Creative Thinking taken in grades 3–5 in 1958–64. Twenty years later the women were asked to describe various aspects of their lives including achievements in high school and subsequent years. The achievements were rated by judges. The women who had scored high on creativity in grade school reported more creative achievement, both of the type publicly recognized and that not publicly recognized.

Creativity and Aging

There is as yet no systematic longitudinal study of creativity after midlife. Some cross-sectional studies of ordinary individuals indicate that creative productivity, thinking, or attitudes decline between 20 and 80. Crosson and Robertson-Tchabo (1983) administered the Barron–Welsh Art Scale to 271 women artists and writers and to a control sample of graduate students not "manifestly creative". The comparison group confirmed the findings of Alpaugh and Birren (1977) that scores of teachers on this measure declined between ages 20 and 70. However, scores of the creative women showed no decline.

Archival research on outstanding creative achievement was revived in the late 1970s, particularly in the work of Simonton (1984). Studies of age and creative achievement have largely confirmed the conclusions of Lehman (1953), discussed earlier, but Simonton has produced a wealth of intriguing findings and hypotheses. His major hypothesis may be considered a cognitive, mathematical, and elaborated version of Beard's theory (1874) which Simonton exposits: Creativity is a function of enthusiasm (originality) and experience (judgment). Enthusiasm tends to peak early in life, but experience gradually increases, and the best equilibrium is attained at about 38–40. Simonton adds to this ideas such as that fields vary in the optimum relation between inspiration and experience, that careers that start late (or early) end late (or early), and that individuals with great creative potential (rich store of originality) last longer than those with less potential.

Psychobiographical Studies
One way in which psychobiographical studies have changed is that they are making use of differentiated conceptions of adulthood, stimulated by the work of Erikson. An example is the account by Elms (1981) of the relation between the writing of *Walden Two* and the periods in Skinner's life, that of an identity crisis at age 21, associated with difficulty in entering the adult world, and a midlife crisis at age 41.

Elms's article also illustrates recent interest in interpreting personality theories through analysis of the lives of the theorists (Atwood and Tomkins 1976; Homans 1979; Stolorow and Atwood 1979). A study of Harry Stack Sullivan (Alexander 1985) seems to me particularly stunning, not only for its effective demonstration of the relation between the life and the theoretical system but also for its instructive contradiction of much of what we say about the "creative personality". As a young adult, Sullivan sometimes performed poorly, and when he lost the social recognition for his intellect that was necessary to his precarious self-esteem, he broke down and "disappeared". According to Alexander, he was terrified of his unconscious, and his techniques are those he had experienced himself as reducing panic and avoiding confrontation.

The Creative Individual and the Milieu

Many psychologists who studied creativity in the 1950s and 1960s were interested in the influence of the environment on the individual. This was particularly true of those concerned with education, business, or research laboratories. For example, Andrews (1965) found that scores on the Remote Associates Test were related to technical contribution in scientists who were influential in the laboratory and able to initiate new activities. Among scientists in less favorable positions, however, the correlation was zero or negative.

Despite a practical interest in the environment and in such facilitating techniques as brainstorming, applied imagination (Osborn 1963), and Synectics (Gordon 1961), no influential conceptualization of the relation between the creative individual and the social milieu distracted personality psychologists from their focus on traits. Stein (1963) sketched a transaction approach to creativity. Perhaps he was ahead of his time. Berlyne (1960), Fiske and Maddi (1961), and others emphasized the importance of environmental stimulation, and this line of experimental research and conceptualization did have continuing influence.

Revival of Archival Studies

By the mid-1970s, however, the social context of creativity was receiving broader theoretical attention. One manifestation was a revival of interest in archival studies of the eminent. Albert (1975) argued that genius is not a pattern of traits but a social judgment, and that the importance of early starts, endurance, and productivity is at least in part that they keep the creative individual in view long enough to be remembered.

The Goertzels (1978) brought out a second book of summarized biographies and autobiographies of the eminent. They noted that some themes, such as the domineering and possessive mother, had declined in importance, and others, such as homosexuality, had become conspicuous.

Martindale (1975, 1984) performed word counts of samples of French and English poetry over a number of historical periods to test a psychological theory of literary change. Changes in art, he argued, are less attributable to social changes than to evolutionary forces endogenous to the social system that produces art. He postulated that the basic problem of the artist is to produce something new. Drawing on concepts from Kris, he proposed that artists can do this in two ways, by changes in inspiration (regression) or by changes in elaboration (i.e. a change in artistic style). He hypothesized that over time works of artists written in the same style would show increasing primary process cognition. When the limits of regression had been reached, there would be a loosening of stylistic rules and a reduction in primary process until the new style was established.

With the aid of his Regressive Imagery Dictionary and other indexes, he offers convincing evidence to support his theory. He said that the poet's role changes from one era to another, and that the theory is only marginally relevant to changes *within* a poet's career. Novelty is an important point of concentration *across* generations. More individuals with pathology are drawn into a field when there is a high level of regression than in other periods. Martindale has conducted similar studies of music and painting.

Simonton (1976, 1984) is less interested in intrapsychic processes than Martindale, but has addressed many issues related to social influences on creativity. He has developed procedures for hypothesis-testing with multivariate and cross-sectional time-series designs in archival research on eminence. As he is a prolific author, I can only illustrate his approach.

In a study of 2,012 Occidental thinkers from 580 B.C. to 1900 A.D. (Simonton 1976), he reports that eminent philosophers were more likely than the less eminent to cover a wide array of philosophical issues, to advocate extremist or minority positions, to combine these positions in unusual ways across issues, to be concerned with the dominant issues of the *previous* rather than the present or following generation, and be more vulnerable to political conditions. Their non-conforming behavior seemed to require the greater freedom of favorable conditions such as political stability and political fragmentation (the number of sovereign political units in the civilization). He found that the availability of role models had a negative effect, presumably because the potential genius was more likely to become a disciple of someone else when there was an abundance of role models.

Other Approaches

What is still needed is perhaps a systematic conceptualization of person-environment relationships and a set of measures with agreed upon validity. Holland's model (1973) of six types of vocational interest, personality, and past and present environments has stimulated a considerable body of research, some of it clearly relevant to creativity.

Helson (1978c), for example, compared 58 writers and 59 critics in children's literature on vocational interests, personality, work style, and past and present environments. Both writers and critics had their highest scores on artistic interests and their lowest scores on conventional interests, but writers scored higher than critics on artistic, investigative and realistic interests, and critics scored higher than writers on the social, enterprising, and conventional. These findings fit Holland's "hexagon" model nicely, and occupations of fathers of the two groups differed along similar lines. The writers and critics differed in personality as Holland would predict: the critics were more goal-oriented, forceful, and efficient, the authors more impulsive and individualistic. They differed also in their work environments: Most critics worked full-time in jobs offering considerable autonomy but requiring an office and formal contacts with others; most authors worked at home. The work style (authors more intuitive and better able to visualize scenes; critics broader and more balanced in critical view, more concerned with needs of child reader) and conditions of life were thus consistent with the roles of the two groups, the writer to create imaginative work, the critic to serve as gate-keeper mediating between writers and the public.

It would be interesting to see similar studies of supervisors and research personnel, with attention to personal interaction between the two groups.

Harrington (1984) espoused an ecological approach to creativity in which individuals are conceived as being in ongoing interaction with others in ecological niches that they make for themselves. Such an approach suggests new questions: What are the particular features of the niches that creative people occupy, and with whom and in what way do they interact to find, maintain, or change it? How does the "eco-system" function in rewarding (or failing to reward) creative achievement, or in adapting and realizing creative ideas?

Some researchers believe that the cultural conditions under which art and science are produced have changed so much that the individual creator and our customary ways of evaluating creative products are anachronistic (Dudek 1984). Others believe that the danger of nuclear war makes it important to summon as much creative awareness as possible from both experts and citizenry in the hope of reducing this threat. Toward this end, Barron and Bradley (1985) have studied the relationship between creative personality characteristics and belief about the "nuclear situation". Creative illumination may be regarded as an energy field intense enough to act simultaneously on many elements (Vargiu 1979). Perhaps when a group is changing a situation, or a social movement is changing society in a way that is new and adaptive, there is a stage of illumination and sense of empowerment in many individuals *at once* that makes the change possible. The work of Barron and

Bradley suggests that individuals who have complexity of outlook, independence of judgment, and originality are better able to imagine a new state of affairs, and thus may take the lead in bringing about an "illumination". How to obtain the power to create and keep it from being "power over" has been the concern of a number of activist thinkers who are interested in feminism, creativity, and the nuclear situation (Pavel 1984). The need to understand such processes can hardly be exaggerated.

DIFFERENTIATION AND INTEGRATION

From Galton to the present, researchers have asked how eminent or creative individuals differ from one field of endeavor to another. Studies agree that scientists are more controlled and less tense than artists, but there has been little empirical study of how the creative process differs in fields that are so different. It seems likely that creativity in art and literature involves the expression of affect in the exploration of the self more often and more centrally than in science, where the individual is usually concerned with effectively neutral problems (Suler 1980). If so, one might expect the creative process of scientists to involve less "regression", and that abilities of imaging and fantasy would be more clearly a part of superior synthetic functioning of the ego (Bush 1969). But there are large individual differences among creative artists and scientists.

Most studies of individual differences in creative style have been conducted within fields of endeavor. Recent examples include comparisons of pure and applied art students by Getzels and Csikszentmihalyi (1976); Katz's (1984) amplification of the patterns of research style (elucidator, artificer, etc.) identified by Gough and Woodworth (1960); and Kirton's (1976) useful distinction between adaptors and innovators.

In a study of work style in men and women mathematicians, Helson (1967, 1983) described two main cluster dimensions, one with the theme of order vs. disorder and emotionality, the other with the theme of assertive inventiveness vs. a preference for the settled and self-contained. She suggested that these themes represent "essential tensions" within the field of mathematics, similar to that of Kuhn (1963) described in science between respect for accumulated knowledge and readiness to depart from it. (Kuhn thought psychologists did not appreciate adequately the respect for tradition and order, with the consequence that their picture of the scientist was that of the inventor, who works largely outside scientific tradition and has little effect upon it.)

On a graph of cluster scores on the two dimensions, Helson found that the less creative Ss were preponderantly in the "orderly and settled" quadrant of work style. These seemed to be individuals who were attracted to mathematics because of its beauty and its clarity about what was right and what was wrong. They experienced little tension between the needs for order and change. Ss is the various quadrants tended to show distinctive "needs" on the ACL, and subgroups of women mathematicians, who were assessed at IPAR, were described distinctively by staff observers, differed in cognitive skills, and had different specialties—"orderly and settled" women having chosen the more traditional fields of mathematics. The subgroups also had distinctive patterns of background and current life-situation.

This study makes an attempt to show both what creative individuals within a field have in common and how they diverge, and to relate patterns of differences in personality and work style to fundamental characteristics of the field. The differentiation of subtypes both within and across fields has been handicapped by the lack of a psychological map of the areas of human endeavor. Holland's (1973) overview has been mentioned, and Gardner (1983) offered another. Gardner considers seven kinds of "intelligence" that meet criteria such as potential isolation by brain damage, the existence of prodigies, and an evolutionary history and plausibility. The seven are linguistic, musical, logical-mathematical, spatial, bodily-kinaesthetic, and two kinds of "personal intelligence". His descriptions of the activities and concerns of people of these different intelligences provides a rich overview of the terrain of human culture.

WHAT WE NEED, WHERE WE'VE BEEN, WHERE WE'RE GOING

Some researchers would like to have a consensus about what creativity is and what the main classes of variables are. Amabile (1983) offered a definition and distinguished domain factors, creativity factors, and task motivational factors in a componential approach to creativity that would seem to be useful for certain experimental studies and applied settings. But, as she says, her schema is not appropriate for the field of creativity as a whole. MacKinnon (1975) believed that creativity was too multifaceted a phenomenon to be precisely defined and organized. It was more like the title of a book, he said—a rubric under which a number of related topics naturally fall.

For this chapter I chose to describe topics that have engaged thinkers and researchers who studied the creative person. We saw that psychoanalysts were especially interested in questions about psychopathology and the relation between conscious and unconscious. Galton and his successors studied the influence of genetics, field, and the early environment. Personality psychologists of the period between 1950 and 1975 were interested in the relation between creativity and intelligence, in personality characteristics of outstanding creative individuals across fields, and in various issues of conceptualization and measurement. In the last 10–15 years there has been more attention to social influence on the creative or potentially creative person, the development of the creative individual over time, and attempts to understand not only similarities but individual differences among creative people within and across fields of endeavor.

The old issues continue to receive attention. The new work may support earlier findings, but it usually adds dimensionality and perspective. With experience, we have learned that ratings and other measures of creativity can be very reliable in some circumstances, but in others it may be advisable to study another sample, devise a different measure, or measure something other than creativity.

The vitality in the field today comes from real-life studies that contextualize, rather than compartmentalize, creative behavior. The use of historical, developmental, and ecological contexts, in combination with our tools for measuring creativity and personality, should enable us to see more clearly which persons are creative, how, where, and why.

REFERENCES

Albert, R. S. 1975: Toward a behavioral definition of genius. *American Psychologist*, *30*, 140–51.
Albert, R. S. (Ed.) 1983: *Genius and eminence*. New York: Pergamon.
Alexander, I. 1985, June: On the life and work of Harry Stack Sullivan: An inquiry into unanswered questions. Paper presented at the meeting of the Society for Personology, Chicago.
Allport, G. W. 1937: *Personality*. New York: Holt, Rinehart & Winston.
Allport, G W., Vernon, P. E. & Lindzey, G. 1960: *Study of values* (3rd. ed.): *Manual of Directions*. Boston: Houghton Mifflin.
Alpaugh, P. K. & Birren, J. E. 1977: Variables affecting creative contributions across the adult life span. *Human Development*, *20*, 240–8.
Amabile, T. M. 1983: The social psychology of creativity: A componential conceptualization. *Journal of Personality and Social Psychology*, *45*, 357–76.
Andrews, F. M. 1965: Factors affecting the manifestation of creative ability by scientists. *Journal of Personality*, *33*, 140–52.

Atwood, G. E. & Tomkins, S. S. 1976: On the subjectivity of personality theory. *Journal of the History of the Behavioral Sciences*, *12*, 166–77.

Bachtold, L. M. & Werner, E. E. 1970: Creative psychologists: Gifted women. *American Psychologists*, *25*, 234–43.

Barron, F. 1963: *Creativity and psychological health*. New York: Van Nostrand.

Barron, F. 1965: The psychology of creativity. In *New directions in psychology II* 1–134. New York: Holt, Rinehart & Winston.

Barron, F. & Bradley, P. 1985: Opposed philosophies in the nuclear gun-pointing situation: Development of a scale and exploration of its meaning. Manuscript submitted for publication.

Barron, F. & Harrington, D. M. 1981: Creativity, intelligence, and personality. *Annual Review of Psychology*, *32*, 439–76.

Barron, F. & Welsh, G. S. 1952: Artistic perception as a factor in personality style: Its measurement by a figure preference test. *Journal of Psychology*, *33*, 199–203.

Beard, G. 1874: *Legal responsibility in old age*. New York: Russell.

Berlyne, D. 1960: *Conflict, arousal, and curiosity*. New York: McGraw-Hill.

Block, J. 1961: *The Q-sort method in personality assessment and psychiatric research*. Springfield: Charles C. Thomas.

Bloom, B. S. 1985: *Developing talent in young people*. New York: Ballantine.

Bush, M. 1969: Psychoanalysis and scientific creativity with special reference to regression in the service of the ego. *Journal of the American Psychoanalytic Association*, *17*, 136–90.

Castle, C. S. 1913: A statistical study of eminent women. *Columbia University contributions to philosophy and psychology*, Vol. 22. New York: Science Press.

Cattell, J. M. 1906: A statistical study of American men of science. II. The measurement of scientific merit. *Science*, *24*, 699–707.

Cattell, R. B. & Butcher, H. J. 1970: Creativity and personality. In P. E. Vernon (Ed.), *Creativity* (pp. 312–326). Middlesex, England: Penguin.

Cattell, R. B. & Drevdahl, J. E. 1955: A comparison of the personality profile (16 PF) of eminent researchers with that of eminent teachers and administrators, and of the general population. *British Journal of Psychology*, *46*, 248–61.

Cox, C. M. 1926: *Genetic studies of genius*, Vol. II. *The early mental traits of three hundred geniuses*. Stanford, Calif.: Stanford University Press.

Crosson, C. W. & Robertson-Tchabo, E. A. 1983: Age and preference for complexity among manifestly creative women. *Human Development*, *26*, 149–55.

Crutchfield, R. S. 1955: Conformity and character. *American Psychologist*, *10*, 191–8.

Dennis, W. 1958: The age decrement in outstanding scientific contributions: Fact or artifact? *American Psychologist*, *13*, 457–60.

Domino, G. 1976: Primary process thinking in dream reports as related to creative achievement. *Journal of Consulting and Clinical Psychology*, *44*, 929–32.

Dudek, S. Z. 1984: Creativity: Neither oblique nor equicrural . . . but all and none at once. *Division 10 Newsletter*, Winter, pp. 1–14.

Ehrenzweig, A. 1953 (1965): *The psychoanalysis of artistic vision and hearing: An introduction to a theory of unconscious perception*. New York: Braziller.

Eiduson, B. T. & Beckman, L. 1973: *Science as a career choice: Theoretical and empirical studies*. New York: Russell Sage Foundation.

Ellis, H. 1904: *A study of British genius*. London: Hurst & Blackett.

Elms, A. 1981: Skinner's dark year and Walden II. *American Psychologist*, *36*, 470–9.

Erikson, E. H. 1958: *Young man Luther*. New York: Norton.

Erikson, E. H. 1969: *Gandhi's truth*. New York: Norton.

Feldman, D. 1980: *Beyond universals in cognitive development*. Norwood, N.J.: Ablex.

Fiske, D. W. & Maddi, S. R. 1961: *Functions of varied experience*. Homewood, Ill.: Dorsey.

Freud, S. 1905: Jokes and their relation to the unconscious. *Standard edition of the complete psychological works of Sigmund Freud*, *8*. London: Hogarth.

Freud, S. 1908: The relation of the poet to day-dreaming. *Collected Papers*, *4*, 173–83. London: Hogarth.

Freud, S. 1910: Leonardo da Vinci and a memory of his childhood. *Standard edition of the complete psychological works of Sigmund Freud*, *12*, 3–82. London: Hogarth.

Galton, F. 1869: *Hereditary genius: An inquiry into its laws and consequences*. New York: Macmillan.

Gardner, H. 1983: *Frames of mind: The theory of multiple intelligences*. New York: Basic Books.

Gedo, J. E. 1983: *Portraits of the artist: Psychoanalysis of creativity and its vicissitudes*. New York: Guilford.

George, A. L. & George, J. L. 1956: *Woodrow Wilson and Colonel House: A personality study*. New York: Dover.

Getzels, J. W. & Csikszentmihalyi, M. 1976: *The creative vision: A longitudinal study of problem finding in art*. New York: Wiley.

Getzels, J. W. & Jackson, P. W. 1962: *Creativity and intelligence*. New York: Wiley.

Gilbert, S. M. & Guber, S. 1979: *The madwoman in the attic: The woman writer and the nineteenth-century literary imagination*. New Haven: Yale University Press.

Golann, S. E. 1963: The psychological study of creativity. *Psychological Bulletin*, *60*, 548–65.

Gordon, W. J. 1961: *Synectics: The development of creative capacity*. New York: Harpers.

Goertzel, V. & Goertzel, M. G. 1962: *Cradles of eminence*. Boston: Little, Brown.

Goertzel, M. G., Goertzel, V. & Goertzel, T. G. 1978: *300 eminent personalities*. San Francisco: Jossey-Bass.

Gough, H. G. 1957/in press: *Manual for the California Psychological Inventory*. Palo Alto, California: Consulting Psychologists Press.

Gough, H. G. 1957: Imagination—undeveloped resource. *Proceedings, First Conference on Research Developments in Personnel Management* (pp. 4–10), Los Angeles: University of California Institute of Industrial Relations.

Gough, H. G. 1960: The Adjective Check List as a personality assessment research technique. *Psychological Reports*, *6*, 107–22.

Gough, H. G. 1976: What happens to creative medical students? *Journal of Medical Education*, *61*, 348–51.

Gough, H. G. 1979: A creative personality scale for the Adjective Check List. *Journal of Personality and Social Psychology*, *37*, 1398–1405.

Gough, H. G. & Heilbrun, A. B., Jr. 1980: *The Adjective Check List manual*. Palo Alto, California: Consulting Psychologists Press.

Gough, H. G. & Woodworth, D. G. 1960: Stylistic variations among professional research scientists. *Journal of Psychology*, *49*, 87–98.

Greenacre, P. 1957: The childhood of the artist: Libidinal phase development and giftedness. *The Psychoanalytic Study of the Child*, Vol. 12. New York: International University Press.

Guilford, J. P. 1950: Creativity. *American Psychologist*, *14*, 469–79.

Hall, W. B. 1972: A technique for assessing aesthetic predispositions: Mosaic construction test. *Journal of Creative Behavior*, *6*, 225–35.

Hathaway, S. R. & McKinley, J. C. 1943: *Manual for the Minnesota Multiphasic Personality Inventory*. Minneapolis: University of Minnesota Press.

Harrington, D. M. 1984, October: An ecological approach to creativity. Colloquium, Institute of Personality Assessment and Research. University of California, Berkeley.

Harrington, D. M. & Andersen, S. M. 1979: Creativity, masculinity, femininity, and three models of androgyny. *Journal of Personality and Social Psychology*, *41*, 744–57.

Harrington, D. M., Block, J. & Block, J. H. 1983: Predicting creativity in pre-adolescence from divergent thinking in early childhood. *Journal of Personality and Social Psychology*, 45, 609–23.

Helson, R. 1967a: Personality characteristics and developmental history of creative college women. *Genetic Psychology Monographs*, 76, 205–56.

Helson, R. 1967b: Sex differences in creative style. *Journal of Personality*, 35, 214–33.

Helson, R. 1970: Sex-specific patterns in creative literary fantasy. *Journal of Personality*, 38, 344–63.

Helson, R. 1971: Women mathematicians and the creative personality. *Journal of Consulting and Clinical Psychology*, 36, 210–20.

Helson, R. 1978a: Creativity in women. In J. Sherman & F. Denmark (Eds.), *The psychology of women: Future directions in research* (pp. 553–604). New York: Psychological Dimensions.

Helson, R. 1978b: Experiences of authors in writing fantasy: Two relationships between creative process and product. *Journal of Altered States of Consciousness*, 3, 310–26.

Helson, R. 1978c: Writers and critics: Two types of vocational consciousness in the art system. *Journal of Vocational Behavior*, 12, 351–63.

Helson, R. 1983: Creative mathematicians. In R. S. Albert (Ed.), *Genius and eminence* (pp. 311–30). New York: Pergamon.

Helson, R. 1985: Which of those young women with creative potential became productive? Personality in college and characteristics of parents. In R. Hogan & W. H. Jones (Eds.), *Perspectives in personality: Theory, measurement, and interpersonal dynamics*. Vol. 1, (pp. 49–80). Greenwich, Connecticut: JAI Press, Inc.

Helson, R. (in press): Which of those young women with creative potential became productive? II. From college to midlife. In R. Hogan & W. H. Jones (Eds.), *Perspectives in personality: Theory, measurement, and interpersonal dynamics*. Greenwich, Connecticut: JAI Press, Inc.

Helson, R. & Crutchfield, R. S. 1970: Creative types in mathematics. *Journal of Personality*, 38, 177–97.

Hocevar, D. 1981: Measurement of creativity: Review and critique. *Journal of Personality Assessment*, 45, 450–64.

Hogan, R. 1980: The gifted adolescent. In J. Adelson (Ed.), *Handbook of adolescent psychology* (pp. 536–59). New York: Wiley.

Holland, J. L. 1973: *Making vocational choices*. Englewood Cliffs, New Jersey: Prentice-Hall.

Holland, J. L. & Astin, A. W. 1962: The prediction of the academic, artistic, scientific, and social achievement of undergraduates of superior scholastic aptitude. *Journal of Educational Psychology*, 53, 132–43.

Holland, J. L. & Nichols, R. 1964: Prediction of academic and extracurricular achievement in college. *Journal of Educational Psychology*, 55, 55–65.

Holt, R. R. 1956: Gauging primary and secondary process in Rorschach responses. *Journal of Projective Techniques*, 20, 14–25.

Holt, R. R. 1968: *Manual for the scoring of primary process manifestations in Rorschach responses*. New York: Research Center for Mental Health, New York University.

Homans, P. 1979: *Jung in context*. Chicago: University of Chicago Press.

Jackson, P. W. & Messick, S. 1965: The person, the product, and the response: Conceptual problems in the assessment of creativity. *Journal of Personality*, 35, 309–29.

Jaques, E. 1965: Death and the midlife crisis. *International Journal of Psychoanalysis*, 46, 502–14.

Jung, C. G. 1930: Psychology and literature. In *Collected Works*, 15, pp. 84–105. New York: Bollingen, 1966.

Kanner, A. D. 1976: Femininity and masculinity: Their relationship to creativity in male architects and their independence from each other. *Journal of Consulting and Clinical Psychology*, 44, 802–5.

Katz, A. N. 1983: Creativity and individual differences in asymmetrical cerebral hemispheric brain functioning. *Empirical Studies of the Arts*, 1, 3–16.

Katz, A. N. 1984: Creative styles: Relating tests of creativity to the work patterns of scientists. *Personality and Individual Differences*, 5, 281–92.

Keller, E. 1985: *Reflections on gender and science*. New York: Knopf.

Kirton, M. 1976: Adaptors and innovators: A description and measure. *Journal of Applied Psychology*, 61, 622–9.

Koestler, A. 1964: *The act of creation*. New York: Macmillan.

Kogan, N. 1973: Creativity and cognitive styles: A life-span perspective. In P. B. Baltes & K. W. Schaie (Eds.), *Life-span developmental psychology: Personality and socialization* (pp. 145–78). New York: Academic Press.

Kris, E. 1952: *Psychoanalytic explorations in art*. New York: International Universities Press.

Kubie, L. S. 1953: Some unsolved problems of the scientific career. *American Scientist*, XLI, 596–613.

Kubie, L. S. 1958: *Neurotic distortion of the creative process*. Lawrence, Kansas: University of Kansas Press.

Kuhn, T. S. 1963: The essential tension: Tradition and innovation in scientific research. In C. W. Taylor & F. Barron (Eds.), *Scientific creativity* (pp. 341–54). New York: Wiley.

Lehman, H. C. 1953: *Age and achievement*. Princeton: Princeton University Press.

Lombroso, C. 1891: *The man of genius*. London: Walter Scott.

MacKinnon, D. W. 1962: The nature and nurture of creative talent. *American Psychologist*, 17, 484–95.

MacKinnon, D. W. 1965: Personality and the realization of creative potential. *American Psychologist*, 20, 273–81.

MacKinnon, D. W. 1975: IPAR's contribution to the conceptualization and study of creativity. In I. A. Taylor & J. W. Getzels (Eds.), *Perspectives in creativity* (pp. 60–89). Chicago: Aldine.

Maddi, S. R. 1965: Motivational aspects of creativity. *Journal of Personality*, 33, 330–47.

Martindale, C. E. 1975: *Romantic progression: The psychology of literary history*. New York: Hemisphere.

Martindale, C. 1981: *Cognition and consciousness*. Homewood, Ill.: Dorsey.

Martindale, C. 1984: The evolution of aesthetic taste. In K. Gergen & M. Gergen (Eds.), *Historical social psychology* (pp. 347–70). Hillsdale, New Jersey: Erlbaum.

Martindale, C., Hines, D., Mitchell, L. & Covello, E. 1984: EEG alpha asymmetry and creativity. *Personality and Individual Differences*, 5, 77–86.

Maslow, A. H. 1959: Creativity in self-actualizing people. In H. H. Anderson (Ed.), *Creativity and its cultivation*. New York: Harper.

May, R. 1975: *The courage to create*. New York: Norton.

Mednick, S. 1962: The associative basis of the creative process. *Psychological Review*, 69, 220–32.

Murray, H. A. 1938: *Explorations in personality*. New York: Oxford University Press.

Murray, H. A. 1943: *Thematic apperception test manual*. Cambridge, Mass.: Harvard University Printing Office.

Myers, I. B. 1962: *Myers-Briggs type indicator manual*. Palo Alto, Calif.: Consulting Psychologists Press.

Nicholls, J. G. 1972: Creativity in the person who will never produce anything original and useful: The concept of creativity as a normally distributed trait. *American Psychologist*, 27, 717–27.

Nochlin, L. 1971: Why are there no great women artists? In V. Gornick & B. K. Moran (Eds.) *Women in sexist society* (pp. 480–510). New York: Basic Books.

Oden, M. 1968: A 40-year follow-up of giftedness: Fulfillment and unfulfillment. In R. S. Albert (Ed.), *Genius and eminence* (pp. 203–13). New York: Pergamon, 1983.

Osborn, A. F. 1963: *Applied imagination.* New York: Scribners.

Parloff, M. B., Datta, L., Klemen, M. & Handlon, J. H. 1968: Personality characteristics which differentiate creative male adolescents and adults. *Journal of Personality, 36,* 530–52.

Pavel, M. 1984: Interview on "womanpower". *Woman of Power Magazine,* 27–38.

Pine, F. 1962: Creativity and primary process: Sample variations. *Journal of Nervous and Mental Disease, 134,* 506–11.

Pine, F. & Holt, R. R. 1960: Creativity and primary process: A study of adaptive regression. *Journal of Abnormal and Social Psychology, 61,* 370–9.

Rank, P. 1945: *Will therapy and truth and reality.* New York: Knopf.

Raskin, E. 1936: A comparison of scientific and literary ability: A biographical study of eminent scientists and men of letters of the nineteenth century. *Journal of Abnormal and Social Psychology, 31,* 20–35.

Rieger, M. 1983: Life patterns and coping strategies in high and low creative women. *Journal for the Education of the Gifted, 6,* 98–110.

Roe, A. 1946: Artists and their work. *Journal of Personality, 15,* 1–40.

Roe, A. 1952: *The making of a scientist.* New York: Dodd, Mead.

Rogers, C. R. 1959: Toward a theory of creativity. In H. H. Anderson (Ed.), *Creativity and its cultivation* (pp. 69–82). New York: Harper.

Rorschach, H. 1942: *Psychodiagnostics.* Bern: Huber.

Rothenberg, A. 1971: The process of Janusian thinking in creativity. *Archives of General Psychiatry, 24,* 195–205.

Rothenberg, A. & Hausman, C. R. 1976: *The creativity question.* Durham, North Carolina: Duke University Press.

Simonton, D. K. 1976: Philosophical eminence, beliefs, and zeitgeist: An individual-generational analysis. *Journal of Personality and Social Psychology, 34,* 630–40.

Simonton, D. K. 1984: *Genius, creativity, and leadership.* Cambridge, Mass.: Harvard University Press.

Stanley, J. C., Keating, D. P. & Fox, L. H. (Eds.) 1974: *Mathematical talent: Discovery, description, and development.* Baltimore: Johns Hopkins University Press.

Stein, M. I. 1963: A transactional approach to creativity. In C. W. Taylor & F. Barron (Eds.), *Scientific creativity* (pp. 217–27). New York: Wiley.

Stein, M. I. & Heinze, S. J. 1960: *Creativity and the individual: Summaries of selected literature in psychology and psychiatry.* Glencoe, Illinois: Free Press.

Stolorow, R. D. & Atwood, G. E. 1979: *Faces in a cloud: Subjectivity in personality theory.* New York: Aronson.

Strong, E. K., Jr. 1938: *Vocational interest blank for men.* Palo Alto: Stanford University Press.

Suler, J. R. 1980: Primary process thinking and creativity. *Psychological Bulletin, 88,* 144–65.

Taylor, C. W., Smith, W. R. & Ghiselin, B. 1963: The creative and other contributions of one sample of research scientists. In C. W. Taylor & F. Barron (Eds.), *Scientific creativity* (pp. 53–76). New York: Wiley.

Terman, L. M. 1925: *Genetic studies of genius.* Vol. I: *Mental and physical traits of a thousand gifted children.* Stanford: Stanford University Press.

Torrance, E. P. 1962: *Guiding creative talent.* Englewood, New Jersey: Prentice-Hall.

Tryk, H. E. 1968: Assessment in the study of creativity. In P. McReynolds (Ed.), *Advances in psychological assessment,* Vol. 1. (pp. 34–54). Palo Alto, Calif.: Science and Behavior Books.

Vargiu, J. 1979: Creativity. *Synthesis, 3–4,* 17–53.

Wallach, M. & Kogan, N. 1965: *Modes of thinking in young children: A study of the creativity-intelligence distinction.* New York: Holt, Rinehart & Winston.

Wallach, M. & Wing, C. W. 1969: *The talented student: A validation of the creativity-intelligence distinction.* New York: Holt, Rinehart & Winston.

Welsh, G. S. 1975: *Personality and intelligence: A personality approach.* Chapel Hill, North Carolina: University of North Carolina Institute for Research in Social Sciences.

White, R. K. 1930: Note on the psychopathology of genius. *Journal of Social Psychology*, *1*, 311–15.

Wild, C. 1965: Creativity and adaptive regression. *Journal of Personality and Social Psychology*, *2*, 161–9.

3

Adaptors and Innovators: Problem Solvers in Organizations

M. J. KIRTON

BACKGROUND

The Adaption–Innovation theory defines and measures *style* of decision-making, problem-solving and, by implication, creativity (Kirton 1976, 1987a). The concept of *level* or *capacity* is explicitly and completely excluded from the definition of style which is the individual's characteristic manner of undertaking problem solving. Style includes the tendency either to accept the problem as consensually perceived or to take it, itself, together sometimes with its own setting (or in other words "the paradigm"), as processable parts of the problem to be solved. Style, therefore, also includes the kind of solution or cognitive production elicited by the problems in relation to the prevailing paradigms—being at the extremes either paradigm-supporting or paradigm-cracking. Care has been taken in validating the measure of the theory (the Kirton Adaption–Innovation theory, or KAI) to test whether, as the theory demands, it is unrelated (or, to be precise, insignificantly related) to measures of level. Two sets of data are available (Kirton 1987b) to demonstrate such orthogonality. The first are the results from four countries in five studies involving ten IQ tests and over 1,600 subjects, which yielded 13 correlations ranging between the magnitudes of −0.14 to +0.12. The second set of results were based on data collected by Torrance and Hrong (1980) in which 18 measures of creativity (including KAI) were intercorrelated. The usual muddled matrix emerged with correlations ranging between small minus to largish plus, without any obvious rationale. Adaption–Innovation theory does supply a rationale: psychometrically good scales of pure level should correlate significantly with each other but insignificantly with psychometrically good scales of pure style. Before factor analyzing the results, therefore, a prediction could be made that two factors would emerge with many scales making heavy (0.3+) contribution to only one factor (pure scales) and some others to both (impure scales). The remaining scales (unreliable or invalid) would be isolated each to their own factor. The results fitted perfectly to the prediction, with KAI appearing as a style measure with such others as Right–Left Brain Hemispheric Preference (see Table 1).

Table 1. Factor analysis Torrance Matrix.

Scales		Factors*	
		One	Two
Left hemisphere style of thinking (C)	Torrance, Reynolds, Bill, and Riegel 1978	0.84	—
Right hemisphere style of thinking (C)		0.76	—
Creative personality (WKPAY)	Khatena and Torrance 1976	0.72	—
KAI	Kirton 1976*	0.66	—
Creative self perception (SAM)	Khatena and Torrance 1976	0.57	—
Creative motivation	Torrance 1971	0.56	0.33
Cue test	Stein 1975	0.42	—
Originality (Rorschach)	Hertz 1946	0.35	—
TTCT fluency	Torrance 1974	—	0.87
TTCT originality		0.35	0.84
TTCT flexibility		0.33	0.69
TTCT elaboration		0.35	0.67
Possible jobs	Gershon and Guilford 1963	—	0.41
Similies	Schaefer 1971	—	0.36
Movement (Rorschach)	Hertz 1946	—	0.31
(Cumulative eigenvalue %)		(50.5)	(67.7)

* Only loadings ≥ 0.30 entered—KAI weakened as blind item scored in error. Analysis on data provided by kind permission of Professor E. Torrance (see Torrance and Horng 1980).
Tests not loading 0.30 on either factor: Integrated Hem. S.T. (heaviest loading on Factor 3); Seeing problems—Guilford 1969 (Factor 4); TTCT Creative strengths checklist (Factor 5).

The separation of the concepts of style and level permits the former to be related or even integrated into personality theory and herein later reference is made to many studies relating KAI to trait measures. So, according to the Adaption–Innovation theory, everyone can be located on a continuum ranging from highly adaptive to highly innovative according to their score on the Kirton

Table 2. Reliabilities.

(A)

Sample	Internal reliability coefficient	N	Country	Author
General	Cronbach alpha = 0.88	562	UK	Kirton 1976
Population	Cronbach alpha = 0.87	835	Italy	Prato Previde 1984
	K-R20 = 0.86	214	USA	Goldsmith 1985
Managers	K-R20 = 0.88	256	USA	Keller and Holland 1978a
	Cronbach alpha = 0.89	203	UK	De Ciantis 1987
	*Cronbach alpha = 0.91	142	UK	McCarthy 1987
	Cronbach alpha = 0.90	221	USA	Keller 1986
Students 1	Cronbach alpha = 0.85	123	USA	Ettlie and O'Keefe 1982
2	Cronbach alpha = 0.84	106	USA	Goldsmith 1984
2	Cronbach alpha = 0.86	98	USA	Goldsmith 1986
2	Cronbach alpha = 0.83	138	USA	Goldsmith and Matherly 1986a
2	Cronbach alpha = 0.87	123	USA	Goldsmith & Matherly 1986b
3	Cronbach alpha = 0.86	412	New Zealand	Kirton 1978b
4	Cronbach alpha = 0.76	374	Ireland	Hammond 1986

* All women, other management samples wholly or predominantly men.
1 Graduates and undergraduates. 2 Undergraduates. 3 Students 17–18 years old. 4 Students 15–19 years old.

(B)

Sample	Time interval	Test-retest coefficient		N	Country	Author
Students (17–18 years old)	7 months	0.82		64	New Zealand	Kirton 1978b
Managers	5–17 months	0.84		106	USA	Gryskiewicz et al. 1987
		Mean				
		(1)	(2)			
Students	4 months	91.2	91.1	121	South Africa	Pottas (unpub.)

Adaption–Innovation Inventory. The range of responses is relatively fixed and stable (see Table 2), and in the general population, approaches the normal curve distribution. For the purpose of clarity, the following descriptions are those that characterize individuals located at the ends of the continuum.

Adaptors characteristically produce a sufficiency of ideas, based closely on, but stretching, existing agreed definitions of the problem and likely solutions. They look at these in detail and proceed within the established mores (theories, policies, practices) of their organizations. Much of their effort when initiating change is in improving the "doing better" (which tends to dominate manage-ment, e.g. Drucker 1969).

Innovators, by contrast, have a taste for producing a proliferation of ideas, are more likely in the pursuit of change to reconstruct the problem, separating it from its enveloping accepted thought, paradigms and customary viewpoints, and emerge with much less expected, and probably less acceptable solutions (see Fig. 1). They are less concerned with "doing things better" than with "doing things differently". [Repeated factor analyses (Kirton 1987a) show that total adaptor–innovator scores are composed of three traits (i.e. characteristic preferences for, but not capacities of): suf-ficiency versus proliferation of originality; (personal) efficiency; and group-rule conformity. They are closely related respectively to Rogers' (1959) creative loner; Weber's (1970); and Merton's (1957) typical bureaucrat and bureaucratic behaviour.]

The development of the A–I theory began with observations made and conclusions reached as a result of a study of management initiative (Kirton 1961). The aim of this study was to investigate the ways in which ideas, which had led to radical changes in the companies studied, were developed and implemented. In each of the examples of initiative studied, the resulting changes had required the co-operation of many managers and others in more than one department.

Numerous examples of successful "corporate" initiative, such as the introduction of a new product or new accounting procedures were examined, and this analysis highlighted the stages through which such initiative passed on the way to becoming part of the accepted routine of the company, i.e. perception of the problem, analysis of the problem, analysis of the solution, agreement of change, acceptance of change, delegation, and finally implemen-tation. The study also looked at what went wrong at these various stages, and how the development of a particular initiative was thus affected. From this, a number of anomalies were thrown up that at the time remained unexplained.

Delays in Introducing Change

★ Despite the assertion of managers that they were collectively both sensitive to the need for changes and willing to embark on them, the time lag between the first public airing of most of the ideas studied, and the date on which an idea was clearly accepted as a possible course of action, was a matter of years—usually two or three. Conversely, a few were accepted almost immediately, with the bare minimum of in-depth analysis. (The size of proposed changes did not much affect this time-scale, although all the changes studied were large.)

Objections to New Ideas

★ All too often, the new idea had been formally blocked by a series of well-argued and reasoned objections which were upheld until some critical event—a "precipitating event"—occurred, so that none of these quondam cogent contrary arguments (lack of need, lack of resource, etc.) was ever heard again. Indeed, it appeared at times as if management had been hit by almost total collective amnesia concerning past objections.

Rejection of Individuals

★ There was a marked tendency for the majority of ideas which encountered opposition and delays to have been put forward by managers who were themselves unacceptable to an "establishment" group, not just before, but also after the ideas they advocated had not only become accepted, but even rated as highly successful. At the same time, other managers putting forward the more palatable ideas were themselves not only initially acceptable, but remained so, even if these ideas were later rejected or failed.

The A–I theory now offers a rational, measured explanation of these findings.

ADAPTORS AND INNOVATORS—TWO DIFFERENT STYLES OF THINKING

Adaptive solutions are those that depend directly and obviously on generally agreed paradigms, are more easily grasped intellectually, and therefore more readily accepted by most—by adaptors as well as the many innovators not so directly involved in the resolution of the problem under scrutiny. The familiar assumptions on which the solution depends are not under attack, and help "butter" the solution advanced, making it more palatable. Such derived ideas, being more readily acceptable, favourably affect the status of their authors, often even when they fail—and the authors of such ideas are much more likely to be themselves adaptors, characterized as being personally more acceptable to the "establishment" with whom they share those underlying familiar assumptions (Kirton 1976). Indeed, almost irrespective of their rank, they are likely to be part of that establishment, which in the past has led innovators to claim somewhat crudely that adaptors owe their success to agreeing with their bosses. However, Kirton (1977) conducted a study in which KAI scores were compared with a scale of Superior/Subordinate Identification (Chapman and Campbell 1957) in a sample of 93 middle managers. No connection was found between KAI scores and tendency to agree with one's boss. Instead a more subtle relationship is suggested, i.e. that those in the upper hierarchy are more likely to accept the same paradigms as their adaptor juniors, and that there is, therefore, a greater chance of agreement between them on broad issues and on approved courses of action. Where they disagree on detail within the accepted paradigm, innovators may be inclined to attach less significance to this and view the broad agreements reached as simple conformity.

It can thus be seen how failure of ideas is less damaging to the adaptor than to the innovator, since any erraneous assumptions upon which the adaptor's ideas were based were also shared with colleagues and other influential people. The consequence is that such failure is more likely to be written off as "bad luck" or due to "unforeseeable events", thereby directing the blame away from the individuals concerned.

In stark contrast to this, innovative ideas, not being as closely related to the group's prevailing, relevant paradigms, and even opposing such consensus views, are more strongly resisted, and their originators are liable to be treated with suspicion and even derision. This rejection of individuals tends to persist even after their ideas are adopted and acknowledged as successful. (It should be noted that both these and the further descriptions to come are put in a rather extreme form (as a heuristic device) and usually therefore occur in a somewhat less dramatic form.)

DIFFERENCES IN BEHAVIOR

Evidence is now accumulating from a number of studies (Beene et al. 1985; Carne and Kirton 1982; Ettlie and O'Keefe 1982; Kirton

1976, 1978a, 1987b; Kirton and De Ciantis 1986; Kirton and Hammond 1980; Goldsmith 1984, 1986a, 1986b, 1986c; Goldsmith and Matherly 1986a, 1986b; Goldsmith et al. 1986; Gryskiewicz 1982; Prato Previde 1984; Prato Previde and Carli 1987) that *personality* is implicated in these characteristic differences between adaptors and innovators.* [Also related is a recently tested measure of learning style (De Ciantis 1987); the same study shows management styles, however, are unrelated.] Indeed it must be so, since the way in which one thinks affects the way in which one behaves, and is seen to behave, in much the same way as there are differences in personality characteristics between those who are left-brain dominated and those who are right-brain dominated—the former being described as giving rise to methodical, planned thinking and the latter to more intuitive thinking (Torrance 1982). (There is a significant correlation between left–right brain preference scores and adaptation–innovation—see Table 1.) The personality characteristics of adaptors and innovators that are part of their cognitive style are described here.

Innovators are generally seen by adaptors as being abrasive and insensitive, despite the former's denial of these traits. This misunderstanding usually occurs because the innovator attacks the adaptor's theories and assumptions, both explicitly when he feels that the adaptor needs a push to hurry him in the right direction or to get him out of his rut, and implicitly by showing a disregard for the rules, conventions, standards of behavior, etc. Irritations can occur in seemingly minor ways, such as the innovator's marked preference to produce a plethora of ideas when "a couple of good ones would do for now". What is even more upsetting for the adaptor is the fact that on many an occasion the innovator does not even seem to be aware of the havoc he is causing. Innovators may also appear abrasive to each other, since neither will show much respect for the other's theories, unless of course their two points of view happen temporarily to coincide. Adaptors can also be viewed pejoratively by innovators, suggesting that the more extreme types are far more likely to disagree than collaborate. Innovators tend to see adaptors as stuffy and unenterprising, wedded to systems, rules and norms which, however useful, are too restricting for their (the innovators') liking. The adaptors' preference towards producing "one or two good, relevant, obviously useful ideas" makes them look pedestrian in innovators' eyes. So often, innovators seem to overlook how much of the smooth

---

* Management styles, however, are unrelated (De Ciantis 1987).

IMPLICATIONS	ADAPTORS	INNOVATORS
For Problem Solving	Tend to take the problem as defined and generate novel, creative ideas aimed at "doing things better". Immediate high efficiency is the keynote of high adaptors.	Tend to redefine generally agreed problems, breaking previously perceived restraints, generating solutions aimed at "doing things differently".
For Solutions	Adaptors generally generate a few wellchosen and relevant solutions, that they generally find sufficient but which sometimes fail to contain ideas needed to break the existing pattern completely.	Innovators produce numerous ideas many of which may not be either obvious or acceptable to others. Such a pool often contains ideas, if they can be identified, that may crack hitherto intractable problems.
For Policies	Prefer well-established, structured situations. Best at incorporating new data or events into existing structures of policies.	Prefer unstructured situations. Use new data as opportunities to set new structures or policies accepting the greater attendant risk.
For Organizational "Fit"	Essential to the ongoing functions, but in times of unexpected changes may have some difficulty moving out of their established role.	Essential in times of change or crisis, but may have some trouble applying themselves to ongoing organizational demands
For Potential Creativity	The Kirton Inventory is a measure of style but not level or capacity of creative problem solving. Adaptors and innovators are both capable of generating original, creative solutions, but which reflect their different overall approaches to problem solving.	
For Collaboration	Adaptors and innovators do not readily get on, especially if they are extreme scorers. Middle scorers have the disadvantage that they do not easily reach the heights of adaption or innovation as do extreme scorers. This, conversely is a positive advantage in a team where they can more easily act as "bridgers", forming the consensus group and getting the best (if skilful) out of clashing extreme scorers.	
For Perceived Behaviour	Seen by Innovators: as sound, conforming, safe, predictable, relevant, inflexible, wedded to the system, intolerant of ambiguity.	Seen by Adaptors: as unsound, impractical, risky, abrasive; often shocking their opposites and creating dissonance.

© M.J. KIRTON 1985

Fig. 1. Characteristics of adaptors and innovators.

running of all around them depends on good adaptiveness (Merton 1957; Weber 1970), but are acutely aware of the less acceptable face of efficient bureaucracy. Disregard of convention when in pursuit of their own ideas has the effect of isolating innovators in a similar way to Rogers' (1959) creative loner.

While innovators find it difficult to combine with others, adaptors find it easier. The latter will more rapidly establish common agreed ground, assumptions, guidelines and accepted practices on which to found their collaboration. Innovators also have to do these things in order to fit at all into a company, but they are less good at doing so, less concerned with finding out the anomalies within a system, and less likely to stick to the patterns they help form. This is at once the innovator's weakness and source of potential advantage.

WHERE ARE THE INNOVATORS AND THE ADAPTORS?

As already noted, much research has been devoted to detailed personality description of these two cognitive styles. More recently, attention has been focused on the issue of how they are distributed and whether any distinctive patterns emerge. It has been found from a large number of studies that KAI scores are by no means haphazardly distributed. Individuals' scores are derived from a 32-item inventory, giving a theoretical range of 32–160, and mean of 96. The observed range is slightly more restricted, 46–146, based on over 1,000 subjects. [General population samples in the U.K.; 2,000 more subjects in Italy and U.S.A. are all in close accord.] The observed mean is near to 95 and the distribution conforms almost exactly to a normal curve. The studies have also shown that variations by identifiable subjects are predictable, their mean shifting from the population mean in accordance with the theory. However, the groups' range of scores is rarely restricted—even smallish groups showing ranges of approximately 70–120—a finding with important implications for change, against the background of differences found at cultural level, at organizational level, between jobs, between departments, and between individuals within departments. This is a somewhat arbitrary grouping since norms of cognitive style can be detected wherever a group of people define themselves as different or distinct from others, by whatever criteria they choose, be it type of work, religion, philosophy, etc. However, while allowing for a certain amount of overlap, the majority of research studies can be classified according to these groupings.

INNOVATORS AND ADAPTORS IN DIFFERENT CULTURES

A considerable amount of research information has been accumulating regarding the extent to which mean scores of different samples shift from culture to culture. The three principal works based on large general population samples (Kirton 1976; Goldsmith 1985; Prato Previde 1984) have produced means within a single point of one another (c. 95) and very similar standard deviations. Other published normative samples collected in Britain (Kirton 1980), U.S.A. (Ettlie and O'Keefe 1982; Gryskiewicz et al. 1987; Keller and Holland 1978a), Canada (Kirton 1980) and New Zealand (Kirton 1978b) have all produced remarkably similar means. When the KAI was validated on a sample of Eastern managers from Singapore and Malaysia (Thomson 1980) their mean score of 95 (s.d. 12.6; N = 145) were compatible with those of their Western counterparts, e.g. U.K. mostly male managerial sample had a mean of 97 (s.d. 16.9; N = 88) compared to male norms in general U.K. samples of 98 (s.d. 17; N = 290).

However, samples of Indian and Iranian managers (Hossaini 1981; Khaneja 1982) yielded more adaptive means, c. 91, with N = 622. Exactly similar adaptive norms were also found in work still in progress on a sample of black South African business students (Pottas, unpublished). Clearly there may be cultural differences in adaptor–innovator norms, but the possibility of select sampling cannot yet be ruled out. There is already evidence accumulating for a further speculation put forward by Kirton (1978c, 1980; also Kirton and Pender 1982) that people who are most willing to cross boundaries of any sort are likely to be more innovative, and the more boundaries there are and the more rigidly they are held, the higher the innovative score should be of those who cross. In the Thomson study, managers in Western-owned companies in Singapore scored higher in innovativeness than either those working for private local companies or those in the Civil Service; those in this last category had the most adaptive scores of the triad. Work on Indian managers (Dewan 1982) found that, as expected, entrepreneurs scored higher on the KAI than non-entrepreneurs (97.9 and 90.5 as opposed to 77.2 for Government Officers), but Indian women [women generally, in all the general population studies, have means about $\frac{1}{2}$ s.d. more adaptive than men] entrepreneurial managers were found to be even more innovative than their male counterparts. They had to cross two boundaries: they broke with tradition by becoming a manager in the first place, and they had succeeded in becoming a manager in a risky entrepreneurial business. A similar phenomenon has been found in Britain, the gap between women personnel managers and male counterparts being the same as that for women and men in the general population. Women engineering managers, however, are significantly more innovative than women generally, and men generally, as well as male engineering managers. The interpretation of these results (McCarthy 1987) is that British culture now accepts women managers in the personnel field, whereas women engineers are still in the realms of paradigm-cracking rarities.

INNOVATORS AND ADAPTORS IN DIFFERENT ORGANIZATIONS

Organizations in general (Mulkay 1972; Whyte 1957) and especially organizations which are large in size and budget (Swatez 1970; Veblen 1928) have a tendency to encourage bureaucracy and adaptation in order to minimize risk. It has been said by Weber (1970), Merton (1957) and Parsons (1951) that the aims of bureaucratic structure are precision, reliability and efficiency, and that the bureaucratic structure exerts constant pressure on officials to be methodical, prudent and disciplined, and to attain an unusual degree of conformity. These are the qualities that the adaptor–innovation theory attributes to the 'adaptor' personality. For the marked adaptor, the longer an institutional practice has existed, the more he [throughout for he, him, his read also she, her, hers] feels it can be taken for granted. So when confronted by the problem, he does not see it as a stimulus to question or change the structure in which the problem is embedded, but seeks a solution within that structure, in ways already tried and understood—ways which are safe, sure, predictable. He can be relied upon to carry out a thorough, disciplined search for ways to eliminate problems by "doing things better" with a minimum of risk and a maximum of continuity and stability. This behavior contrasts strongly with that of the marked innovator. The latter's solution, because it is less understood and its assumptions untested, appears more risky, less sound, involves more 'ripple-effect' changes in areas less obviously needing to be affected; in short, it brings about changes with outcomes that cannot be envisaged so precisely. This diminution of predictive certainty is unsettling and not to be undertaken lightly, if at all, by most people—but particularly by adaptors, who feel not only more loyal to consensus policy but less willing to jeopardize the integrity of the system (or even the institution). The innovator, in contrast to the adaptor, is liable to be less respectful to the views of others, more abrasive in the presentation of his solution, more at home in a turbulent environment, seen initially as less relevant in his thinking towards company needs (since his perceptions may differ as to what is needed), less concerned with people in the pursuit of his goals than adaptors readily tolerate. Tolerance of the innovator is thinnest when adaptors feel under pressure from the need for imminent radical change. Yet the innovators' very disadvantages to institutions make them as necessary as the adaptors' virtues in turn make them.

Every organization has its own particular "cognitive climate", and at any given time most of its key individuals reflect the particular mode (say that exhibited by those $\frac{1}{2}$ s.d. on either side of the group mean [$\frac{1}{2}$ s.d. = 8–9 points; an amount large enough to be "visible" (Kirton and McCarthy 1985)]). Sufficient evidence has been collected to enable predictions to be made not only about the direction of, but also the extent to which these shifts in KAI mean will occur from organization to organization. For example, it has been suggested by Kirton (1980) that the mean scores of managers who work in a particularly stable environment will incline more towards adaption, while the mean scores of those whose environment could be described as turbulent will tend towards innovation. This hypothesis was supported by a number of studies of which those of Dewan (1982) and Khanaja (1982) have already been noted. Thomson (1980), whose study has also already been noted, showed that a Singapore sample of middle-ranking Civil Servants were markedly adaptor-inclined (mean = 89, s.d. 10.5), whereas the means of a sample of managers in multinational companies were just as markedly innovator-inclined (mean = 107, s.d. 11.4). Both Gryskiewicz et al. (1987) and Holland (1986) suggest that bank employees are inclined to be adaptors; so are Local Government employees (Hayward and Everett 1983) and accountants (Gul 1986). Employees of R&D oriented companies, however, show the opposite inclination (Keller and Holland 1978b; Lowe and Taylor 1986). Two of these studies support and refine the hypothesis that, given time, the mean KAI score of a group will reflect its ethos. Both studies found that groups of new recruits had means away from those of the established group they were joining, however, within three (Holland) or at most five (Hayward and Everett) years, as a result of staff changes, the gaps between the means of the new groups and the established groups narrowed sharply. New work in progress by Holland and Bowskill (unpublished) is providing further support.

If there are predictable variations between companies, wherever selection has been allowed to operate for a sufficient length of time, then variations may be expected within a company as adaptors and innovators are placed in the part of the organization which suits them best. It is unlikely (as well as undesirable) that any organization is so monolithic in its structure and in the "demands" on its personnel that it produces a total conformity of personality profiles. This hypothesis was tested and supported by Kirton (1980) when adaptors were found to be more at home in departments of a company that must concentrate on solving problems which mainly emanate from within their departmental system (e.g. production) and innovators tend to be more numerous in departments that act as interfaces (e.g. sales, progress chasing). Studies by Keller & Holland (1978b) in American R&D departments found that adaptors and innovators had different roles in internal company communications; adaptors being more valued for communications on the workings of the company and innovators being more valued for communications on advanced technological information. Kirton (1980) also found that managers who tend to select themselves to go on courses (i.e. self-selected) will have significantly different mean KAI scores from the managers on courses who were just sent as part of the general scheme (i.e. personally unselected), the former being innovator-inclined. Members of three groups of courses were tested: one British "unselected", one British "selected" and one Canadian "selected". The results showed that the unselected managers scored significantly more adaptively than the selected groups. Among the Canadian sample of managers, there was sufficient information on their job titles to be able to divide them into two groups of occupations: those liable to be found in adaptor-oriented departments (e.g. line manager) and those liable to be found in innovator-oriented departments (e.g. personnel consultant). The latter group were found to be significantly more innovative than the former, having a mean of 116.4 for non-line managers as opposed to a mean of 100.14 for line managers. [Because of the nature of this course and selection system, both groups' means were displaced towards innovativeness; however, they retain their distance vis-à-vis each other.] These findings later led to a full-scale study (Kirton and Pender 1982) in which data on 2,375 subjects collected in 15 independent studies were cross-tabulated with reference to different occupational types and varying degrees of selection to course. Engineering instructors and apprentices were studied as examples of occupations involving a narrow range of paradigms, thorough rigid training and a closely structured environment, while research and development personnel were examined as examples of occupations involving a number of flexible paradigms and a relatively unstructured environment. The differences were large, significant and in the expected direction.

These variations which exist between companies and between occupational groups are also found within the relatively narrow boundaries of the job itself. For example, work in progress suggests that within a job there may be clear subsets whose tasks differ and whose cognitive styles differ, e.g. an examination of the job of quality control workers for a Local Government body revealed that the job contained two major aspects. One was the vital task of monitoring, and one was the task of solving anomalies which were thrown up in the system from time to time. The first of these tasks was carried out by an adaptively-inclined group, and the second by an innovative one. Gryskiewicz et al. (1987) have found that subsets of bank managers whose tasks made them interfaces between the bank and its clients had significantly more innovative means (110 s.d. 15) compared with their U.S. norm group (91 s.d. 14) and Holland's U.K. norm group (91 s.d. 17).

Foxall (1986) has found similar subsets in a number of groups, separating Cost Accountants, Technical Managers and Administrative General Managers from Financial Accounts, Managing Engineers and Directive General Managers. Gul (1986) found marked relationships between preferences between sub-specialisms in accounting and adaption–innovation scores. Such knowledge about jobs and who is inclined to do them could eventually lead to better integration of adaptors and innovators within a company. It is always leading to an examination of the role of cognitive style in the concept of organisational climate (Clapp & De Ciantis 1987, Kirton & McCarthy 1988).

BUYING NEW IDEAS AND PRODUCTS

Recent research has explored the relevance of adaptive–innovative cognitive styles to the adoption of new ideas and products. Mudd and McGrath (1985) investigated the adoption of teaching innovations. They found an encouraging relationship between adaption–innovation and the number of innovations adopted, although significant relationships were not found for the adoption of any particular new teaching device or method. Presumably too many other general and unique factors also intervened. Path-breaking work on the adoption of consumer innovations was at the same time being undertaken by Foxall (1987), who has since made important contributions to both empirical knowledge and the theoretical framework for the analysis of consumer innovativeness.

Although there are well-established reasons for believing consumer innovators (i.e. the first buyers of new products) to differ in personality terms from later adopters, the empirical evidence for such a relationship is extremely tenuous (note again Mudd and McGrath's finding for single ideas). Although hundreds of investigations have been undertaken, the correlations they have produced have been so low as to be "questionable or even meaningless" (Kassarjian and Sheffet 1982). Foxall argues that this has been due to researchers having chosen an inappropriate level of analysis, their insensitive use of tests validated in areas other than consumer behavior, and their lack of consideration of the degree of continuity (similarity) or discontinuity (radical difference) represented by different innovative brands and products.

His research indicates that consumers who buy discontinuously innovative brands and products are more innovative (in KAI terms) than purchasers of continuous items. However, purchasers of the largest *number* of new brands and products tend to be adaptive. This finding suggests, in line with KAI theory, that committed purchasers of a given brand or product class tend to be assiduous seekers-out of some particular class of novelty. The characteristic of assiduous (systematic) activity probably related to a paradigm (e.g. healthy eating), is more likely to be found in adaptors than innovators. The research which has been published (Foxall and Haskins 1986, 1987) suggests that models of consumer innovativeness which have recently appeared in the marketing literature (Midgley and Dowling 1975; Hirschman 1980) be modified to take full account of adaptive and innovative cognitive styles. Such a conclusion, when added to the pilot work of Mudd and McGrath, suggests that a theoretical structure can be formalized that is wider and more general than that contained in any of the three individual studies; this has already been suggested by Foxall and Haskins.

WHO ARE THE AGENTS FOR CHANGE?

It has already been noted that the mean adaptor–innovator score of a group may shift quite considerably depending on the population in question, whilst the range remains relatively stable. This suggests that many a person is part of a group whose mean adaptor-innovator score is markedly different from his own. There are three possible reasons why these individuals should be caught up in this potentially stressful situation: (a) they are in transit, for example, under training schemes; (b) they are trapped, unhappy, and may soon leave; [Hayward & Everitt, 1983; Lindsay, 1985] (c) they have found a niche which suits them and have developed a particular role identity. (These three categories should be regarded as fluid, since given a change in the individual's peer group, boss, department or even organizational outlook, he may well find himself shifting from one category to another.)

It is the identification of the third category which will most repay further investigation since it contains refinements of the A–I theory which have considerable practical implications, though these are as yet speculations and work is currently being undertaken to explore their ramifications more fully.

The individual who can successfully accept and be accepted into an environment alien to his own cognitive style must have particular survival characteristics, and it is those characteristics which make him a potential agent for change within that particular group. In order to effect a change an individual must first have job "know-how" which is also an important quality keeping him functioning as a valuable group member when major changes are not needed. He must also be able to gain the respect of his colleagues and superiors, and with this comes commensurate status, which is essential if he wants his ideas to be recognized. Lastly, if a person is embarked on a course of action for change, he will of course require the general capacity, e.g. leadership, management qualities, to carry out such a task. His different cognitive style gives him a powerful advantage over his colleagues in being able to anticipate events which others may not see (since due to their cognitive styles, they may not think to look in that direction).

Therefore, the agent for change can be seen as a competent individual who has enough skill to be successful in a particular environment (which he may in fact have made easier by selecting or being selected for tasks within the unit less alien to his cognitive style). At this point he plays a supportive role to the main thrust of the group with its contrasting cognitive style. Given a "precipitating event" however, (particularly if he has anticipated and prepared for it), the individual becomes at once a potential leader in a new situation. In order to be able to take advantage of this position, he must have personal qualities to bring to bear, management must have the insight to recognize the position, and management development must have also played its part. However, this may need to be reinforced by individual and group discussion or even counselling (e.g. Lindsay 1985) which makes use of an understanding of Adaption–Innovation theory.

It should be emphasized here that the change agent can be either an adaptor or an innovator, and this is solely determined by the group composition, so that if it is an innovator group, the change agent will be an adaptor, and vice versa. This discovery challenges traditional assumptions that heralding and initiating change is the innovator's perogative because a precipitating event could demand either an adaptive or innovative solution, depending on the original orientation of the group and the work. An example in which an adaptor is the change agent in a team of innovators might be where the precipitating event takes the form of a bank's refusal to give further financial support to a new business enterprise. At this stage the change agent (who may have been anticipating this event for months) is at hand with the facts, figures and a cost cutting contingency plan all neatly worked out. It is now that the personal qualities of know-how, respect, status and ability will be crucial for success. All this assumes that many groups will have means away from the centre. It seems likely that the more the mean is displaced in either direction, the harder it will be, the bigger the precipitating event needs to be, to pull the group back to the middle, which may be unfortunate both for the group and the change agent. However, an "unbalanced" team is what may be required at any particular time. To hold such a position and yet to be capable of flexibility is a key task of management to which this theory may make a contribution.

CONCLUSION

A convenient way of concluding is by a brief evaluation of the theory and the data it has generated, both at a level close to the individual and on a wider scale. For the individual, a better understanding of what contribution to problem solving he and his immediate colleagues make, each in their own characteristic way, is the basis of more mutual tolerance, especially in times of stress. This in turn may lead to better collaboration which may well be reinforced by greater success. The training needed to hasten that acquisition and the reinforcing of such insights are the work of particular people and departments supported by significant leaders. Much experience is developing at the Centre for Creative Leadership (Gryskiewicz et al. 1987) and elsewhere (see McHale and Flegg 1985, 1986) on ways of putting these insights across and in seeing them in operation. Training is helped by the knowledge (Kirton and McCarthy 1985) that people can readily be brought to give good estimates of their own and others relative positions on the KAI continuum. Widely heterogeneous groups are intrinsically more difficult to manage than homogeneous ones; wide differences in cognitive style between women managers and their (often) male colleagues lead to more reported stress (McCarthy 1987). Some heterogeneous groups may manage well enough over long periods, however, (Clapp and De Ciantis 1987) and education has a role to play in promoting reasonable harmony in the presence of useful diversity.

In a wider context, it is hoped that the Adaption–Innovation theory will offer an insight into the interactions between the individual, the organization and widespread change. By using the theory as an additional informational resource when forward planning, it may also be possible to anticipate and retain control of changes brought about by extraneous factors. Thoughts are being pursued (v.d. Molen 1985) which may throw light on how such changes might take place amid less imbalance and confusion, thereby rendering them more effective.

REFERENCES

Beene, J. M. Helton, J. R., Zelhart, P. F. & Markley, R. P. 1985: *Self-actualization and anxiety.* Symposium Presented at the 31st Annual Convention of the South Western Psychological Association.

Carne, J. C. & Kirton, M. J. 1982: Styles of creativity: Test score correlations between the Kirton Adaption–Innovation Inventory and the Myers-Briggs Type Indicator. *Psychological Reports, 50,* 31–6.

Chapman, L. J. & Campbell, D. T. 1957: An attempt to predict the performance of three-man teams from attitude measures. *Journal of Social Psychology, 46,* 277–86.

Clapp, R. G. & De Ciantis, S. M. 1987: The influence of organisational climate on the relationship between cognitive style and observed behaviour (in press).

De Ciantis, S. M. 1987: *The relationship of cognitive style to managerial behaviour and the learning of skills in 3-D managerial effectiveness training.* PhD thesis, the Hatfield Polytechnic.

Dewan, S. 1982: *Personality characteristics of entrepreneurs.* Ph.D. Thesis, Institute of Technology, Delhi.

Drucker, P. F. 1969: Management's new role. *Harvard Business Review, 47,* 49–54.

Ettlie, J. E. & O'Keefe, R. D. 1982: Innovative attitudes, values and intentions in organizations. *Journal of Management Studies, 19,* 163–82.

Foxall, G. R. 1986: Managers in transition: An empirical test of Kirton's Adaption–Innovation theory and its implications for the mid-career M.B.A. *Technovation, 4,* 219–32.

Foxall, G. R. 1987: Consumer innovativeness: Novelty-seeking, creativity and cognitive style. *Research in Consumer Behaviour, 3.*

Foxall, G. R. & Haskins, C. G. 1986: Cognitive style and consumer innovativeness an empirical test of Kirton's Adaption–Innovation Theory in the context of food purchasing. *European Journal of Marketing, 20,* (in press).

Foxall, G. R. & Haskins, C. G. 1987: Cognitive style and discontinuous consumption. *Food Marketing, 3.*

Gershon, A. & Guilford, J. P. 1963: *Possible jobs: scoring guide.* Orange, CA: Sheridan Psychological Services.

Goldsmith, R. E. 1985: A factorial composition of the KAI Inventory. *Educational and Psychological Measurement, 45,* 245–50.

Goldsmith, R. E. 1986a: Adaption–Innovation and cognitive complexity. *Journal of Psychology, 119,* 461–7.

Goldsmith, R. E. 1986b: Convergent validity of four innovativeness scales. *Educational and Psychological Measurement, 46,* 81–7.

Goldsmith, R. E. 1986c: Personality and Adaptive–Innovative problem solving. *Journal of Personality and Social Behaviour*, *1*, 95–106.

Goldsmith, R. E. 1987: Personality characteristics: association with Adaption–Innovation. *Journal of Psychology* (forthcoming).

Goldsmith, R. E. & Matherly, T. A. 1986a: The Kirton Adaption–Innovation Inventory: Faking and social desirability. A replication and extension. *Psychological Reports*, *58*, 269–70.

Goldsmith, R. E. & Matherley, T. A. 1986b: Seeking simple solutions. Assimilators and explorers, adaptors and innovators. *Journal of Psychology*, *120*, 149–55.

Goldsmith, R. E., Matherly, T. A. & Wheatley, W. J. 1986: Yeasaying and the Kirton Adaption–Innovation Inventory. *Educational and Psychological Measurement*, *46*, 433–6.

Gryskiewicz, S. S. 1982: *Creative leadership development and the Kirton Adaption–Innovation Inventory.* Paper at British Psychological Society Conference.

Gryskiewicz, S. S., Hills, D. W., Holt, K. & Hills, K. 1987: *Understanding managerial creativity: The Kirton Adaption–Innovation Inventory and other assessment measures.* Technical Report, Center for Creative Leadership, Greensboro, N.C.

Gul, F. A. 1986: Differences between Adaptors and Innovators attending accountancy courses on their preferences in work and curricula. *Journal of Accounting Education*, *4*, 203–9.

Hammond, S. M. 1986: Some pitfalls in the use of factor scores: the case of the Kirton Adaption–Innovation Inventory. *Personality and Individual Differences.*

Hayward, G. & Everett, C. 1983: Adaptors and Innovators: data from the Kirton Adaption–Innovation Inventory in a local authority setting. *Journal of Occupational Psychology*, *56*, 339–42.

Hertz, M. R. 1946: *Frequency tables to be used in scoring responses to the Rorschach Ink Blot Test.* (3rd Ed.). Cleveland, OH: Western Reserve University.

Hirchsman, E. C. 1980: Innovativeness, novelty seeking and consumer creativity. *Journal of Consumer Research*, *7*, 283–95.

Holland, P. A. 1987: Adaptors and Innovators: Application of the Kirton Adaption–Innovation Inventory to bank employees. *Psychological Reports*, *60*, 263–70.

Holland, P. A. & Bowskill, I. 1987: Unpublished correspondence. Loughborough University, England.

Hossaini, H. R. 1981: *Leadership effectiveness and cognitive style among Iranian and Indian middle managers.* Ph.D. Thesis, Institute of Technology, Delhi.

Kassarjian, H. H. & Sheffet, M. J. 1982: Personality and consumer behaviour: An update. In H. H. Kassarjian & T. S. Robertson, (Eds.) *Perspectives in Consumer Behaviour.* Illinois: Scott Foresman & Co.

Keller, R. T. 1986: Predictors of project group performance in research and development organisations. *Academy of Management Journal*, *29*, 715–26.

Keller, R. T. & Holland, W. E. 1978a: A cross-validation study of the Kirton Adaption–Innovation Inventory in three research and development organizations. *Applied Psychological Measurement*, *2*, 563–70.

Keller, R. T. & Holland, W. E. 1978b: Individual characteristics of innovativeness and communication in research and development organizations. *Journal of Applied Psychology*, *63*, 759–62.

Khaneja, D. 1982: *Relationship of the Adaption–Innovation continuum to achievement orientation in entrepreneurs and non-entrepreneurs.* Ph.D. Thesis, Institute of Technology, Delhi.

Khatena, J. & Torrance, E. P. 1976: *Manual for Khatena–Torrance creative perception inventory.* Chicago: Steolting Co.

Kirton, M. J. 1961: *Management initiative.* London: Acton Society Trust.

Kirton, M. J. 1976: Adaptors and innovators: a description and measure. *Journal of Applied Psychology*, *61*, 622–9.

Kirton, M. J. 1977: Adaptors and innovators and superior-subordinate identification. *Psychological Reports*, *41*, 289–90.

Kirton, M. J. 1978a: Field dependence and Adaption–Innovation theories. *Perceptual and Motor Skills*, *47*, 1239–45.

Kirton, M. J. 1978b: Have adaptors and innovators equal levels of creativity? *Psychological Reports*, *42*, 695–8.

Kirton, M. J. 1978c: Adaptors and innovators in culture clash. *Current Anthropology*, *19*, 611–12.

Kirton, M. J. 1980: Adaptors and innovators in organizations. *Human Relations*, *3*, 213–24.

Kirton, M. J. 1987a: *Manual of the Kirton Adaption–Innovation Inventory.* (2nd ed.). London: Occupational Research Centre.

Kirton, M. J. 1987b: Adaptors and innovators: cognitive style and personality. In S. G. Isaksen, *Frontiers of creativity research: beyond the basics*, Buffalo, New York: Bearly Limited.

Kirton, M. J. & De Ciantis, S. M. 1986: Cognitive style and personality: The Kirton Adaption–Innovation and Cattell's Sixteen Personality Factor Inventories. *Personality and Individual Differences*, *7*, 141–6.

Kirton, M. J. & Hammond, S. 1980: Levels of self actualization of adaptors and innovators. *Psychological Reports*, *46*, 1321–2.

Kirton, M. J. & McCarthy, R. 1985: Personal and group estimates of the Kirton inventory scores. *Psychological Reports*, *57*, 1067–70.

Kirton, M. J. & McCarthy, R. 1988: Cognitive climate and organization. *Journal of Occupational Psychology* (forthcoming).

Kirton, M. J. & Pender, S. R. 1982: The Adaption–Innovation continuum: occupational type and course selection. *Psychological Reports*, *51*, 883–6.

Lindsay, P. R. 1985: Counselling to resolve a clash of cognitive styles. *Technovation*, *3*, 57–67.

Lowe, E. A. & Taylor, W. G. K. 1986: The management of research in the life sciences: The characteristics of researchers. *R&D Management*, *16*, 45–61.

McCarthy, R. 1987: *The relationship of cognitive style with coping as a member of a minority group at work.* Ph.D. Thesis, Hatfield Polytechnic.

McHale, J. & Flegg, D. 1985: How Calamity Jane was put in her place. *Transition*, November, 14–16.

McHale, J. & Flegg, D. 1986: Innovators rule OK—or do they? *Training and Development Journal*, Oct. 10–13.

Merton, R. K. (Ed.) 1957: *Bureaucratic structure and personality*. In *Social Theory and Social Structure*. New York: Free Press of Glencoe.

van der Molen, P. P. 1985: The "compliance vs self-will" trait, incrowd-outcast selection and population cycles. Paper presented at the fifth meeting of the European Sociobiological Society, Oxford.

Midgley, D. F. & Dowling, G. R. 1978: Innovativeness: the concept and its measurement. *Journal of Consumer Research*, *4*, 229–42.

Mudd, S. & McGrath, R. A. 1984: *The validity of the Kirton Adaption–Innovation Inventory in predicting the adoption of teaching innovations.* Paper presented to KAI Users Group, Hatfield Polytechnic.

Mulkay, M. S. 1972: *The social process of innovation.* London: Macmillan.

Parsons, T. 1951: *The social system.* Glencoe: Free Press.

Prato Previde, G. 1984: Adattatori ed Innovatori: i risultati della standardizzazione italiana del KAI. *Ricerche di Psicologia*, *4*, 81–134.

Prato Previde, G. & Carli, M. 1987: Adaption–innovation typology and right–left hemispheric preferences. *Journal of Personality and Individual Differences.*

Pottas, C. D. 1985: Unpublished correspondence, University of Pretoria, South Africa.

Rogers, C. R. 1959: *Towards a theory of creativity.* In H. H. Anderson (Ed.), *Creativity and its cultivation.* New York: Harper.

Schaefer, C. E. 1971: *Similies test manual.* Goshen, New York: Research Psychologist Press.

Stein, M. I. 1975: *Manual: Physiognomic cue test.* New York: Behavioral Publications.

Swatez, G. M. 1970: The social organization of a university laboratory. *Minerva: A Review of Science Learning & Policy, VIII,* 36–58.

Thomson, D. 1980: Adaptors and innovators: A replication study on managers in Singapore and Malaysia. *Psychological Reports, 47,* 383–7.

Torrance, E. P. 1971: *Technical norms manual for the creative motivation scale.* Athens GA: Georgia Studies of Creative Behavior, University of Georgia.

Torrance, E. P. 1974: *Norms technical manual: Torrance tests of creative thinking.* Bensenville, Ill.: Scholastic Testing Service.

Torrance, E. P. 1982: Hemisphericity and creative functioning. *Journal of Research and Development in Education, 15,* 29–37.

Torrance, E. P. & Horng, R. Y. 1980: Creativity, style of learning and thinking characteristics of adaptors and innovators. *The Creative Child Adult Quarterly, V,* 80–5.

Torrance, E. P., Reynolds, C. R., Ball, O. E. & Riegel, T. R. 1978: *Revised norms—technical manual for your style of learning and thinking.* Athens, GA: Georgia Studies of Creative Behavior, University of Georgia.

Veblen, T. 1928: *The theory of the leisure class.* New York: Vanguard Press.

Weber, M. 1970: In H. H. Gerth & C. W. Mills (Eds. & trans.), *From Max Weber: essays in sociology,* London: Routledge & Kegan Paul.

Whyte, W. H. 1957: *The organization man.* London: Cape.

4

Problem Solving and Creativity

GEIR KAUFMANN

A NEW DEAL FOR PROBLEM SOLVING

Over the past two decades a sweeping shift of paradigm for research programs in psychology has come about (e.g. Dennett 1981). The "bottom up" psychology of Behaviorism, with its strategy of extrapolating from studies of microprocesses of learning to the complexities of human problem solving, has been supplanted by the "top down" psychology of Cognitivism where higher-order cognitive control processes are given primacy.

The basic research goal of the new school is to describe and explain the structure, dynamics, and capacities of the human information processing system (e.g. Boden 1979). Doing psychology from the vantage point of man as an active information processor naturally moves problem solving to a focal spot on the research map. Since creativity is an integral part of human problem solving, we may consequently expect a reawakening of interest in the experimental psychology of creative thinking which has largely been dormant since the pioneering research undertaken by Gestalt psychologists (e.g. Duncker 1945; Luchins 1942; Scheerer 1963; Wertheimer 1959).

It is certainly true that human problem solving is now a major preoccupation in contemporary cognitive psychology (e.g. Anderson 1985; Greeno and Simon 1985). However, the field of creative thinking has not enjoyed the beneficial spin-offs one might have expected. In fact, it seems that creativity has received more acts of courtesy than acts of courtship by cognitive psychologists (e.g. Simon 1978). A possible reason for this state of affairs may lie in the intrinsic complexity and elusive nature of the phenomenon. Considerable basic knowledge obtained under well-defined and well-controlled conditions may be required before the more ill-defined area of creative thinking becomes available for manageable treatment. However, the psychology of creative thinking was not born yesterday, and the fact that the field has been experimentally mapped in productive ways by psychologists of the previous generation makes one wonder if the issue of creative thinking may represent a deviant case for the computer inspired paradigm currently dominating contemporary research (e.g. Weizenbaum 1976). Nourishment for this suspicion is given by a recent general introductory text on problem solving written from an information processing point of view (Kahney 1986). Here, the creativity piece of the pie has been cut out completely, and no reference at all is made to the valuable research that exists in the area.

Consequently, the aim of the present treatment is two-fold: We want to provide a representative state-of-the-art review of concepts, theories, and research in the field of human problem solving with particular emphasis on the creativity region. Since this area is still underdeveloped, our ambitions also include an attempt to expand on the conceptual framework and general theory of problem solving in order to facilitate direction for research in the creativity domain of problem solving.

WHAT IS A PROBLEM?

Raaheim (1974) has pointed out that there is a considerable degree of consensus between the different attempts in the literature to define what is to be meant by a "problem". In all definitions it is emphasized that the individual has a problem when he has a goal, but is uncertain as to what series of action he should perform to reach it (e.g. Dewey 1910; Johnson 1955; Køhler 1927; Newell and Simon 1972; Woodworth and Schlosberg 1955).

In such a definition it is normally implied that a problem arises when the individual is confronted with a difficulty. We have previously argued (Kaufmann 1980) that this kind of definition is too narrow in limiting problems to the situation where the individual is set over against a presented difficulty. Particularly in the creativity domain it may often be the case that the difficulty is a result of comparing an existing situation with a *future, imagined* state of affairs that constitutes a desirable goal for problem solving (e.g. Wertheimer 1959). We have therefore argued (Kaufmann 1984a) that a satisfactory definition is to regard a problem as *a discrepancy between an existing situation and a desired state of affairs* (see also Pounds 1969). With this kind of definition at hand we are not limiting problem solving to tasks that present themselves with a "gap".

Aspects of Difficulty

In the literature, problems are frequently described as varying on a continuum from "well structured" to "ill-structured" (ISP), which are defined in terms of degree of definition in important respects (Simon 1973, p. 181). Mintzberg et al. (1976) define an ISP as a task calling for "decision processes that have not been encountered in quite the same form and for which no predetermined and explicit set of ordered responses exists . . ." (p. 246). However, the concept of an ISP itself seems to be ill-structured and in need of being unpacked into its basic constituents. More specifically, we will argue that it is important to distinguish between different determinants of ill-structuredness which may be quite distinct. We may here distinguish between at least three conceptually distinct stimulus conditions responsible for turning a task into an ISP. These are *novelty, complexity* and *ambiguity*. The justification for nuancing between these stimulus conditions as distinct sources of an ISP may become more apparent when we consider the example of what makes a jig-saw puzzle difficult (unstructured). One source of difficulty may be located in the unfamiliarity of the goal structure that is to be attained (novelty). A quite different ISP-producing condition would be the number of pieces that are to be put together to make up the puzzle (complexity). A third condition of difficulty would surface if the task is indeterminate in the sense that quite different goal structures may be visualized and it is hard to see which is the correct one (ambiguity). It is easy to see that these dimensions can be varied systematically and *independently* of each other. The important point that needs to be underscored here is that novelty, complexity, and ambiguity may call for the use of quite different capacities and strategies on the part of the problem solver. When operating with an indifferentiated concept of an ISP, important differentiations in the problem-solving domain may be lost in the blur. Turning to the creativity aspect in particular, it is reasonable to argue that the novelty component of difficulty is of primary importance and should be specifically focused in creativity research.

Still another source of difficulty that is not entirely reducible to the above-mentioned task determinants, and that may be of special interest to map in the creativity domain, may materialize in what we will term "deceptive problems". In many tasks the problem may lie in "apparent familiarity", when the presentation (or representation) of the task is such that a conventional, well-programed line of attack is suggested which is, however, not adequately tailored to the solution, and a new twist is necessary to meet the goal specifications. This problem condition seems well illustrated in the so-called Hatrack Problem—one of the pet laboratory tasks employed in the experimental study of thinking. In its original version, the Hatrack Problem confronts the subject with the concrete task of making a construction for hanging up a coat by means of two sticks and a C-clamp. The construction is to be built in the center of the room, and no other objects are allowed for the task. The correct solution consists of clamping the sticks together and wedging them between the floor and the ceiling. The clamp handle functions as the hook on which to hang the coat. On the basis of detailed observations, Raaheim (1974) argues that the problem of the experimental subjects may aptly be described as trying to reconstruct the ordinary floor-standing rack—a "direction" that is suggested by the task as presented which is, however, not a possible solution strategy with the means available. A new twist is necessary to construct a solution arrangement that fully meets the goal specifications.

Although a novel response is required, there is an important distinction to be made between handling a situation that is novel, and recognized as such, and a situation which looks familiar, but deceptively directs the problem solving to look for conventional solutions where a novel one is really requested. In the former case, search for new information and creative use of past experience is required. In the latter case, successful problem solving requires the individual to shake off a misleading representation and to realize that solutions along conventional lines do not adequately meet the goal specifications.

Maneuvering from an undifferentiated concept of ISP may, thus, entail the danger of glossing over important distinctions that must be made in the problem-solving domain. This confounded point of departure may also tempt researchers to make unwarranted generalizations from one kind of task domain to another with quite different properties, where conditions for success relate to a different set of processes and capacities. Scientific scrutiny of creative thinking therefore presupposes a clear conception of the properties that characterize that special class of problem solving.

In the following section an attempt will be made to point out the most central vital measures of this body of problem solving.

THE ANATOMY OF CREATIVITY

In our discussion above we implied that creativity is most intimately linked to problem solving that results in high-novelty solutions.

However, novelty in thought product does not constitute a sufficient condition for defining creativity. The weird ideas of a psychotic person may rank high in originality and novelty, but we would hardly regard them as creative. To justify the use of the term "creative thinking", a thought product also has to satisfy the criterion of having some *use* or *value*. This requirement may be fulfilled in the way of functional use, as in technical inventions, or in aesthetic value, as in artistic productions. In a lucid discussion of the concept of creative thinking, Newell, Shaw, and Simon (1979) suggest some additional criteria which may provide an opportunity for a more precise delimiting of the concept. These are:

★ Creative thinking is *unconventional* by requiring the modification or rejection of previously accepted ideas.

This defining characteristic is important to include since it legitimately brings into the line of creative thinking the kind of thinking required in what we have termed "deceptive tasks". Here the rejection of conventional lines of thinking is of primary importance for a successful solution to the problem.

★ Creative thinking normally feeds on high motivation and persistence and takes place either over a relatively long period of time—continuously or intermittently—or at high-level intensity.

The criterion suggested here seems well motivated. It is interesting to note that creative thinking is now placed in the category of "hot cognition" (e.g. Janis and Mann 1977). Including the dynamic dimension in the very definition of creativity may serve to raise questions that are normally absent in contemporary, computer-inspired research on problem solving (e.g. Kahney 1986).

★ Finally, Newell, Shaw, and Simon (1979) argue that a problem that requires creative thinking is normally vague and ill-defined, and that part of the task is to formulate the problem itself.

This points suggests that creative thinking is seen most clearly in the category of "constructed problems", where the main share of creativity may lie in the formulation of the problem itself. The criterion of creativity also applies, of course, to the process of identifying a productive problem definition in a presented problem.

The defining characteristics presented above seem to place creative thinking in its proper kinship relation in the family of problem solving and to safeguard against too straightforward generalization from the general body of problem-solving research to the special domain of creative thinking. In view of their careful consideration of the concept of creativity, it seems almost paradoxical when Newell, Shaw, and Simon (1979) fall prey to exactly this line of hazardous reasoning when they extrapolate from research findings on problems where the difficulty is mainly that of complexity (see p. 162), to the domain of creativity where task (and thought)/ *novelty* is the cardinal characteristic. Their main concern is with the issue of *reducing* complex problem spaces and what kind of heuristics are used for this purpose. As we will see, however, the problem is often the exact *opposite* in the case of creative thinking, where good solutions are blocked by searching in a too *narrow* space. The implicit assumption underlying such dubious generalizations as those made by Newell, Shaw, and Simon seems to be that ISP's constitute a unitary class of phenomena. While there may be important regularities pertaining to the general level of ISP's irrespective of differences in basic constituents, these are probably of a highly general nature and may not throw the kind of specific light on the sub-category of creative thinking that we are aiming at.

We will now move one step forward to consider the prevailing general theory of human problem solving. A major point in our present discussion is that the theory as presently worked out is not adequately developed to deal with the creativity aspect of problem solving. Particular emphasis will therefore be placed on expanding theory in order to further research in this domain.

THEORY OF PROBLEM SOLVING

The theory that dominates and directs research in the problem-solving field is worked out from an explicit information processing perspective, where the computer-metaphor is the cornerstone of the edifice (e.g. Kahney 1986; Mayer 1983; Simon 1978).

Components of the Theory

The basic point of departure in the information processing approach is given in the thesis that problem solving can be fruitfully understood as a sequence of information processes, where "operators" are applied to "problem states" to bridge a gap between an "initial state" and a "goal state". The "initial state" is the problem as given, where the starting conditions are the amount of information available, resources at hand, constraints, etc. The "goal state" designates the final situation to be attained with relevant goal specifications attached to it. Intermediate "problem states" are problem situations that are generated through processing between the initial and the goal state. "Operators" consist of all transformations that are regarded as legal by the problem solver, and that serve to change one state to another state. All possible problem states that result from every possible sequence of operator-application constitute a "basic problem space" which represents an omniscient observer's view of the structure of a problem. A major point in the theory of problem solving is that the "state action three" that arises from the basic problem space is—even in trivial problems—too large to handle for the problem solver. Thus, the individual has to create his own "representation" or "model" of the problem that reduces the amount of search to a manageable size limited to regions that look promising for the purpose of finding a workable solution. The term "problem space" is used to designate the area of search that is defined by the individual's representation of the problem. As will be seen, this aspect of problem solving is regarded as particularly sensitive, and much research has been directed at illuminating the problem representation process and its importance for subsequent problem-solving behavior. Another major source of research input has been allocated to the task of uncovering the major kinds of strategies used by the problem solver for the purpose of "pruning" state-action trees. Schematically we can describe the problem-solving strategies as shown in Figure 1. Input is given in the initial statement of the problem, which is translated to a "model" of the problem through the individual internal representation of it. The solution is reached through the application of suitable problem-solving strategies.

Information-Processing System

In contemporary information processing theories of problem solving, considerable emphasis is placed on the properties and workings of the memory system as the basic machinery involved in thinking and problem solving (e.g. Greeno 1973). The three main components of particular relevance for problem solving are:

Short-term memory (STM)—the component where external description of the problem is received.
Long-term memory (LTM)—past experience relevant to solving problems is stored here. Included are *operational* knowledge (i.e. rules, algorithms, and general strategies) and *semantic* knowledge (facts, knowledge about *kinds* of problems).

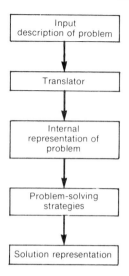

Fig. 1. The problem-solving process.

Working memory (WM). Information from STM and LTM is interchanged here and solution alternatives are generated and tested. This kind of model is schematically presented in Figure 2.

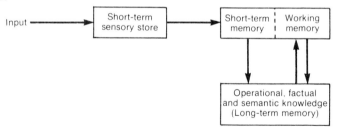

Fig. 2. Information processing system in problem solving. Adapted from Mayer (1983)

The information-processing picture of the problem-solving process involves a cyclus where a description of the problem containing information about the initial state, the goal state, and operators enters working memory through short-term memory. Stored knowledge about how to solve the problem enters working memory from LTM. The arrows from WM and STM to LTM suggest that more information from the outside world may be required in the course of problem solving, and the arrow from LTM to WM illustrates that WM may need more information from previous experiences as fuel for its search activity.

The reader may legitimately feel that somehow the head is lost in this model. Little attention has been paid to the translation process in this description. In cognitive theories (e.g. Anderson 1985), higher-order control operations are assumed to be performed through *executive functions*. Recently, Sternberg (1979, 1985) has filled this somewhat elusive concept with more specific substance as it relates to problem solving. Executive functions are meta-components in the information-processing system involving considerations and decisions about problem representation (i.e. how to understand the nature of the problem), choice of strategies, allocation of attentional resources, and monitoring of solution process (e.g. keep track of what one already has done).

The theory as briefly sketched here is completely general and resembles more a rather loosely formulated conceptual framework that can be used to think systematically about problem solving and to guide and describe empirical research. Certainly, no particular implications for gaining insight into the domain of creative thinking is provided. In the sequel we will make some effort at expanding this theoretical scheme with the aim of improving the point of departure for understanding human problem solving at a somewhat more differentiated level with a particular eye to the case of creative thinking.

THE CONCEPTUAL GEOGRAPHY OF PROBLEM SOLVING

In the discussion above we have argued that care must be taken not to operate with an undifferentiated concept of ISP. Problem spaces may not only be too large, but also too narrow and inappropriate. Executive functions must also be explicitly brought into the theory.

In Figure 3 we present a schematic picture of an expanded information processing model of human problem solving, where these considerations are worked in. The model represents an attempt to "slice" the problem-solving event into its main components and to capture the essential, general properties that have to be considered. Locus refers to the particular point of consideration in the various phases of problem solving, where attention may be given to the environment, the mediator of the problem (others or self), and the individual seen as an information processing system.

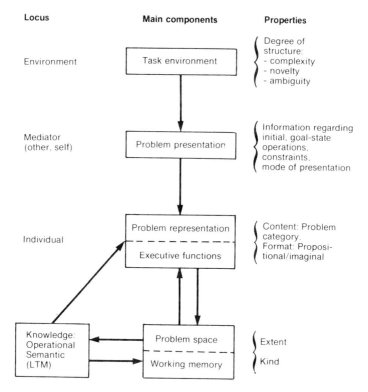

Fig. 3. A conceptual model of problem solving.

Main components in the corresponding order are task environment, problem presentation, problem representation, and problem space. Within the individual, the three main components of executive functions, working memory, and long-term memory are singled out as the most central ones in the problem-solving context. Executive functions are responsible mainly for meta-cognitive considerations related to problem representation, which in turn gives rise to a problem space that is the search field defined through the individual's subjective conception of the nature of the problem. Working memory operates at the operational level, where different solution alternatives are generated and tested. There is an interplay between the different components in the information processing system as pictured in Figure 3, which also allows for redefining of the problem (arrow from working memory to executive functions).

Most central to our discussion are the considerations on properties. As is seen, the concept of an ISP is differentiated into the basic constituents of complexity, novelty, and ambiguity for the reasons given above. On the mediator level, information is presented regarding initial, goal state, operators, and constraints (boundaries of problem, restrictions regarding resources to be used, etc.). The problem may also be presented in different modes, e.g. in abstract or concrete terms, which may influence the subsequent problem solving process to a considerable degree. The translation of the problem as given to a particular problem representation has to do with content, i.e. how the individual classifies the problem (i.e. what *kind* of problem it is). Also choice of mode of representation is relevant here. The problem solver may opt to deal with the problem in a linguistic/propositional format (words, mathematical formulas) or in an imaginal format (visualized, graphic representation). The conditions that may trigger the choice of different forms of representation will be discussed later on specifically related to the case of creative thinking. Of particular relevance for getting a somewhat more focused research aim on the creativity aspect is the distinction made with regard to the nature of problem spaces. Here we have distinguished between "extent" and "kind" of problem space. As maintained above there is a striking tendency in contemporary problem solving literature to hold forth the case of the *too large* problem space as the only interesting one. This aspect is, of course, prominent when the difficulty of the task lies in its *complexity*. However, it is not as obvious when the problem is one of novelty or ambiguity, or in the case of the "deceptive problem". Quite the contrary, the difficulties observed in these tasks may often be aptly described as related to a *too narrow* problem space, where the problem solver has to enlarge the space and see new possibilities in order to succeed in solving the problem. There is a clear tendency for human problem solvers, to be more completely discussed later on, to stick too closely to established lines of procedures when the problem requires new lines of attack. With ambiguity, the problem is rather one of choosing between different *kinds* of problem spaces that are conflicting alternatives. In the case of the "deceptive problem", an adequate description of the difficulties facing the problem solver is that the task directs the individual into the *wrong* problem space, where a solution is not possible to locate. We believe that these differentiations on the nature of the problem space and the difficulties that arise from it are important and may give a better focus for directing research on human problem solving that may lead to a clearer picture than the one presently available. In particular we feel that one reason why creative thinking has been so poorly treated by contemporary information processing oriented psychologists is the misguided over-emphasis on large problem spaces as the single most interesting case in problem solving. While complexity may certainly be involved in creative thinking problems, there is good reason to believe that this aspect is not the most interesting one, and that attempts to generalize to the creativity domain from evidence gathered on this kind of problem situation is not very illuminating, and may even be misguided in an important way.

PHASES OF PROBLEM SOLVING

The theory of problem solving has also been developed to account for the phases involved in the total problem solving event. The central questions to be answered here are whether the process of problem solving can be divided into distinct phases which exist across a wide variety of different tasks, and whether they follow a simple and orderly sequence. The answer to the first question seems to be in the affirmative, while the answer to the second is essentially negative. It seems to be possible, then, to identify lawful regularities in the form of distinct phases in problem solving, but these phases do not seem to follow a simple and straightforward sequence.

There is striking agreement in the literature describing the phases of a problem solving event. Normally, three major phases are identified. Johnson (1955) distinguishes between *preparation*—understanding and identifying the problem; *production*—development of different solution alternatives; and *judgment*—which involves choice of the best solution. In a series of experiments, Johnson and his collaborators have provided evidence that suggests that these three phases are empirically distinguishable and independent of each other in important ways. In one experiment (Johnson and Jennings 1963), the subjects were given a story-problem. Their task was to read the story to understand its plot ("preparation"), to suggest five titles for the story ("production"), and to choose the best title for the story ("judgment"). Time was recorded separately for each stage, and Johnson and Jennings examined the interrelationship between individual differences in performance during the different stages. The results showed no correlation between performance in the different stages, and thus indicate that the different stages are independent of each other. More evidence is needed, however, to examine whether this pattern is found in other types of tasks and if stage performance is correlated across tasks. An interesting implication of the Johnson and Jennings findings that deserves further inquiry is that there are important individual differences in profiles of problem solving ability. High ability in one phase does not seem to imply success in other phases of problem solving. Simon (1965, 1977) has suggested a trichotomy that is essentially commensurate with the Johnson formulation. "Intelligence" describes the phase of identifying the nature of the problem, "design" involves "inventing, developing and analyzing possible courses of action" (p. 41). The third phase deals with the selection of a particular course of action from those available. (When dealing with problems in a practical management context, Simon argues that we may distinguish a fourth phase termed "review", which involves evaluating past choices.) The finest and most extensive research on phases has probably been done by Mintzberg et al. (1976). On the basis of comprehensive studies of real-life problem solving, Mintzberg et al. were able to confirm the trichotomy theory of phases in problem solving. Mintzberg et al. distinguish the three major phases under the headings of "identification", "development", and "selection" and go on to give a detailed picture of the microstructure of the problem solving process by identifying seven recurring central "routines" within the tripartite structure.

The three phases seem to be related logically to form a strict sequence with "identification" first, followed by "development", ending in "selection". However, the logic of the situation only requires that *some* identification has to be made before development, and a certain minimal level of development has to precede selection. The evidence presented by Mintzberg et al. seems to demonstrate very clearly that a simple, straightforward sequence is the rare case. Normally, the cycle of phases is a lot more complex, and a high degree of overlapping occurs with lots of commuting between the different phases.

Apart from the importance of having a detailed descriptive model of the phases of problem solving, research in this category also stages the important question of where the most narrow bottlenecks in problem solving are located. The answer to these questions will probably vary a lot according to the particular aim of the problem solver (i.e. finding one workable solution vs. finding the best one possible vs. finding several good alternatives for an adequate solution, etc.). With the creativity aspect specifically in mind, the research that has been done in the context of these questions seems to point to some interesting answers of a more general nature.

According to Hayes (1978), what sets an ill-defined problem apart from a well-defined one is that "ill-defined problems require problem solvers to contribute to the *problem definition*" (p. 212, italics ours). Newell, Shaw, and Simon (1979) make the very same point when they claim that problems that require creative thinking are typically vague and ill-defined when initially posed and that an important part of the task is to formulate the problem itself. In a recent discussion of creative thinking, Perkins (1981) argues that the danger of "premature closure" is a major obstacle to creativity. Premature closure means that the problem is encapsulated in a too narrow perspective, which in turn prevents the consideration of alternative solution routes that may lead to high-quality, creative solutions to a problem. What these views have in common is the singling out of problem identification and definition as a particularly delicate phase in problem solving.

The view that choice of representation can make an important difference in our problem-solving performance has attracted quite a bit of research interest in contemporary experimental studies of thinking. To exemplify the point, we may consider the problem of

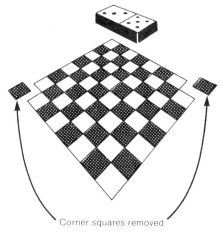

Fig. 4. The mutilated Checkerboard problem.

the mutilated checkerboard. As illustrated in Figure 2 the task consists of an ordinary checkerboard with 64 squares and a set of 32 rectangular dominoes. Each domino covers 2 checkerboard squares. Thus, the 32 dominoes can be arranged to cover the complete board. The problem is the following one: Suppose that two black squares are cut from opposite corners of the board, as shown in Figure 4, is it possible to cover the remaining 62 squares of the checkerboard by using exactly 31 dominoes?

Put this way, the problem is exceedingly difficult and very few can solve it (Hayes 1978). The correct answer is "No", for the following reason: Since each domino covers 1 white square and 1 black square, 31 dominoes will cover 31 white squares and 31 black squares. But the two corners that have been removed are both black. The mutilated checkerboard therefore has 32 white squares and 30 black ones. Thus it cannot be covered by 31 dominoes. It is easily seen that a correct identification of the problem gives the correct solution immediately. Hayes contrasts the Mutilated Checkerboard problem with a so-called problem-isomorph, i.e. the same problem described in a different context: "In a small but very proper Russian village, there were 32 bachelors and 32 unmarried women. Through timeless efforts, the village matchmaker succeeded in arranging 32 highly satisfactory marriages. The village was proud and happy. Then one Saturday night of drinking, two bachelors, in a test of strength, stuffed each other with pirogies and died. Can the matchmaker, through some quick arrangements, come up with 31 satisfactory marriages among the 62 survivors?" (Hayes 1978, p. 182).

This Matchmaker problem is an exact parallel to the Mutilated Checkerboard problem and is trivially easy to solve. This example rather dramatically illustrates that the same problem can differ vastly in difficulty when presented in different ways, thus underscoring the importance of problem definition for success in subsequent problem solving.

The example also illustrates the research strategy that has been used to study the impact of problem representation on performance. The same base problem is presented in various isomorphic versions and the effect on performance is measured. By using this experimental technique, Hayes and Simon (1974, 1976, 1977) have demonstrated that changes in problem representation may cause a large difference in performance (by a factor of 2). As Hayes (1978) points out, the effect is particularly prone to occur in "discovery tasks", where search activity is often of minimal importance after the "correct" (or productive) representation of the problem has been achieved (see also Wickelgren 1974).

The way research evidence stands today, then, it is reasonable to claim that the identification phase of problem solving is a particularly sensitive area and a potentially important bottleneck in the process of finding creative solutions to problems.

MAIN AREAS OF RESEARCH ON HUMAN PROBLEM SOLVING

We will now set our feet more firmly on empirical ground and consider the major areas of contemporary research on human problem solving with particular emphasis on the creativity aspect. Five areas of research will be singled out for special review and discussion. First we will discuss the implications of research on *capacity* limitations in the human information processing system for understanding the way people approach problems. A major focus of research has here been on the *strategies* people use to cope with ill-structured problems. As will be shown, research on the use of *symbolic tools* (e.g. verbal/imaginal) connects naturally with the strategy studies. More recently, a strong research emphasis has been placed on the knowledge base of problem solvers and the requirements for *expertise* in problem solving. Finally, we will provide a selective review of less system-oriented, empirical research on *antecedent and situational conditions* that either facilitate or inhibit the process of creative thinking in particular.

Capacity limitations and problem solving

A major source of direction for contemporary research on human problem solving is given in the theory of *bounded rationality* developed by Simon (1945, 1957, 1969, 1979, 1983). The theory has grown out of an attack on the classical Subjective Expected Utility (SEU) theory. The SEU theory assumes among other things that the individual will make decisions from an exhaustive set of alternative strategies, probability calculations of scenarios for the future associated with each strategy, and a policy of maximizing expected utility. Apart from the unrealistic assumptions of complete knowledge, Simon has also argued that the cognitive capacity assumptions implied in the SEU model are widely at odds with observed realities. As an alternative to the super-rationality assumed in SEU theory, he formulated the principle of bounded rationality: "The capacity of the human mind for formulating and solving complex problems is very small compared with the size of the problems whose solution is required for objectively rational behavior in the real world—or even for a reasonable approximation to such objective rationality" (1957, p. 198). Rather than maximizing, the individual will follow a strategy of *satisficing*, directed by a realistic aspiration level of what is "good enough". More precisely, the thinker is assumed to select the first alternative he encounters which meets some minimum standard of satisfaction.

Substituting satisficing for maximizing greatly reduces the demands upon the computational capabilities of the thinker.

Research on capacity limitations in the human information processing system has clearly corroborated the basic thesis in Simon's theory. There are important bottlenecks in the cognitive system that easily impose cognitive strain on the individual when confronted with a non-trivial task. Available evidence seems to show that only a small number of such limits exist, but these seem to be constant across a wide variety of tasks, and are of crucial importance for the processing capabilities involved in solving problems. Thus, insight into cognitive capacity limitations may have great implications for our understanding of basic features of problem-solving behavior.

Schiffrin (1978) has argued that many of the computational limitations observed in cognitive tasks may be traced back to basic limitations in short-term memory (including working memory). 7 ± 2 limits of information has been given as a "magical number" (Miller 1956) of the storage capacity of short-term memory. Recent research by Simon (1979) suggests that this is an overestimation, and that the true number is in the order of 5 units. Roughly, we may define a unit of information as the information contained in a unitary concept. The size of the unit may vary (letter, word, sentence) and it is interesting to note that the parameter seems to be constant across size of unit. To increase memory capacity the individual has to organize information in higher-order, unitary "packages" called "chunks of information". The letters U, C, B represent three units of information. Drawn together they may mean University of California, Berkeley, which is one unit. Such capacity-increasing "chunking" may be of great importance in tasks of memory and problem solving (e.g. Anderson 1985). Another bottleneck in the human cognitive system is given in the time it takes to transfer information from short- to long-term memory. Research reported by Simon (1979) indicates that it takes 8–10 seconds to transfer a new unit of information from short- to long-term memory. Again it is interesting to note that this parameter seems to be invariant across different types of information.

The important implication for problem solving is that the threshold for "information leak" is rather low. The following illustration will give the reader a taste of this problem. Try to solve the following task:

The task is to substitute numbers for letters, and perform the calculation correctly. One is certain to experience rather quickly that there are "many balls in the air", and that solution of the task is dependent on ability to mobilize efficient problem solving strategies that may guard against information leak.

According to information processing theory of problem solving, most non-trivial tasks contain an enormous number of paths to the goal. A basic difficulty in problem solving is thus that a problem solver with strongly limited computational capacities is confronted with a task that contains a vast number of possible solution paths. According to the theory of bounded rationality this makes it necessary for the individual to construct strongly simplified models of reality. The subjectively constructed model then serves the basis for the individual's problem solving activity (see also Johnson-Laird 1983). For most non-trivial problems, then, only a limited amount of search is possible within the realities of the problem-solver's world.

The critical problem is therefore to identify promising sequences of operations. Here the so-called heuristic methods come into play. The basic aim of heuristics is to aid the problem solver to simplify the problem by serving to identify the most likely solution paths.

With this point of departure it is easy to understand why the major share of contemporary research on problem solving has been devoted to uncover heuristic strategies used by problem solvers in their transactions with ill-structured task environments.

Strategies in Problem Solving

Newell (1969) has dealt with the issue of strategies in human problem solving from a particularly systematic point of view, and suggests a distinction between "weak" and "strong" methods be made. A strong method has high power in the sense of guaranteeing a solution to a problem. The generality of a strong method is low, however, and the need for information about the specific problem is high. Mathematical formulas for solving specific problems are good examples of this category of method. In an ill-structured problem, it is implied by definition that the information about the problem is low, and here the individual has to fall back on weak methods. A weak method makes very weak demands on the environment for information. The power of delivering a solution is low compared to strong methods, but it can be used across a wide variety of different tasks. Thus there seems to be an inverse relation between power and generality of problem solving methods, as illustrated in figure 5.

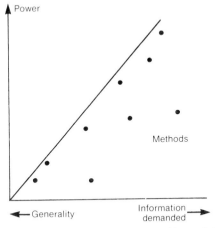

Fig. 5. Relation between power and generality of problem-solving methods. Source: A. Newell (1969).

As Simon (1981) points out, a science of problem solving should deal with the process at a high level of generality and concentrate on invariant (or relatively invariant) features. In line with such a dictum, research on strategies is aimed at uncovering methods of high generality that are regularly used over a wide specter of tasks.

Inspired by the theory of bounded rationality, research has been heavily focused on the heuristics used for the purpose of simplifying problems and confine search activity to the most promising solution paths. Here we have room only for a brief glimpse of some of the major strategies identified. Excellent reviews are found in Anderson (1975), Hayes (1978), Newell and Simon (1972), Wickelgren (1974), and Winston (1977).

Heuristics of Simplification
Hayes (1978) distinguishes between three varieties of search aimed at short-cutting the problem. These are: proximity methods, pattern matching, and planning.

Proximity Methods
The Hot and Cold Strategy. Problem solvers often behave on the principle of the "Hot and Cold" child's game, and look for "hot" signals to move closer to, and "cold" signs to move away from, a

search route. An everyday example of use of this strategy would be the location of another person in a large house by maneuvering on the basis of the loudness of the other person's voice.

Hill Climbing. This strategy derives its name from the analogy of climbing a hill under foggy conditions, where the climber chooses one step at a time that increases the altitude. A practical example of the hill-climbing strategy would be to improve the performance of a soccer team by taking one step at a time and evaluating its consequences. Altitude here corresponds to winning performance and standing in the league. We can take a single step at a time—for instance increase physical exercise—while others are kept constant. Operating along these lines, we can continue to improve the team by implementing the steps that help and rejecting those that do not. By a hill-climbining method, as with other heuristics, we should arrive at good solutions to problems although we may not get the best possible solutions.

Means–end Analysis. Means–end analysis is a method with considerable problem solving power and generality. It differs from those considered above by offering a choice among different *means* of approaching the goal. In means–end analysis, search is guided by attempts to *reduce the difference* between the initial state and the goal state, selecting from a set of available means.

Making a travel plan may exemplify the use of means–end analysis. If the goal is to go from Oslo to New York, the difference between the initial state and the goal is a large distance. This large distance can be eliminated by using an aeroplane, which is the first operator. But the precondition for using the first operator is having a ticket and getting to the airport. We now have two sub-problems. The problem of the ticket can be solved by using a travel agent, but then there is the problem of choosing the agent who will give the best deal. A new sub-problem then arises, and the process will continue until the solving of the subsidiary problems results in the solution of the overall problem.

This example also illustrates the important heuristic of the *sub-goal strategy*. Rather than attacking the problem head on, as a whole it is useful to follow the strategy of breaking it up into subsets of subsidiary sub-goals which are more well-structured than the total problem. The general structure of the means-end strategy is illustrated in Figure 6. The main purpose of the means-end analysis is to solve an ill-structured problem by reducing it to a series of smaller well-structured problems. Newell and Simon (1972) report

Goal: Transform object A into object B

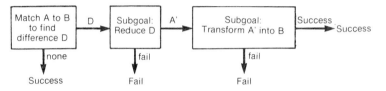

Goal: Reduce difference D between object A and object B

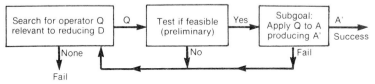

Goal: Apply operator Q to object A

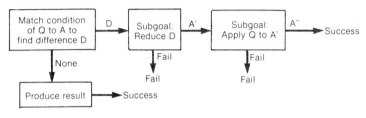

Fig. 6. Flow diagram of General Problem Solver methods. Source: A. Newell and H. A. Simon (1972).

that this is a procedure that is very frequently resorted to by their subjects presented with a variety of different problems.

Planning Methods

Another class of heuristics that may be used to confine search in complex task environments is the planning methods. Hayes (1978) distinguishes between three types of planning strategies.

Planning by Modeling. To guide complex activities in a rational way, it is useful to construct a simplified model of the situation to avoid going in wrong directions and making errors. Before starting on the job of making a suit the tailor needs to draw a model of it.

Planning by Analogy. Plans may be formed by analogy where the solution of one problem is used as a platform for solving another problem. An example would be the problem of exploiting the energy in waves by way of the principle of a lens.

Planning by Abstraction. Plans are formed by abstraction when the original problem is simplified to a related but easier one. The solution of the simpler problem can then be used as a plan for solving the original more complex problem.

Writing a book for the first time may be approached by building on the successful solution to the problem of writing a paper on a related topic.

Working Backwards

The ordinary directional strategy in problem solving is to work forwards from initial state to goal state. However, when the goal is less connected than the givens, it may be more useful to work backwards from the goal to the givens. The reason is that fewer paths lead to the solution from the goal than from the givens. Such a situation is often present in mathematical problems involving formal proof. When this strategy was implemented in a computer program, Newell, Shaw and Simon (1958) even discovered a shorter and more elegant proof to one of the theorems in Whitehead and Russell's *Principia Mathematica*.

The general problem solving strategies considered so far all aim at simplifying problems. The research directed at identifying this bundle of heuristics is clearly driven from the theory of bounded rationality, where the issue of the too comprehensive problem space is focused.

This is perhaps an example of how students of problem solving themselves may fall into the trap of illegitimately narrowing the problem space due to a certain problem representation. Heuristics of simplification are particularly relevant when the task is one of *complexity*. As argued above, however, in creative thinking tasks, where *novelty* is the dominating determinant of difficulty, the problem is often that of operating within a too restricted problem space, or—as in "deceptive problems"—in the wrong space. The scope for investigating strategy use thus clearly stands in need of expansion to cover these realities of problem solving. This kind of lacuna has intuitively been spotted by Anderson (1975). He considers a different class of strategies that he coins "heuristics of stimulus variation". This category contains exactly the kind of strategies that would be expected to be useful in typical creative thinking tasks.

Heuristics of Variation

Anderson (1975) considers three kinds of strategies within this category.

Adding Stimuli. New stimuli may be needed to expand a too narrow problem perspective, and to set up new patterns of activity in the representational system. One way of achieving this aim is to present the problem in *different terms.* An example given by Anderson (1975) is the problem of proving whether the hypotenuse of a right-angled triangle is longer or shorter than the sum of the two sides. If expressed in concrete terms by asking the student whether it would be shorter to take a path that cuts across a vacant lot or to take a sidewalk that goes around it, the answer would probably be much easier to see. In this connection it is interesting to note the observation that creative thinkers often shift between concrete and abstract ways of representing the problem (Kaufmann 1980).

Thinking in analogies is another way of adding stimuli, Gordon (1961) has reported numerous examples of using this strategy. In one case the problem was to construct a roof that would absorb heat under cold conditions and reflect heat under hot conditions. The analogy of a flounder that can change color according to variations in light conditions was used. The problem was solved on the basis of this analogy by constructing a roof impregnated with white plastic balls that would contract under low temperatures rendering the roof black, and expand under high temperatures, thus reflecting heat.

Removing Stimuli. Particularly relevant to the case of the "deceptive problem" is the presence of irrelevant or misleading stimuli that may lure the problem solver into the wrong space. Adamson and Taylor (1954) have shown that fixations in problem solving may decay over a period of some days. Theoretically, such an effect would be expected from the theory of selective forgetting (Anderson 1975, p. 284). Weaker items are forgotten more rapidly than stronger ones. It is reasonable to believe that in most cases incorrect items are weaker than stronger ones. When used deliberately as a strategy, the problem is of course to decide which stimuli are the irrelevant ones. Anderson (1975) suggests that the problem solver should ask himself exactly what the solution requires. In our example of the deceptive Hatrack Problem, it is seen that such a strategy could prove useful. The rack is to hold a heavy coat and has to be very stable. The false lead of constructing a conventional rack could very well be shaken off by carefully considering this crucial feature of the solution.

Rearranging Stimuli. A final way of achieving the aim of restructuring a problem space by deliberate, strategical means is to rearrange stimuli. Temporal orderings may here play a significant role. Given the task of crossing out the word that does not fit into the sequence *skyscraper*, *temple*, *cathedral*, *prayer*, a subject will normally choose *prayer*. When presented in the reverse order, the word *skyscraper* is most likely to be crossed out (Anderson 1975 p. 285). Spatial arrangements are also important. The chessboard looks differently from the opponent's view. A useful strategy in solving a chess problem could therefore be to imagine the chessboard from the opponent's perspective.

As is seen, heuristics of stimulus variation comprise a much more loosely defined category than heuristics of simplification. The reason is not that the former are less important. Rather, it seems they fall outside the scope of the prevailing paradigm, which is largely driven from the basic ideas derived from the concept of bounded rationality. Our main point here is that this theory tends to confine the area of search mainly to the aspect of difficulty defined by complexity. A more differentiated view of what constitutes an ISP may serve the fruitful purpose of widening the scope for research on human problem solving. Also it may provide a more systematic basis for the description and classification of heuristics than is now available. In particular, we believe that such expansion of perspective will throw new light on and revitalize specifically the area of creative thinking. This is strongly needed in current research on human problem solving.

Symbolic Activity in Problem Solving

Closely related to research on heuristics is the issue of the usefulness of different kinds of symbolic strategies in problem solving. Research in this category has concentrated on the basic question of the nature of the symbolic systems at disposal in cognition. One group of theorists have argued that all information is represented in a common underlying format of a conceptual/propositional format (single code). Their opponents have argued for a multiple-code theory, where there are several different symbolic codes (verbal-propositional and perceptual analogs in imagery) that have different functional properties in information storing and processing (see e.g. Kaufmann 1980, 1986; Kosslyn 1980; Paivio 1971, 1986, and Shepard 1978).

On the basis of extensive research, it now seems possible to claim that the general position defined by multiple-code theory has won out (Anderson 1985). Representations in verbal and imagery based structures seem not to be entirely reducible to a basic representation system consisting in abstract meanings (propositions). It seems clear, then, that human beings have developed different types of codes that are assigned to different information processing functions.

The spotlight of research has been cast specifically on the use of imagery as a symbolic medium. From the perspective of our discussion this is particularly interesting. Reports pertaining to major inventions and scientific discoveries suggest that the inventors were visualizing complex situations when their revealing "flash of insight" took place (Kaufmann 1980; Shepard 1978). Such informal evidence suggests that somehow imagery is of particular service to creative thought. Corroborating this hypothesis, reviews of the experimental literature suggest quite strongly that imagery gains increasing importance in direct proportion to the degree of ill-structuredness of the task. With high novelty, complexity or ambiguity the subject seems to switch from a linguistic-propositional code to an imagery based representation (Kaufmann 1984b). Why is this so?

With reference to the general theory of strong and weak methods in problem solving considered above, we have suggested that imagery is a back-up system that gives access to a set of simpler cognitive processes of a perceptual kind. Such simpler processes may be needed in an ISP where computational processes in the form of rule-governed inferences are difficult or impossible to perform (Kaufmann 1987). The theory is presented in Table 1.

Table 1. Elements and structure of the theory.

CONSCIOUS REPRESENTATIONS	VERBAL		IMAGINAL	
MODE OF OPERATION	COMPUTATIONAL TRANSFORMATIONS (Rule governed inferences)		PERCEPTUAL SIMULATIONS (Mental modelling)	
Main information processing categories	Deductive reasoning	Inductive reasoning	Perceptual comparisons	Perceptual anticipations
UNDERLYING REPRESENTATIONS	PROPOSITIONAL		ANALOG	

The basic thesis of the theory is as follows: A linguistic-propositional representational format is a strong one in the sense that great precision may be achieved in the form of explicit descriptions. It is easily and quickly manipulated and contains the full range of computational operations within its potential. In contrast, imagery is more ambiguous, sluggish and less easily manipulated and only realizes simple cognitive operations of a perceptual kind, like anticipations and comparisons. This may be useful and even necessary in ill-structured task environments, where computational operations in the sense of rule-governed inferences are difficult or impossible to perform. More specifically, limitations of the possibility of using computational operations may be due to lack of experience with the task at hand (either factual or strategic), where computational processes break down due to lack of rule-based information to feed on (novelty). Limitations of computational operations may also result from strain on working memory due to high information load (complexity). Finally, uncertainty as to which rule or procedure should be applied may lead to computational dysfunctions (ambiguity). Images thus may be best described as perceptual-like *mental models* (see also Sanford 1985). Such perceptual mental models allow a translation from computational to perceptual operations. More specifically, we have suggested that deductive operations may be translated into simple, quasi-perceptual comparisons, where certainty of judgment may be reached. The imagery parallel to inductive operations may be quasi-perceptual anticipation, where a future state of affairs may be imagined on the basis of a previous sequence of events.

The theory offers an explanation of the evidence that links the usefulness of imagery to ill-structuredness in the task environment and to creative thinking specifically. However, the research on symbolic representation in problem solving is not as precise and advanced as in the fields of memory and learning. Clearly, more precise experimental evidence is necessary to subject our theory formulation to a more critical test.

Expert Performance

So far we have considered the use of very general strategies and capacities that may facilitate performance in ill-structured task environments. This picture needs to be supplemented with research evidence that relates to the significance of domain specific knowledge for effective and creative problem solving.

In his seminal work on human intelligence, Spearman (1904, 1927) arrived at the conclusion that successful problem solving is determined by two factors: general intelligence and knowledge specifically bearing on the task in question. The factor of general intelligence has in recent years been unpacked by Sternberg (1985). He claims that the realities behind the term "general intelligence" consist of the efficient use of highly general problem solving strategies of the kind we have described above. The condition of "positive manifold" (i.e. positive correlations between most problem solving tasks) is, according to Sternberg, due to the fact that the same general strategies are used across a wide variety of different tasks.

This general picture of the basic conditions underlying successful problem solving performance has been corroborated in interesting ways by recent experimental studies. In addition to efficient use of general problem solving strategies, it seems that expertise in problem solving is highly dependent on the availability of extensive, well-organized domain specific knowledge. The major strategy that has been employed in these studies is to compare the performance of experts and novices in solving problems with the aim of uncovering the conditions that affect performance differences.

Chase and Simon (1973a, b) reported a series of experiments where they examined the ability of chess players of different skill levels to reconstruct positions on a game board after viewing the board for a very brief period of time. The Master's performance was vastly superior to that of a novice. However, the differences were not observed when the pieces were placed in random positions. The Master's superior performance does not, then, seem to depend on greater visual memory ability. Rather, highly skilled players outperform novices due to their ability to perceive *relations* between pieces. Chase and Simon conclude that the Master's superior ability is due to the possession of a vast number of stored patterns in long-term memory which allow them to match perceived board positions with stored memory representations. Thus, the Master can organize his information in a more efficient way. Chase and Simon estimated that a Chess Master has stored in memory 50,000–100,000 such "chunks" of information. It is the advantage of possessing this extensive and well-organized knowledge that makes for superior problem solving ability.

McKeithen et al. (1981) have corroborated these findings in a study where computer programmers at different levels of skill were asked to recall computer programs. Following the logic of the Chase and Simon experiments, performance was observed under meaningful and non-meaningful conditions (i.e. meaningful programs vs. randomly scrambled instructions). Expert programmers were clearly superior in recalling meaningful programs, but no better than others in recalling scrambled instructions. By using the technique of asking the subjects to recall key-words, McKeithen et al. were able to confirm the hypothesis that the superior performance of experts was probably due to their better organized domain specific knowledge.

In several other experiments of similar design, where novices and experts in other domains like physics and architecture have been studied, the same results are found (Kahney 1986).

Differences in the strategies employed by experts and novices have also been studied. Bhaskar and Simon (1977) report that experts tend to work forwards when searching for a solution to a problem, while novices more typically use the "weak" strategy of working backwards or resort to general means–end strategies. With specific address to the question of problem representation, Chi et al. (1981) have shown that novices try to understand a problem by working from its surface features, paying attention to the kinds of objects mentioned in the problem statement, while experts focus on the deeper, underlying principles.

The results of the research on expertise, then, seem to show that the possession of extensive, well-organized domain specific knowledge is a crucial factor in expert performance. The point here is not only the rather trivial one that experts outperform novices because they have more factual information about the task. Rather the important point is that a higher level of organized, domain specific knowledge gives the expert access to more powerful problem solving methods. Thus, high-level cognitive abilities should not be seen as existing aside of knowledge. Rather, powerful cognitive operations seem only to materialize in a system of well-organized, extensive knowledge. This is a new and important insight in cognitive psychology.

Creative problem solving may be seen as expert performance that produces new insights. We may now expect that high-level creativity is crucially dependent on a large amount of well-organized domain specific knowledge. Before the fruits of real creativity can be reaped, then, we may expect that a long history of building up domain specific knowledge and skills must precede.

Hayes (reported in Simon 1983) has examined this question by way of biographical evidence on famous chess masters, composers, painters, and mathematicians. Hayes concludes that for top performance, ten years of work in the field seems to be a magic number. Almost none of the individuals studied had produced world-class performance without first having invested at least ten years of intensive learning and practice. It is interesting to note that the folklore surrounding child prodigees seems to be distorted. For instance, by the standard Hayes used (five appearances in the

Swann catalog), he found no world-class work of Mozart before the age of seventeen.

On the basis of Hayes' extensive and systematic observations, Simon concludes: "A *sine qua non* for outstanding work is diligent attention to the field over a decade or more" (p. 28).

In general, expertise and high-level creativity seem to require an extensive base of well-organized domain specific knowledge that takes a long time to acquire. The point is not, of course, that having extensive knowledge in itself guarantees creativity of high rank. Rather, it seems to be a *necessary condition* for high-level performance. Thus, Wallach seems to be entirely justified in his main conclusion in the opening chapter of this volume, where he argues that there has been a narrow perspective in the field of creativity research that has focused mainly on very general capacities underlying creativity at the expense of the existensive domain specific knowledge and skills that have to be present for high-level creativity to unfold. The results of recent cognitive research on expertise and highly talented work in a number of different fields here seem definitely to have balanced the scales in a proper way.

Antecedent and Situational Determinants of Problem Solving

So far our discussion has revolved around issues defined by general theoretical formulations on the nature of human problem solving. In this section we will give a selective review of some of the major research evidence stemming from more straightforward empirical research on problem solving. The aim has been to identify antecedent and situational conditions that affect problem solving performance. We may group these into two categories: conditions that *facilitate* and those that *inhibit* productive problem solving.

Facilitating Conditions

(a) Exploring the Problem Situation
In our discussion above, we presented evidence to the effect that the initial *problem identification* is a most sensitive part of the problem-solving process. Differences in *problem representation* may strongly affect the success of solving a problem. It is to be expected that this relationship holds true especially in the area of creative thinking. Here we are often dealing with "discovery tasks", where formulation of the problem is of critical importance for success, as is seen in the problem of the Mutilated Checkerboard presented above.

Raaheim (1964) performed an interesting experiment that addresses this issue directly. The subjects were given a typical "insight problem" to solve. The task was to get hold of a ring at the bottom of a three metres' deep shaft by a rapid and safe method. The following objects were put at their disposal: A glass tube three metres long with the same diameter as the ring, a nail, a knife, a point of pliers, and a piece of wood. The best solution is considered to be the following: By means of the knife, a small wooden plug is made. The nail is fastened to the plug and bent into a hook with the pliers. The plug is then fitted into the glass tube. By way of this tool, the ring may quickly and safely be brought up from the shaft.

The experimental group was carefully instructed to find out exactly what was missing, and then to replace it with the object available before starting on solving the problem. The control group was given no such instructions.

The results showed an impressive 50 percent increase in correct solutions in the experimental group compared with the control group.

Again it seems that spending time in working out a suitable problem representation may greatly facilitate insightful problem-solving performance.

Along similar lines, Maier (1963) has advocated the strategy of exploring the problem by *locating multiple obstacles*, and has presented evidence that this approach will increase productive problem solving. However, the evidence in question is largely based on informal observations, and more precise and controlled experimentation is necessary before we can assess the potential benefits of such a strategy. The thesis is, however, soundly anchored in the general body of theory and research, and opens the avenue for interesting and important experiments on the conditions that affect creative problem solving.

(b) Turning a Choice Situation into a Problem Situation

Addressing the same issue, Maier and Hoffman (1970) have argued that "there is a tendency to seek solutions before the problem is understood" (p. 369). When the aim is for creative solutions to a problem, it is therefore important that an attitude of "problem-mindedness" replaces the attitude of "solution-mindedness". Maier and Hoffman submitted this hypothesis to experimental test by having the subjects obtain a second solution after a first one had been produced. The problem given was the so-called Change of Work Procedure, where the task is to find a productive solution to an assembly-line problem. The assembly operation was divided into three positions and the workers adopted a system of hourly rotation among the three jobs. Each worker had a best position, and the issue was to find a new method that would increase productivity. The interesting feature of this problem is that there are several possible solutions that may be graded on a scale of creativity. The creative solutions are called *integrative*. Here individual differences in ability are exploited, while the unfavorable effects of monotony are avoided.

Maier and Hoffman obtained results that confirmed their hypothesis by showing that, whereas only 16 percent of the subjects produced integrative solutions on the first try, 52 percent of the second solutions were of this type.

Such findings underscore the importance of sufficient problem exploration for creative problem solving and fall nicely in line with the general body of results that point to the critical importance of problem representation for finding high-quality solutions to a problem.

(c) Separate Idea Generation and Idea Evaluation

Osborn (1963) has argued that a major block in creative thinking is the tendency to premature evaluation of ideas. Consequently, Osborn argues for a strategy of "brainstorming" in problem solving where creativity is a major requirement. The basic idea of brainstorming is to separate idea generation from idea evaluation.

Much experimental research has gone into testing the soundness of this principle. The general design has been to determine the effects of brainstorming instructions on originality and productivity in problem solving (see Stein [1975] for an extensive review). When used as a group problem solving method, the results have been rather depressing for the brainstorming thesis. Brainstorming instructions often turn out to have a *detrimental* effect on the quality and productivity of problem solving, as compared to the effect of neutral instructions. However, Parnes (1963) has reminded us that the brainstorming technique is not inherently a group technique. Anxiety over negative evaluations from others when suggesting "wild ideas" may, indeed, block the productivity of the individual performing in a group setting. Research on the deferment-of-judgment principle used in individual problem solving tends to show a positive effect on performance. Furthermore, encouraging the subjects to "free wheel" and not resist wild ideas is not a necessary feature of the principle. Maier (1963), on the basis of several experiments, reports that the principle of separating idea generation and idea evaluation in itself seems to be sound policy for promoting productive problem solving.

Rickards and Freedman (1979) presented interesting evidence to the effect that a greater *variance* in quality of ideas is obtained under the deferment-of-judgment principle. This means that a greater number of poor solutions *and* high-quality solutions are obtained. These findings suggest that there is a special place for the deferment-of-judgment principle in idea-deficient situations, where the requirements for creativity are particularly strong.

(d) Conflictual Thinking

In the literature on problem solving and creativity, the conditions of a *conflict* between opposing ideas are seen as a potentially powerful spur for productive thinking (e.g. Duncker 1945). The general idea is that such a conflict puts pressure on the problem solver which may be resolved by finding a new idea containing elements from the two opposing base ideas.

Hoffman (1961) has spelled out this theory and subjected it to an experimental test. According to Hoffman, four conditions must be fulfilled for a conflict to have a productive function in problem solving: (1) Opposing, but compatible cognitions must be present at the same time. (2) At least two opposing cognitions must have a higher value than a certain minimum for their being accepted as interesting. These should have approximately the same value of attractiveness. (3) Problem solving must take place in a situation where there is pressure towards finding the best possible solution. (4) The cognitive components that give rise to the value of each solution should be identified and clarified.

The theory was put to the test in a series of experiments where the Change of Work Problem was employed in order to get a measure of the creativity of solutions. Conditions were manipulated to create the right "conflict atmosphere". The result clearly supported the theory. More creative solutions were obtained under the conflict conditions. Thus, a cognitive conflict seems to have the potential to facilitate creative restructuring in problem solving.

More recently, Rothenberg (1976) has defined a process termed "Janusian thinking" that is held to be characteristic of creative thought. Janusian, or "oppositional thinking", is "the capacity to conceive and utilize two or more opposite or contrary ideas, concepts, or images simultaneously" (Rothenberg 1976, p. 313).

Rothenberg points out that highly creative works of art (paintings, compositions, poetry, etc.) often rest on a simultaneous conception of opposites. This feature is also found in great scientific discoveries (the "double helix" springs out of the notion of identical chains running in opposite directions). Rothenberg has pursued the issue by gathering clinical and experimental evidence on the importance of thinking in conflictual opposites in creativity. In his clinical studies, highly acclaimed writers (winners of Pulitzer prizes, etc.), novice writers of high creative potential (rated by literary critics and teachers) have been compared with "non-creative" persons who try to write a work of fiction or poetry for financial reward. According to the results, Janusian thinking figures frequently in the works of the creative writers, but never in non-creative persons. Experimental evidence has been produced by studying word association responding to the Kent–Rosanoff (K–R) test. Rothenberg finds a high tendency to rapid opposite responding ("white" to "black", "health" to "sickness", etc.) in creatively oriented groups of male college students. Special association tasks as well as the K–R test were also given to prominent and novice creative writers. The results show a high tendency to rapid oppositional associations in these groups as well.

Taken together, the evidence tends to support the general idea that thinking in a context of cognitive conflict will promote creativity in solving problems.

(e) High Motivation and Persistence

We have seen that an attitude of persistence is mentioned as a defining criterion of creativity in problem solving. There are several good reasons for accepting this "hot element" in the otherwise consistently "cold" descriptions of creative problem solving.

Simon (1966) has pointed out that high-level creativity is a rare event in scientific discovery, and a theory of creativity also needs to account for this rarity of occurrence. The idea that high-level creativity requires extraordinarily high persistence fits nicely into this observation. Creativity often involves "going against the tides" and a lot of resistance to change has to be overcome.

Anderson (1980) also points to the possibility that high persistence may have to do with the openness that high-level creativity requires. He posits that highly creative individuals may be willing to continue working because they are less willing than others to accept the many conflicting facts that are present on the route to creativity.

Several findings suggest that high motivation and persistence are indeed vital ingredients in the creative process. Roe (1953) examined characteristics of a group of 64 American physicists, biologists, and social scientists selected for the importance of their contributions to their fields. The only trait that Roe found to be common to her subjects was the *willingness to work extremely hard.* MacKinnon (1962), in his studies of creative architects, reports that his highly creative subjects had developed a "healthy obsession" with their problems.

Hyman (1964) has approached the issue along more direct experimental lines. He made the interesting observation that when people were asked to continue working on a problem beyond the point where they thought they had come up with their best effort, they frequently are able to produce ideas that are even more creative than their previous ones.

This area of research is obviously important both from a theoretical and practical point of view, and more research in this category would therefore be particularly welcome. Obviously, there are limitations to a one-sided framework of "cold cognition" in the experimental study of human problem solving.

Inhibiting Conditions

According to the theory of bounded rationality, a basic motive governing the cognitive behavior of the individual is to preserved *cognitive economy.* A strategy of satisficing is chosen to reduce the strain on the computational capabilities of the thinker. This is a rational strategy that makes for good adjustment given the cognitive limitations of the human information processor. It may be argued, however, that preserving cognitive economy means to keep variation and changes to a minimum. Thus, we may expect dysfunctional consequences of this orientation where restructuring and creative change is required. To relinquish established perspectives and standard operating procedures that are associated with safety and predictability may conflict with the individual's striving for cognitive economy and meet with resistance to change. Thus, a rational orientation to solving problems, guided by the motive to preserve cognitive economy, may entail a danger of rigidity and stereotypi when faced with a situation where a restructuring of established conceptions and lines of procedure is required.

The psychological literature on human problem solving confirms this expectation through many examples of rigidity, stereotypi and dysfunctional resistance to change. The Gestalt psychologists in particular have been active in producing many striking demonstrations of this "other side of the coin" of human cognitive adjustment. A major orientation in the experimental psychology of creative thinking has, indeed, been to investigate the conditions that *inhibit* creativity in problem solving. In this section we will describe some of the major phenomena in the category of fixations and resistance to change in problem solving.

(a) The Einstellung Effect

According to the theory of satisficing, the problem solver is assumed to select the first alternative he encounters which meets some minimum standard of satisfaction. Luchins (1942) has shown how such a strategy, under certain conditions, may produce fixation and stereotypi in problem solving behavior. The so-called "Einstellung" effect shows itself under conditions where the individual has discovered a strategy that initially functions well in solving certain tasks, but later on blocks the realization of new and simpler solutions to similar problems. Luchins has investigated this phenomenon in a series of experiments and shown it to be a reliable and robust one.

In one experiment the subject was presented with three jars (A, B, and C) that would contain a certain volume of fluid. By using these jars, the subject should attain a given amount of fluid. Table 2 shows the design in Luchin's original experiment. Problem 1 is employed for instructional purposes. Problems 2–6 are the so-called "Einstellung" tasks. These can all be solved according to the formula B–A–2C. In problem 2, the task is solved by first filling up B, then empty 21 liters in A and then 2×3 liters in C. We then have 100 liters left in B.

The remainder of the Einstellung tasks will all be solved according to the B–A–2C formula. The following two tasks may be solved according to the standard formula, but also through far simpler methods (A–C, A+C). Problem 9 is not possible to solve by way of the standard formula, but is solved very easily by the formula A–C. Problems 10 and 11 are similar to 7 and 8.

In the experiment, the Experimental group solves all the problems, while a Control group only solves tasks 1, 7, 8, 9, 10, and 11. This procedure allows a test of the effect of establishing a standard procedure on subsequent solving of the critical problems.

The general conclusion from the water jar experiments is that the fixation in problem solving is dramatically strong. While just about all of the subjects in the Control group solve the test problems by way of the simplest formulas, as many as 80 percent in the Experimental group use the complicated standard procedure on problems 7 and 8. About 60 percent are not at all able to solve problem 9, where the formula breaks down.

The Einstellung phenomenon also shows up in other types of tasks, like anagrams (letter combinations), concept formation

Table 2. A water jar problem.

| Problem | Given jars of the following sizes | | | Obtain the amount |
	A	B	C	
1.	29	3		20
2. Einstellung 1	21	127	3	100
3. Einstellung 2	14	163	25	99
4. Einstellung 3	18	43	10	5
5. Einstellung 4	9	42	6	21
6. Einstellung 5	20	59	4	31
7. Critical 1	23	49	3	20
8. Critical 2	15	39	3	18
9.	28	76	3	25
10. Critical 3	18	48	4	22
11. Critical 4	14	36	8	6

Possible answers for critical problems (7, 8, 10, 11)

Problem	Einstellung solution	Direct solution
7	$49 - 23 - 3 - 3 = 20$	$23 - 3 = 20$
8	$39 - 15 - 3 - 3 = 18$	$15 + 3 = 18$
10	$48 - 18 - 4 - 4 = 22$	$18 + 4 = 22$
11	$36 - 14 - 8 - 8 = 6$	$14 - 8 = 6$

Performance of typical subjects on critical problems

Group	Einstellung solution	Direct solution	No solution
Control (children)	1%	89%	10%
Experimental (children)	72%	24%	4%
Control (adults)	0%	100%	0%
Experimental (adults	74%	26%	0%

Source: R. E. Mayer (1983).

tasks, and geometry tasks. Thus, it seems to reflect a dysfunctional consequence of the normal, rational way of approaching problems that now may block the establishment of a new perspective and more appropriate lines of procedure in task environments that resemble those encountered before. It is interesting to note that Cyert and March (1963) have observed similar behavior among managers in real-life contexts. Typical managerial search is seen as "simple minded", and as overemphasizing previous experience, by selectively searching in regions close to where previous solutions have been found.

(b) Functional Fixedness
Duncker (1945) has also investigated how past experience may block productive problem solving through a different experimental

set-up. Duncker coined the term "functional fixedness" to refer to a block against using an object in a new way that is required to solve a problem. Duncker devised a series of experiments to explore this phenomenon. In the so-called Box Problem (see Figure

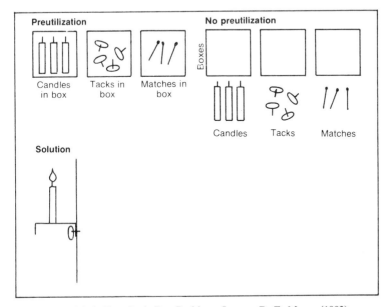

Fig. 7. Materials in Duncker's Box Problem. Source: R. E. Mayer (1983).

7), the subject was given three cardboard boxes, matches, thumb tacks, and candles. The task was to mount a candle vertically on a screen nearby to serve as a lamp. The Experimental group was given a box containing matches, a second box holding candles, and a third one that contained tacks. This was called the "preutilization" condition. A "no preutilization" condition was administered to the Control group. The same supplies were given, but with the matches, tacks, and candles outside the boxes. The solution to the problem is to mount a candle on the top of a box. This may be achieved by melting wax onto the box, sticking the candle to it, and then tacking the box to the screen. This was much harder for the Experimental group. Similar effects are found in other tasks, where the same general set-up was used.

Duncker takes the results of his experiments as a demonstration of how previous experience may have dysfunctional consequences in problem solving by blocking new insights and necessary restructuring in the fact of the requirements of the task. Later experiments have replicated and extended Duncker's original findings in interesting ways (see e.g. Adamson 1952; Adamson and Taylor 1954; Birch and Rabinowitz 1951; Glucksberg and Danks 1968; Glucksberg and Weisberg 1966; Weisberg and Suls 1973).

Some interesting examples of how the factor of functional fixedness may operate in real-life contexts and seriously hamper the process of technical invention are given by Weizenbaum (1984). According to Weizenbaum the steam engine had been in use for a hundred years to pump water out of mines before Thretwik got the idea of using it as a source of locomotive power. Another example is the computer, that for a long time was seen just as a calculator before its potential as a general symbol manipulator was conceived.

(c) Hidden Assumptions

A similar effect of dysfunctional effect of a mental set is due to the "hidden assumption". Fixation in problem solving may be caused by certain assumptions of how a problem has to be solved which delimits the search for a productive solution.

An illustrative experimental demonstration is provided by Scheerer (1963) through the case of the so-called Nine Dot Problem

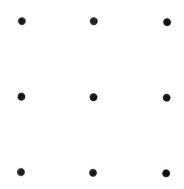

Fig. 8. The Nine Dot Problem. Source: B. F. Anderson (1980).

(see Figure 8). The Nine Dot Problem presents the problem solver with a 3×3 grid of nine dots. The task is to draw four straight lines through all nine dots without lifting the pencil from the paper. The problem is difficult to solve, and the main reason for the difficulties seems to be the hidden assumption that one has to stay within the initial configuration of a square. Once this assumption

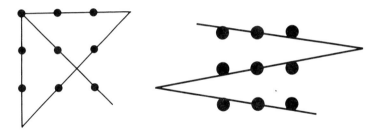

Fig. 9. Solutions to the Nine Dot Problem. Source: B. F. Anderson (1980).

is broken, several possible solutions are possible (see Figure 9). The difficulties involved in solving the Nine Dot Problem also illustrate the delicate nature of problem representation, and how it may affect success in problem solving.

(d) Confirmation Bias

Going back to our thesis about possible adverse consequences of the striving for cognitive economy, we may now posit that the individual will show resistance to relinquishing established hypotheses, and be generally reluctant to actively seeking disconfirming evidence.

A number of experiments seem indeed to show that people have a natural propensity to seek confirming evidence and to avoid disconfirmation or discard disconfirming evidence when it is present.

Wason (1960, 1968) observed a marked confirmation bias in a situation where the subjects were to discover an experimenter defined rule by seeking evidence to evaluate their hypotheses. The subjects in these experiments were given an initial three-digit sequence, 2, 4, 6, which was compatible with the rule and instructed to generate their own test sequences. Following each sequence, the subjects were informed as to the appropriateness of their example. The majority of subjects did not actively seek disconfirming evidence and tended to ignore it when it occurred.

The same conclusion was reached by Einhorn and Hogarth (1978) who observed that confidence in a hypothesis generally increases more following positive feedback than it decreases following negative feedback.

Mynatt, Doherty and Tweney (1977) investigated hypothesis-testing behavior in a simulated research environment. The subjects were presented with moving objects on a screen and their task was to discover the principle that could account for the particular movements. Seventy percent of the subjects chose screens that could only *confirm* their hypothesis. In a follow-up, Doherty et al. (1979) found that many people do not consult observations relevant to an alternative hypothesis, even when such observations were readily available.

(e) Conservatism in Hypothesis Testing

Closely related to the experiments described above are a series of experiments purporting to demonstrate a marked bias in hypothesis testing behavior. Based on normative prescriptions derived from Bayes' theorem, people have been observed to manifest a *conservative bias* in hypothesis testing consisting of a reluctance to reduce their confidence in a decision following disconfirmation.

Phillips and Edwards (1966) investigated the effects on posterior probability estimates of (1) prior probabilities, amount of data, and diagnostic impact of data; (2) payoffs; and (3) response modes. The subjects were given the task of imagining ten bags, where each bag contained 100 poker chips, with red chips predominating in r of the bags, and blue chips predominating in the remaining 10-r bags. They were shown a bag and told that the experimenter had chosen it from the 10 bags, where each of the 10 bags was equally likely to have been chosen. The subjects were asked to make estimates of the probabilities that a predominantly red or a predominantly blue bag was the chosen one. In all the experiments the subjects typically behaved conservatively. This was manifested by the fact that the difference between their prior and posterior probability estimates was less than that prescribed by Bayes' theorem.

Pitz, Downing and Rheinhold (1967) likewise observed striking departures from predictions of Bayes' theorem. Under conditions where revisions toward uncertainty should be identical to revisions toward certainty, the former was much less than the latter. Pitz (1969) has argued that commitment is an important cause of this "inertia effect" (see also Pitz [1975] for a review of experiments on conservatism in hypothesis testing).

Taken together, the experiments on hypothesis-testing behavior seem to suggest that there is a basic, natural tendency to resist change and restructuring in problem solving. Together with the experiments on Einstellung and fixations they point to dysfunctional cognitive tendencies that we may expect to block creativity in problem solving.

ARE THERE "CREATIVITY SPECIALS"?

In spite of our attempts to focus specifically on the case of creative thinking in our previous discussion, it may be argued that we have not been able to capture its unique flavor. According to such a view, the domain of creative thinking harbors special features of cognition that are not easily described or explained within the framework of a rational information processing model of the kind that have guided our treatment of the subject. Creative thinking, it will be argued, is driven largely by more irrational processes, and this is why the field is so badly understood within the context of a purely rational model of human cognition. The processes that are the dominant driving forces behind creative thinking are held to be largely unconscious, and stand in need of being described and explained within the framework of a totally different conceptual scheme. The processes referred to here are mainly those of "intuition" and "incubation"—both held to depend on the workings of the unconscious mind and to be driven by illogical, irrational processes. These phenomena are indeed often described as being intimately related to the process of high-level creativity in prominent scientists and artists (e.g. Patrick 1938), and present a challenge to the information-processing approach.

Much of the discussion of these phenomena is of a rather popular and/or informal nature. We therefore need to address the question of how real these purportedly special features of the mind are before. Only then can we say something meaningful about what *kind* of phenomena we are dealing with, with the special purpose of deciding whether they fall outside the scope of an information-processing system as defined by current cognitive models.

Intuition—ESP or IPS?

In the popular writings, intuition is often described in rather occult terms as a kind of indefinable gift that only the highly-creative person possesses to a full degree. On this view, intuition is not an ordinary, natural part of an information-processing system (IPS). Rather, it is a kind of extra-rational, ESP-like ability that allows the individual who possesses it to point out the correct directions in problem solving without knowing why. It is like having a friendly homunculus in the back of the head that whispers sweet answers to the searching mind. Often such views are coupled with naive and oversimplified ideas about the human brain. According to the popular theory (e.g. Blakesley 1980), the left brain is the site of logic and language and is dry and editorial, while the right brain harbors all the goodies of intuition, imagination and imagery, and is the engine of creativity. Such a picture of the division of labor in the human brain seems, however, to be totally misguided. In a recent review of the research available, Gazzaniga (1983)—a prominent researcher in the field—even argues that the right brain as an isolated system functions at the level of intelligence found in chimpanzees. Even if this assessment should turn out to be a bit overdrawn, we are not likely to find the magic sources of creativity exclusively in the right brain. Moreover, the issue of location in the brain is largely irrelevant to our inquiry. It would obviously not make much of a difference for the advance of the psychology of problem solving to know that creativity is to the left or to the right, or up or down in the brain for that matter. What we see here is confusion in levels of explanation. The *psychological* questions can only be answered by way of functional explanations at the "software" and behavioral levels.

With specific address to the case of intuition, Simon (1983) has thrown some cold water into the veins of the most ardent proponents of the extra-rational model. According to Simon, intuition is something that we all have, and it is not dependent on a mysterious, indescribable process. Intuition has to do with making a correct judgment without conscious awareness of the process behind it. Such a capacity is not the exclusive property of the chosen few. Rather it is a common ingredient in everyday, cognitive functioning. The skilled chess player will be able to make a move in a midgame situation in just a few seconds without exactly knowing why, and often it turns out to be a correct move. The explanation of such good intuition is well known to cognitive psychologists. It depends on the availability of a large, well-organized knowledge base and a corresponding elaborate discrimination net that makes for quick and accurate judgments. According to Simon, it is no more mysterious than the ordinary ability to recognize immediately one of your friends when walking on the street. The recognition ability here is likewise dependent on considerable experience with a large number of "friends". Thus, we have an elaborate, well-organized sorting net that allows us to perform the correct judgment very quickly.

We can now see that the phenomenon of intuition may depend on elaborate sorting nets derived from extensive experience. Since the knowledge is well organized in "chunks", processing will occur at very high speed, largely outside the conscious awareness of the individual. Thus, it is not necessary to postulate fancy and elusive processes behind the power of human intuition. The phenomenon is real and available to every human information processor that possesses elaborate and well-organized discrimination nets derived from extensive experience. Thus, intuition can be seen as a natural IPS component rather than an occult ESP phenomenon.

Incubation—Fact or Fiction?

A closely related phenomenon that is held to be a "creativity special" is the process of incubation. Incubation is said to occur when the individual sets the problem aside for a while and does something else that is unrelated to the problem. At a later occasion—often very suddenly—the correct solution springs to mind. Many dramatic examples of the phenomenon have been described in the literature (e.g. Ghiselin 1952; Patrick 1938). A good example has been provided by Amy Lowell, the poet (in Ghiselin 1952):

> I registered the horses as a good subject for a poem: and having so registered them, I consciously thought no more about the matter. But what I had really done was to drop my subject into the subconscious, much as one drops a letter into a mail-box. Six months later, the word of the poem began to come into my head, the poem—to use my private language—was "there" (p. 110).

According to the popular theory, the incubation runs on active unconscious processes that go on in the interval. This is, however, not the only possible explanation. It may be due to the breaking of an unproductive set, stress reduction, reduction of fatigue, selective forgetting, and facilitating effects of incidental stimuli (Olton and Johnson 1976).

Informal descriptions are, of course, not sufficient to establish the reality of the phenomenon, nor the eventual mechanism behind it. Thus, several controlled experiments have been performed to study incubation at a more stringent level. Olton (1980) has provided a good review of the literature on experimental studies of incubation. In all the studies that deserve serious attention, incubation was examined by comparing the performance of an Experimental group and a Control group in solving problems that require creative insight. The Experimental and the Control groups worked for an equal amount of time on the actual problem, but the Experimental group spent an additional intermittent time working on unrelated tasks. In some studies, an incubation effect was present, but in others no difference was observed between the two groups. It is difficult to draw a definite conclusion from these experiments. Olton argues that a possible explanation for the difficulty in demonstrating the incubation effect could be the artificial and restricted nature of the laboratory situation. He then went on to examine the effect of an incubation period in a more ecologically valid situation. Here the subjects worked on an engaging mid-game chess problem. However, no incubation effect was observed in this study either.

The existence of the phenomenon thus seems difficult to prove in a controlled setting, even in situations closely resembling a real-life context. This does not, of course, prove that the phenomenon is a pure fiction. Some experiments have produced an incubation effect, and many suggestive informal descriptions of the phenomenon exist. We now turn to the next question: Given its existence, can we give a convincing explanation of the phenomenon in pure information processing terms, or do we have to admit the elusive explanation in terms of active unconscious processes?

According to Hayes (1978), selective forgetting is the most likely explanation for incubation effects. On the principle of selective forgetting it seems that incubation may be well understood as a natural ingredient of ordinary human information processing. Simon (1966) has given the following detailed explanation of the possible mechanisms behind incubation:

★ In the early stages of problem solving, the individual forms a solution plan to guide his attempts at solving the problem. Solution plans are normally stored in working memory rather than in long-term memory.

★ During attempts to execute the plan, the individual learns a great deal more about the problem (constraints, etc.). Thus, new information groups may be formed. Individual problem solving steps may be formed into sub-routines. This sort of information may be stored in long-term memory.

★ During a delay the initial plan is forgotten, or suffers more forgetting than the newly acquired information about the nature of the problem.

★ When the problem is approached again after a delay, the old plan is forgotten, and a new one has to be formed. The new plan is based on better information about the nature of the problem than the old one and is likely to be a better plan. The break therefore should have the effects of increasing the probability of solution, and this is exactly what is supposed to happen during incubation.

The sudden realization of a solution may occur when we are dealing with discovery tasks, where the solution to the problem is largely dependent on a correct representation, as in the example of the Mutilated Checkerboard.

This account does not, of course, prove that this is what happens during incubation. It does prove, however, that a purely rational, information-processing explanation can be expanded to cover also the phenomenon of incubation in a very satisfactory and convincing way.

It seems, then, that we do not have to borrow ingredients from extra-rational theories to account for the creativity domain of human problem solving. This is also the conclusion reached by Reitman (1965) and Simon & Sumner (1968) in a series of investigations of the processes involved in musical composition. In none of these investigations of creative problem solving was evidence obtained that made it necessary to posit processes that occur in creative acts only.

Is there, then, no limitations present in a computer oriented information-process account of problem solving that is to include the processes involved in creativity?

We have already pointed to some potential lacunas above, particularly those that have to do with the nature of the problem space. It seems, however, that these difficulties can be remedied by expanding the conceptual scheme appropriately. More difficult is perhaps a potential limitation in the scope of problems that the IPS model is fit to deal with. We will wind up our discussion by raising some critical comments on this particular issue.

Three Kinds of Problems and IPS—From Problem Solving to Problem Finding

When Hayes (1978) claims that "Problems are presented to us by the outside world" (p. 177), he voices well the spirit of the traditional conception of the nature of problems that has guided our discussion so far. Problems are thrusted upon us, and this categorization of problems is exhaustive.

There is reason to believe, however, that this is an unduly myopic view of the world of problems that seems to reflect the traditional view of humans as basically reactive and responding to "stimuli" or "inputs". To cover the full territory of problem solving, we may need a broader view. We will here suggest that there are three broad kinds of problems that a theory of human problem solving has to deal with in a satisfactory way.

First, there are *presented problems*. The individual is faced with a difficulty that has to be handled. Such a situation may be well structured (initial conditions, goal conditions, and operators are clearly definable). At the other pole is the unstructured problem situation, where the number of unknowns is at a maximum. But these are not the only problems people deal with. There is also a class of problems that may be called *foreseen problems*; that is, the individual anticipates that a problem (serious pollution, a massive traffic jam, etc.) will result if present developmental trends continue. Evasive action may then be taken. Even more interesting in the context of the present discussion is the class of problems that may be called *constructed problems*. The initial condition may here be a consistently reinforcing, satisfactory state of affairs. Nevertheless, a problem may arise when an individual compares the existing situation to a future, hypothetical state of affairs that could represent an improvement over the present situation. An example would be the present TV technology which may be said to be quite satisfactory. Yet, an individual might see a problem here in that there is no TV set with an adjustable-sized monitor unit.

Wertheimer (1959) linked this aspect of problem solving explicitly to creativity when he claimed that "The process starts, as in some creative processes in art and music, by envisaging some feature in an S2 that is to be created" (197–198).

Recently, Getzels and Csikszentmyhalyi (1978) argue a similar point and present evidence to the effect that problem *finding* is more intimately related to creativity than is problem *solving* in the traditional sense. However, Getzels and Csikszentmyhalyi consistently blur a subtle, but important distinction between "discovered" and "constructed" problems. In the former case the problem finder is able to ". . . articulate out of vague tensions the significant problems" (p. 251). But this is a case of an "indeterminate situation" (p. 172), where the more creative individual is able to discern a problem on weak signals. When Getzels and Csikszentmyhalyi speak of presented problems vs. discovered problems they often mean explicit vs. inexplicit problems. Conceptually, these situations both belong to the category of presented problems since the stimulus for problem solving is a "tension" (clear or vague). In the case of constructed problems there is no "tension" inherent in the situation. Rather, the first step in the problem-solving process is here to *create* a tension between an existing situation and a new, desirable future situation in order to propel the process of innovation. In the case of "discovered problems" success is dependent on problem sensitivity, whereas with "constructed problems" the driving force seems to be "innovation-orientation" or "opportunity seeking".

It is interesting to note that this important aspect of problem solving has been clearly seen by researchers observing managerial problem solving in real-life contexts. Mintzberg et al. (1976) identify an important class of problems as "opportunity decisions" which are "initiated on a purely voluntary basis, to improve an already secure situation, such as the introduction of a new product to enlarge an already secure market share" (p. 251). These are to be contrasted with "crisis decisions", where individuals and organizations "respond to intense pressures" (p. 251).

The narrow, presented problem perspective is also found in keynote organizational theories of problem solving of the information-processing type (e.g. Braybrooke and Lindblom 1963; Cyert and March 1963). In the traditional line of reasoning, Cyert and March (1963) argue that innovation occurs in the face of adversity. It is interesting to note that this is one of their major hypotheses that gains no support from the evidence obtained in their empirical studies of problem solving in organizations. Thus, Mintzberg et al. (1976) seem entirely justified when they blow their whistle on this perspective and claim that ". . . there is the need to reassess the increasingly popular point of view in the literature that organizations tend to react to problems and avoid uncertainty rather than seek risky opportunities" (p. 254).

Working in "discovered" and "constructed" problem environments, thus, seems to be intimately related to the creativity aspect of problem solving. It is possible that the popular computer-metaphor in this respect has put a blinder on students of problem solving. Computers, as we know them today, are *problem solving systems* par excellence and work on given, explicit problem descriptions. As Simon et al. (1986) points out, development is under way in computer-aided design that may present new opportunities to provide human designers with computer-generated representations of their problems. However, these new features relate to problem representation in the narrow sense of clarifying different ways of interpreting a given problem. They do not capture the human processes involved in identification of problems on weak signals, and definitely not the creation of "tensions" and discrepancies between existing and imagined situations. It would be foolish to forestall a development of these features in future computer technologies, but it is important to point out discrepancies between human and computer capabilities that may block perspectives and narrow our views on problem solving to those aspects that are assimilable to a computer-metaphor.

REFERENCES

Adamson, R. E. 1952: Functional fixedness as related to problem solving: A repetition of three experiments. *Journal of Experimental Psychology*, *44*, 288–91.

Adamson, R.W. & Taylor, D. W. 1954: Functional fixedness as related to elapsed time and to set. *Journal of Experimental Psychology*, *47*, 122–6.

Anderson, B. F. 1975: *Cognitive psychology*. New York: Academic Press.

Anderson, B. F. 1980: *The complete thinker*. Englewood Cliffs, N.J.: Prentice Hall.

Anderson, F. R. 1985: *Cognitive psychology and its implications*. San Francisco: Freeman.

Birch, H. G. & Rabinowitz, H. S. 1951: The negative effect of previous experience on productive thinking. *Journal of Experimental Psychology*, *41*, 121–5.

Bhaskar, R. & Simon, H. A. 1977: Problem solving in semantically rich domains: an example of engineering thermodynamics. *Cognitive Science*, *1*, 193–215.

Blakesley, T. R. 1980: *The right brain*. New York: Doubleday & Company.

Boden, M. A. 1979: The computational metaphor in psychology. In N. Bolton (Ed.), *Philosophical problems in psychology*. London: Methuen.

Braybrooke, D. & Lindblom, C. E. 1963: *A strategy of decision*. New York: Free Press.

Chase, W. G. & Simon, H. A. 1973a: Perception in chess. *Cognitive Psychology*, *4*, 55–81.

Chase, W. G. & Simon, H. A. 1973b: The mind's eye in chess. In W. G. Chase (Ed.), *Visual information processing*. New York: Academic Press.

Chi, M. T. H., Feltovich, P. J. & Glaser, R. 1981: Categorization and representation of physics problem by experts and novices. *Cognitive Science*, *5*, 121–52.

Cyert, R. M. & March, J. G. 1963: *A behavioral theory of the firm*. Englewood Cliff, N.J.: Prentice Hall.

Dennett, D. C. 1981: *Brainstorms*. Brighton, Sussex: Harvester Press.

Dewey, J. 1910: *How we think*. Boston: D. C. Heath Company.

Doherty, M. E., Mynatt, C. R., Tweney, R. D. & Schiaro, M. D. 1979: Pseudodiagnosticity. *Acta Psychologica*, *43*, 111–21.

Duncker, K. 1945: On problem solving. *Psychological Monographs*, *58*, No. 5 (Whole No. 270).

Einhorn, H. J. & Hogarth, R. M. 1978: Confidence in judgement: Persistence of the illusion of validity. *Psychological Review*, *85*, 395–416.

Gazzaniga, M. S. 1983: Right hemisphere language following brain bisection. A 20-year perspective. *American Psychologist*, May, 525–537.

Getzells, F. & Csikszentmyhalyi. 1976: *The creative vision: A longitudinal study of problem solving in art*. New York: Wiley.

Ghiselin, B. 1952: *The creative process*. Berkeley, CA: University of California Press.

Glucksberg, S. & Danks, J. 1968: Effects of discriminative labels and nonsense labels upon availability of novel function. *Journal of Verbal Learning and Verbal Behavior*, *7*, 659–664.

Glucksberg, S. & Weisberg, R. W. 1966: Verbal behavior and problem solving: Some effects of labeling in a functional fixedness problem. *Journal of Experimental Psychology*, *71*, 659–64.

Gordon, W. J. 1961: *Synectics*. New York: Harper & Row.

Greeno, J. G. 1973: The structure of memory and the process of solving problems. In R. Solso (Ed.), *Contemporary issues in Cognitive psychology: The Loyola Symposium*. Washington D.C.: Winston.

Greeno, J. & Simon, H. A. 1985: Problem solving and reasoning. In R. C. Atkinson, R. Herrnstein, Q. Lindzey & R. D. Luce (Eds.), *Stevens handbook of experimental psychology*. New York: Wiley.

Hayes, J. R. 1978: *Cognitive psychology. Thinking and creating*. Homewood, Ill.: Dorsey Press.

Hayes, J. R. & Simon, H. A. 1974: Understanding written problem instructions. In L. W. Gregg (Ed.), *Knowledge and cognition*. Potomac, Md.: Erlbaum.

Hayes, J. R. & Simon, H. A. 1976: Psychological differences among problem isomorphs. In N. J. Castellan, D. B. Pisoni & G. R. Potts (Eds.), *Cognitive theory*. Vol. 2. Hillsdale, N.J.: Erlbaum.

Hayes, J. R. & Simon, H. A. 1977: Psychological differences among problem isomorphs. In N. J. Castellan, D. B. Pisoni, and G. R. Potts (Eds. 1), *Cognitive theory*, Vol. 2. Hillsdale, N. J.: Erlbaum.

Hoffman, L. R. 1961: Conditions for creative problem solving. *Journal of Psychology*, 52, 429–44.

Hyman, R. 1964: Creativity and the prepared mind: The role of information and induced attitudes. In C. W. Taylor (Ed.), *Widening horizons in creativity*. New York: Wiley.

Janis, I. L. & Mann, L. 1977: *Decision making*. New York: The Free Press.

Johnson, D. M. 1955: *The psychology of thought*. New York: Harper & Row.

Johnson, D. M. & Jennings, J. W. 1963: Serial analysis of three problem solving processes. *Journal of Psychology*, 56, 43–52.

Johnson-Laird, P. N. 1983: *Mental models*. Cambridge: Harvard University Press.

Kahney, H. 1986: *Problem solving: A cognitive approach*. Milton Keynes: Open University Press.

Kaufmann, G. 1980: *Imagery, language and cognition*. Oslo/Bergen/Tromsø: Norwegian Universities Press.

Kaufmann, G. 1984a: Can Skinner define a problem? *The Behavioral and Brain Sciences*, 7, 599.

Kaufmann, G. 1984b: Mental imagery in problem solving. *International Review of Mental Imagery*, 23–55.

Kaufmann, G. 1986: The conceptual basis of cognitive imagery models: A critique and a theory. In D. Marks (Ed.), *Theories of image formation*. New York: Brandon House.

Kaufmann, G. 1987: Mental imagery and problem solving. In M. Denis (Ed.), *Imagery and cognitive processes*. Amsterdam: Martinus Nijhoff.

Kosslyn, S. M. 1980: *Image and mind*. Cambridge: Harvard University Press.

Køhler, W. 1927: *The mentality of apes*. New York: Harcourt, Brace & World.

Luchins, A. S. 1942: Mechanization in problem solving: The effect of Einstellung. *Psychological Monographs*, 54, (Whole No. 248).

MacKinnon, D. W. 1962: The nature and nurture of creative talent. *American Psychologist*, 17, 484–95.

Maier, N. R. F. 1963: *Problem solving discussions and conferences*. New York: McGraw-Hill.

Maier, N. R. F. & Hoffman, L. R. 1970: Quality of first and second solutions in group problem solving. In N. R. F. Maier (Ed.), *Problem solving and creativity*. Belmont, Ca.: Brooks/Cole.

Mayer, R. E. 1983: *Thinking, problem solving and cognition*. San Francisco: Freeman.

McKeithen, K. B., Raitman, J. S., Ruchter, H. H. & Hirtle, S. C. 1981: Knowledge organization and skill differences in computer-programmers. *Cognitive Psychology*, 13, 307–25.

Miller, G. A. 1956: The magical number seven, plus or minus two: some limits on our capacity for processing information. *Psychological Review*, *63*, 81–97.

Mintzberg, H., Duru, R. & Theoret, A. 1976: The structure of unstructured decision processes. *Administrative Science Quarterly*, *21*, 246–75.

Mynatt, C. R., Doherty, M. E. & Tweney, R. D. 1977: Confirmation bias in a simulated research environment: An experimental study of scientific inference. *Quarterly Journal of Experimental Psychology*, *29*, 85–95.

Newell, A. 1969: Heuristic programming: Ill-structured problems. In J. Aronsky (Ed.), *Progress in operations research*. Vol. 3. New York: Wiley.

Newell, A. & Simon, H. A. 1972: *Human problem solving*. Englewood Cliffs, N.J.: Prentice Hall.

Newell, A., Shaw, J. C. & Simon, H. A. 1958: Elements of a theory of human problem solving. *Psychological Review*, *65*, 151–66.

Newell, A., Shaw, J. C. & Simon, H. A. 1979: The processes of creative thinking. In H. A. Simon (Ed.), *Models of thought*. New Haven: Yale University Press.

Olton, R. M. 1980: Experimental studies of incubation: Searching for the elusive. *Journal of Creative Behavior*, *13*, 9–22.

Olton, R. M. & Johnson, D. M. 1976: Mechanisms of incubation in problem solving. *American Journal of Psychology*, *89*, 617–30.

Osborn, A. F. 1963: *Applied Imagination*. New York: Scribners.

Paivio, A. 1971: *Imagery and verbal processes*. New York: Holt, Rinehart & Winston.

Paivio, A. 1986: *Mental representations*. Oxford: Oxford University Press.

Patrick, C. 1938: Scientific thought. *Journal of Psychology*, 55–83.

Parnes, S. J. 1963: The deferment of judgment principle: A clarification of the literature. *Psychological Reports*, *52*, 117–22.

Perkins, D. N. 1981: *The mind's best work*. Cambridge, Mass.: Harvard University Press.

Phillips, L. & Edwards, W. 1966: Conservatism in a simple probability inference task. *Journal of Experimental Psychology*, *72*, 346–54.

Pitz, G. F. 1969: An inertia effect (resistance to change) in the revision of opinion. *Canadian Journal of Psychology*, *23*, 24–33.

Pitz, G. F. 1975: Bayes' Theorem: Can a theory of judgment and inference do without it? In F. R. Restle, R. M. Schiffrin, N. J. Castellan, H. R. Lindman & D. B. Pisoni (Eds.), *Cognitive Theory*. Vol. 1. Hillsdale, N.J.: Erlbaum.

Pitz, G. F., Downing, L., & Rheinhold, H. 1967: Sequential effects in the revision of subjective probabilities. *Canadian Journal of Psychology*, *21*, 381–93.

Pounds, W. 1969: The process of problem finding. *Industrial and Management Review*, *11*, 1–19.

Raaheim, K. 1964: Analysis of the missing part in problem solving. *Scandinavian Journal of Psychology*, *5*, 149–52.

Raaheim, K. 1974: *Problem solving and intelligence*. Oslo/Bergen/Tromsø: Norwegian Universities Press.

Reitman, W. R. 1965: *Cognition and thought*. New York: Wiley.

Rickards, T. & Freedman, B. L. 1978: Procedures for management in idea-deficient situations: An examination of brainstorming approaches. *The Journal of Management Studies*, 43–55.

Roe, A. 1953: A psychological study of eminent psychologists, and a comparison with biological and physical scientists. *Psychological Monographs*, *67*, No. 2 (Whole No. 352).

Rothenberg, A. 1976: The process of Janusian thinking. In A. Rothenberg (Ed.), *The creativity question*. Durham, N.C.: Duke University Press.

Sanford, A. J. 1985: *Cognition and cognitive psychology*. London: Weidenfeld & Nicholson.

Scheerer, M. 1963: Problem solving. *Scientific American*, *208*, 118–28.

Schiffrin, R. M. 1978: Capacity limitations in information processing, attention and memory. In W. K. Estes (Ed.), *Handbook of learning and cognitive processes*. New York: Wiley.

Simon, H. A. 1945: *Administrative behavior*. New York: Macmillan.

Simon, H. A. 1957: *Models of man*. New York: Wiley.

Simon, H. A. 1965: *The shape of automation*. New York: Harper & Row.

Simon, H. A. 1966: Scientific discovery and the psychology of problem solving. In R. G. Colodny (Ed.), *Mind and cosmos: Essays in contemporary science and philosophy*. Pittsburgh: University of Pittsburgh Press.

Simon, H. A. 1969: *The sciences of the artificial*. Mass.: M.I.T. Press.

Simon, H. A. 1973: The structure of ill structured problems. *Artificial Intelligence*, 4, 181–201.

Simon, H. A. 1977: *The new science of management decision*. Englewood Cliffs, N.J.: Prentice-Hall.

Simon, H. A. 1978: Information processing theory of human problem solving. In W. K. Estes (Ed.), *Handbook of learning and cognitive processes*. New York: Wiley.

Simon, H. A. 1979: *Models of thought*. New Haven: Yale University Press.

Simon, H. A. 1981: Cognitive science: The newest science of the artificial. In D. Norman (Ed.), *Perspectives on cognitive science*. Hillsdale, N.J.: Erlbaum.

Simon, H. A. 1983: *Reason in human affairs*. Oxford: Basil Blackwell.

Simon, H. A. & Sumner, R. K. 1968: Pattern in music. In B. Kleinmuntz (Ed.), *Formal representation of human judgement*. New York: Wiley.

Simon, H. A. et al. 1986: Report of research briefing panel on decision making and problem solving. *Research Briefings 1986*. Washington D.C.: National Academy Press.

Shepard, R. N. 1978: The mental image. *American Psychologists*, 125–37.

Spearman, C. 1904: General intelligence objectively determined and measured. *American Journal of Psychology*, 15, 201–93.

Spearman, C. 1927: *The abilities of man*. New York: Macmillan.

Stein, M. I. 1975: *Stimulating creativity. Vol. 2. Group procedures*. New York: Academic Press.

Sternberg, R. J. 1979: The nature of mental abilities. *American Psychologist*, 34, 214–30.

Sternberg, R. J. 1985: *Beyond IQ*. Cambridge: Cambridge University Press.

Wason, P. C. 1960: On the failure to eliminate hypotheses in a conceptual task. *Journal of Experimental Psychology*, 12, 129–40.

Wason, P. C. 1968: "On the failure to eliminate hypotheses . . ."—a second look. In P. C. Wason & P. Johnson-Laird (Eds.), *Thinking and reasoning*. Baltimore: Penguin.

Weisberg, R. W. & Suls, J. 1973: An information processing model of Duncker's candle problem. *Cognitive psychology*, 4, 255–76.

Weizenbaum, J. 1984: *Computer power and human reason*. Harmondsworth: Penguin.

Wertheimer, M. I. 1959: *Productive thinking*. New York: Harper & Row.

Wickelgren, W. A. 1974: *How to solve problems: Elements of a theory of problems and problem solving*. San Francisco: Freeman.

Winston, P. H. 1977: *Artificial intelligence*. Reading, Mass.: Addison-Wesley.

Woodworth, R. S. & Schlosberg, H. 1955: *Experimental psychology*. London: Methuen.

5

From Individual Creativity to Organizational Innovation

TERESA M. AMABILE

What is the creative process like when it occurs within an organization? What are the important influences on that process? The best way to begin examining creativity within an organization may be to simply ask the people who work there. That is precisely the approach that Stan Gryskiewicz and I have taken over the past two years. We have conducted interviews with 120 scientists working in Research and Development laboratories within a wide variety of corporations from around the world. Because we wanted to discover the specific stimulants and obstacles to creativity in the corporate environment (and, particularly, within the R&D environment), we asked a very specific question. We asked each scientist to tell us about two events from his or her work experience—one that stood out as an example of high creativity and one that stood out as an example of low creativity (defining creativity in whatever manner they wished). We told them to describe the problem or task, the context, and anything about the persons involved or the work environment that distinguished that event from others.

These comments were typical of those describing environmental factors present in the high creativity events (Amabile and Gryskiewicz 1985):

> For the event I chose, the thing I want to focus on is the highly creative environment that allowed things to occur. The project was 3 people who were put together in sort of a "do or die" situation, which was, Show us if you can make this work, or tell us if you can't—we will give you whatever resources are necessary. They then put three people from three different divisions in one area, so that we were essentially sharing one office. From then on, we went whatever way we felt we wanted to go. They set some broad objectives at the beginning, but other

than that, it was up to us to define how that objective was to be met. Because of the freedom we were given, we could go into any area where before there would have been communication problems.

*   *   *

The people were in a group environment; they all knew what it meant to the company, so everybody felt part of it and everybody contributed. The leader, Bob, had a very unique way of keeping the people informed. Each week, and sometimes more, we would have a group meeting to update the situation. Bob could communicate very well with all the people. Every week, everyone knew the direction that had to be taken and what was happening.

*   *   *

Something very important was the support provided. There was an allocation of people and capabilities throughout the organization. It was possible to get lots of diverse information.

*   *   *

We wrote up a proposal to use this technique. It was submitted to the management in our division, where it got support. Then we reported about this work in meetings with other divisions, and there was a lot of interest. People wanted us to start right away.

*   *   *

Management was very open-minded in allowing people to do what they wanted to do. Their attitude was: Here is a new invention; let's see where we can make it fit. Management was so fired up that they took us, as lower-level individuals, all the way up to the vice president.

*   *   *

What makes a difference is the atmosphere where you are recognized as a creative person and allowed to pursue the creativity rather than being treated as a production worker.

*   *   *

It turned out that two things needed to be invented. We suggested ways of doing both of those things, but unfortunately we met with very little enthusiasm in the research environment that was responsible for the product. But through the efforts of our innovation office (to whom we had submitted a memo on our ideas), some strong interest was quickly generated in a key marketing area. That marketing area contacted the research group responsible for producing the material, the key connection was made, and the work began to be done.

*   *   *

Having the other scientists to talk to is important. When it came time to introduce the product, he had support from different areas such as marketing, which enhanced its coming into the marketplace.

There were just as many comments about environmental factors that served as obstacles in the stories of low creativity. These are typical:

The thing that is sometimes hard to understand here is my manager, saying, "If you do that you will be rewarded". He shouldn't even say that. I had the motivation to do it in the first place. I don't know if I really got as much credit for the project as I should have, but that's kind of immaterial—I have a tremendous amount of self-satisfaction from it and that's what I'm looking for instead of some monetary situation. I wouldn't reject it, but it is a measure of success—not a motivator. Here, they think that in order to get someone to work they have to say, "You do that and we will reward you".

*   *   *

That was a low creativity situation because they wanted me to follow a particular path without adding any of my input into it. If you want to use your creativity, you can't be told the exact way something should be done.

*   *   *

There is pressure from management to get results, which contributes to a conservative approach. If you stay safe and make little improvements, it will show on your appraisal form that you are moving in the right direction. If you take a chance and it bombs, they will recognize that you did interesting work, but record that you did not come up with a product. You will get marked down in that category because it did not make any money, even though it was a "nice piece of work".

*   *   *

The key problem is that the managers, being young, still want to do research, and they compete with the technical staff.

*   *   *

The people in management change their minds a lot, shifting goals. Sometimes you don't want to focus on a problem because you know it will be changed.

*   *   *

The inputs were changing with time, so there was not a well-defined goal for the people involved.

*   *   *

Lack of communication was the biggest thing that made this fall over.

*   *   *

Marketing dictated a particular format and everybody jumped in to accomplish marketing's goal. Eleven years later, it is still not a success. They still have format constraints on them. Anytime they try anything else, they are told that there isn't a market for it. Instead of building a technology base and trying to solve problems one at a time, marketing is saying this is where the need is, and anything else is not related to this project.

*   *   *

It is very difficult to develop something and then have support for it and have it move forward—one of the difficulties of a large company.

*   *   *

From the beginning, the concept really lacked any innovative thought. It was a defensive strategy; instead of looking for a concept that would be one step beyond what was currently available, we settled for coming out with something that was a "me too" product.

When we did a detailed content analysis of the transcripts of these interviews, we identified nearly 1,350 comments about personal qualities of the scientists involved in the events (talents, experience, personality traits, etc.) and about factors in the work environment. Interestingly, about two-thirds of all comments were about the work environment. This suggests that, though personal qualities of workers certainly played a role, factors in the work environment seemed to have had more of an impact on the creative outcome of the projects described by our interviewees. Certainly, the talents and working styles of individual scientists must contribute heavily to a project's success. But it is quite likely that, in the high-powered organizations we studied, only very talented and highly skilled scientists are hired in the first place. Given this high baseline level of personal qualities, then, environmental factors might well account for a larger percentage of the variance in creative success.

Among these environmental factors, obstacles and stimulants were mentioned with about equal frequency. Within both major themes, the factors mentioned fell into three major categories: (1) factors of organizational climate or corporate culture, such as attitudes towards innovation and risk-taking, organizational structures, evaluation systems, communication channels, and reward procedures; (2) factors of management style, both at the level of the organization or division and at the level of the individual project; and (3) resources, including resources of materials, money, people, and time.

A consideration of the most frequent comments made by these interviewees, within the context of literature on the creative process, leads to three general conclusions. First, an individual's creativity within an organization is influenced by the same factors that influence creativity in other contexts. Repeatedly, as illustrated in the quotes presented earlier, our interviewees mentioned certain specific features of management style that appeared to inhibit individual creativity: evaluation pressure and feelings of constant performance appraisal, reward structures that tie specific rewards closely to specific tasks, competition, and—most importantly—a constraint of choice in methods for carrying out a task (Amabile 1984). All of these agree well with the growing body of experimental evidence on factors that undermine different types of creativity in different contexts and for different subject populations. These experimentally-studied factors include evaluation expectation (Amabile 1979; Amabile, Goldfarb, and Brackfield 1982); surveillance (Amabile, Goldfarb, and Brackfield 1982); reward (Amabile, Hennessey, and Grossman in press; Kruglanski, Friedman, and Zeevi 1971; McGraw and McCullers 1979); competition (Amabile 1982a); restriction of choice (Amabile and Gitomer 1984); and a general orientation toward extrinsic constraints rather than toward the intrinsic properties of a task (Amabile 1985).

The second general conclusion that can be made from these interviews is that an individual's creativity within an organization is also influenced by a number of factors that are *specific* to the organizational context (Amabile 1984). Many of these factors are illustrated in the interview excerpts quoted earlier: goal-setting by management, communication between different areas of an organization, the availability of material resources and people resources within the organization, management attitudes toward innovation, the presence or absence of interest and cooperation from different areas of the organization, and the presence or absence of constructive feedback.

The third general conclusion is that the process of individual creativity is a crucial part of the entire process of innovation within an organization, but it is only *one* part. Other elements are also essential—both before and after the individual generates creative ideas. Some of these other elements were illustrated in the quoted interview excerpts: a mechanism for hearing and disseminating new ideas; practical follow-up on ideas from various areas of the organization; a broad rather than a narrow corporate approach to achieving organizational goals; and an offensive rather than a defensive strategy toward doing the business of the organization (Amabile 1984).

In the remainder of this chapter, I will use each of these three observations to establish a more complete understanding of the causes and consequences of creativity and innovation in organizations. Specifically, I will discuss the process of individual creativity and then explore the intricate connections between individual creativity and organizational innovation, with the aim of moving toward a comprehensive theory of innovation within organizations.

DEFINITIONS

Because I will be drawing distinctions between them, it is important to begin with at least some general definitions of the terms *creativity* and *innovation*. Most current definitions of creativity within the psychological literature include the concepts of novelty and appropriateness as the two main ingredients. In order to be considered creative, a product or idea must be different from what came before it. But it cannot be merely bizarre; it must be in some way appropriate to the requirements of the task at hand. For reasons that will be discussed later, I believe it is important to constrain the definition of creativity by limiting its application to a certain type of task. Simply put, it does not make sense to talk of a creative response to a problem where the path to solution is very clear and straightforward. Thus, this is the definition of creativity that I have adopted (cf. Amabile 1983a): *A product or idea is creative to the extent that it is both a novel and appropriate response to an open-ended task*. In this context, an open-ended task is one that does not have a clear and straightforward path to solution.

This definition, however, like most psychological definitions of creativity, is conceptual rather than operational. The conceptualizations have not been translated into actual assessment criteria; we have no clear, objective way of measuring novelty and appropriateness in products or ideas. One solution that I have suggested to this dilemma (Amabile 1983) is the adoption of two complementary definitions of creativity—a conceptual definition that can be used in building theoretical formulations of the creative process, and an operational definition that is readily applicable to empirical research. For a variety of reasons, it appears most reasonable to adopt an operational definition that is based on the subjective assessment of products by experts. As long as there is consensus in experts' ratings of products on creativity, we can reasonably accept those ratings as valid assessments. They should be more valid, in fact, than any explicit definition of creativity that we, the researchers, could provide to creativity judges (assuming that no creativity researcher could be considered an "expert" in all fields of endeavor). We have found that, although it is quite difficult for experts in any given field to articulate exactly what they mean by "creativity", they do agree quite well in their subjective ratings of the degree of creativity in products from that field. This is the basis of the consensual definition of creativity (Amabile 1982b): *A product or response is creative to the extent that appropriate observers independently agree it is creative. Appropriate observers are those familiar with the domain in which the product was done or the response articulated.*

Clearly, this definition cannot be used effectively at the frontiers of any field. It is well known that the most creative work in any field is often ignored or ridiculed until enough time has passed that people can understand it. There is often no consensual agreement on such products. This is because, in effect, there are no experts suitable to judge these works; the works essentially define their own new field. Only with the passage of time can such pioneering products or ideas be judged on creativity. Thus, the consensual definition of creativity will not work at the frontiers of any field. But neither will any other operational definition of creativity.

In most studies of creativity, however, we are not examining the very highest levels. So the consensual approach to creativity assessment has worked well in the experimental studies of creativity cited earlier, and it is, in fact, the approach taken in the interview study of R&D scientists. As noted earlier, we asked participants in that study to use their own subjective definitions of creativity in identifying high-creativity and low-creativity events from their work experience. Although organizational innovation certainly depends on individual creativity, the two cannot be considered identical. For one thing, the term "innovation", when applied to an organization, implies more than simple creative thinking in a single individual. It suggests a concerted effort by an aggregate of individuals, directed toward doing something novel and appropriate in their business. Also, creative ideas can fail to become innovations not only through a lack of creativity at various stages of implementation, but also through a simple lack of resources or inappropriate timing. Thus, it seems best to define innovation in this way: *Innovation is the successful implementation of creative ideas about products or processes within an organization.* In this definition, "products" and "processes" are very broadly conceived. Products can include anything produced by an organization, from automobiles to theatrical productions, and processes can include any methods of production, methods of management, methods of doing business, or services offered by the organization.

Just as with the conceptual definition of creativity, this conceptual definition of innovation does not readily translate into assessment operations. There is no straightforward way to objectively measure how successfully a creative idea has been implemented. Nonetheless, it is reasonable to assume that, as with assessments of creativity, experts in a domain can agree to a satisfactory degree on the level of innovation exhibited. For that reason, a separate operational definition of innovation will be useful for empirical research. Not only will it be necessary to have consensual agreement about the level of creativity in ideas, but it will also be necessary to have consensual agreement about the degree to which those creative ideas have become successful innovations. This, then, is the operational definition of innovation: *A product or process is innovative to the extent that appropriate observers independently agree it is innovative. Appropriate observers are those familiar with the domain in which the product or process was introduced.*

THE PROCESS OF INDIVIDUAL CREATIVITY

A basic premise of the model I propose for organizational innovation is that individual creativity is a necessary but not sufficient condition of innovation. Because individual creativity plays a crucial role, it is important to consider this process and the factors that influence it. In an earlier paper, I presented a componential model of the individual creative process (Amabile 1983a). I will review that model here, and will then proceed to a discussion of implications for the broader model of organizational innovation.

This conceptualization of creativity relies on a number of assumptions about creative production, assumptions based in both formal and information observation. Most fundamentally, this model relies on some assumptions about the nature of creativity. I assume that there is a continuum of creativity from the lower levels of "garden variety" creativity observed in everyday life to the high level of creativity involved in historically significant advances in any field of endeavor. In contrast to views that creativity is an all-or-nothing entity, this perspective suggests that it is at least theoretically possible for anyone with normal cognitive abilities to produce creative work in some domain. In addition, I assume that there can be different degrees of creativity within a single individual's work within a single domain. For example, it seems reasonable to assume that a particular scientist will produce some less creative work and some more creative work throughout the course of his or her career, and even perhaps within the course of a brief time period. Furthermore, although different individuals may be quite distinct in their potential for creative performance in a given domain (due to unique talents and stable traits), it does appear to be possible to increase creativity to some extent (Stein 1974, 1975).

The model of individual creativity must take into account a number of observations that researchers have made about creative behavior. First, at least for high levels of creativity, there often seems to be a "match" between individuals and domains (Feldman 1980). There appears to be a particularly good fit, for example, between one individual and mathematics or between another individual and musical composition. Such matches seem due to both innate talent and interest patterns. Second, education and training in cognitive skills are essential, but for high levels of creativity they do not appear to be sufficient by themselves. Third, particular clusters of personality traits may correlate fairly well with consistent creativity in individuals (see Stein 1974), but again, they are not sufficient in and of themselves. Certainly, although a given individual might be accurately described by a particular constellation of "creative" personality traits, that person is not creative at all times in all domains—even if he or she does sometimes produce notably creative work in some domains. Fourth, innate abilities ("talents") in a given domain do appear to be important for noteworthy levels of creativity, but education seems also to be essential in most outstanding creative achievements (Feldman 1980). Fifth, although an eagerness to work diligently appears to be an essential component of high levels of creativity (Golann 1963), and although a number of introspective accounts describe the phenomenology of creativity as marked by deep involvement in the activity at hand, these accounts also stress the importance of intellectual playfulness and freedom from external constraints (e.g. Einstein 1949).

THE COMPONENTS OF INDIVIDUAL CREATIVITY

The componential model of creativity was designed to account for those well-established phenomena: the importance of talents, education, cognitive skills, interest patterns, and personality dispositions, all functioning interactively to influence creative behavior, as well as the apparently contradictory importance of both "work" and "play" in the motivation for creative behavior.

Figure 1 outlines the three major components necessary for individual creativity. Although I have proposed (Amabile 1983a) that the three main components constitute a complete set of the general factors necessary for creativity, the listing of elements within each component can only be completed gradually, as progress is made in creativity research. The elements included within each of the components in Figure 1 are examples of the kind of elements that each component contains.

1 DOMAIN-RELEVANT SKILLS	2 CREATIVITY-RELEVANT SKILLS	3 TASK MOTIVATION
Includes:	Includes:	Includes:
- Knowledge about the domain - Technical skills required - Special domain-relevant "talent"	- Appropriate cognitive style - Implicit or explicit knowledge of heuristics for generating novel ideas - Conducive work style	- Attitudes toward the task - Perceptions of own motivation for undertaking the task
Depends on:	Depends on:	Depends on:
- Innate cognitive abilities - Innate perceptual and motor skills - Formal and informal education	- Training - Experience in idea generation - Personality characteristics	- Initial level of intrinsic motivation toward the task - Presence or absence of salient extrinsic constraints in the social environment - Individual ability to cognitively minimize extrinsic constraints

Fig. 1. Components of creative performance.

Within the componential framework, domain-relevant skills can be considered as the basis from which any performance must proceed. They include factual knowledge, technical skills, and special talents in the domain in question. Creativity-relevant skills include a cognitive style favorable to taking new perspectives on problems, an application of heuristics for the exploration of new cognitive pathways, and working style conducive to persistent, energetic pursuit of one's work. Task motivation represents the motivational variables that determine an individual's approach to a given task.

The three components appear to operate at different levels of specificity. Creativity-relevant skills operate at the most general level; they may influence responses in any content domain. Thus, some highly creative individuals do indeed appear to be creative "types", in the sense that they produce unusual responses in many domains of behavior. Domain-relevant skills operate at an intermediate level of specificity. This component includes all skills relevant to a general domain, such as mathematical reasoning, rather than skills relevant to only a specific task within a domain, such as devising an equation to describe the motion of a certain space satellite. Obviously, within a particular domain, skills relevant to any given specific task will overlap with skills relevant to any other task. Finally, task motivation operates at the most specific level. In terms of impact on creativity, motivation may be very specific to particular tasks within domains, and may even vary over time for a particular task. For example, an artist may have a high level of motivation for exploring new painting techniques within her own studio, but may have a low level of motivation for a commercial assignment to paint advertising copy illustrations.

Figure 1 includes several elements within each of the three components. Although all three components appear to be necessary for some recognizable level of creativity, it might not be necessary to have all possible elements within each component. For example, it might be possible for an engineer to design a creative new testing instrument without having any special talent for visually imagining his designs. Of course, it *could* be that such talent is essential for even moderate levels of creativity in such a task. The point is that only future research can indicate which elements constitute a complete set within any one of the components, and which elements are indeed essential.

Domain-relevant Skills

This component comprises the individual's complete set of response possibilities—response possibilities from which the new response is to be synthesized and information against which the new response is to be judged. This component can be viewed as the set of cognitive pathways for solving a given problem or doing a given task. Some of the pathways are more common, well-practiced, or obvious than others, and the set of pathways may be large or small. The larger the set, the more numerous the alternatives available for producing something new, for developing a new combination of steps. As Newell and Simon (1972) poetically described it, this set can be considered the problem solver's "network of possible wanderings" (p. 82). (I am using problem-solving terminology here in a very general sense, where problem solving refers to doing virtually any task that requires cognitive involvement aimed at an observable goal-directed response.)

This component includes familiarity with and factual knowledge of the domain in question: facts, principles, attitudes toward various issues in the domain, knowledge of paradigms, performance "scripts" for solving problems in the domain (Schank and Abelson 1977), and aesthetic criteria. Each of these elements is very broadly conceived. For example, "paradigms" may include anything from formal scientific paradigms to traditional modes of operation in personnel management or standard techniques of advertising. And "aesthetic criteria" may include anything from standards for artistic merit to notions of what constitutes an elegant marketing strategy.

Domain-relevant skills constitute the individual's "raw materials" for creative productivity. Certainly, it is impossible to be creative in planning financial strategy unless one knows something (and probably a great deal) about the stock market, money markets, and current economic trends. In addition to basic knowledge, the component includes technical skills that may be required by a given domain, such as laboratory techniques or techniques for making etchings, and special domain-relevant talents such as an engineer's ability to visually imagine his designs. Domain-relevant skills appear to depend on innate cognitive, perceptual, and motor abilities, as well as on formal and informal education in the domain of endeavor.

The nature of the domain-relevant information and the manner in which it is stored can make an important difference in creative production. Wickelgren (1979) has argued that "the more we concentrate on . . . heavily contextualized (specific) concepts and propositions, the less capacity we will have available to learn general principles and questions that crosscut different areas and perspectives" (p. 382). In other words, knowledge organized according to general principles is of greater utility than specific, narrowly applicable collections of facts. Likewise, performance information organized according to general approaches to problems rather than blind response algorithms should be more likely to contribute to high levels of creativity. This perspective directly contradicts the popular notion that creativity will decrease as a person gains more knowledge in a domain. In general, an increase in domain-relevant skills can only lead to an increase in creativity—if the domain-relevant information is organized appropriately. Inappropriate organization of information would be marked by a memorization and rote acceptance of established algorithms in a field; appropriate organization would be marked by a learning of basic principles and a critical examination of established algorithms. This proposition fits well with the assertion of previous theorists (e.g. Campbell 1960) that larger stores of properly coded knowledge increase the probability of outstanding responses. In other words, although it is possible to have too many algorithms, it is not possible to have too much knowledge.

Creativity-relevant Skills
Creativity-relevant skills determine the way in which problem-solving proceeds and, to a large extent, the final level of performance attained. Herein lies the "something extra" of creative performance. Assuming that an individual has some incentive to

perform an activity, performance will be "good" or "adequate" or "acceptable" if the requisite domain-relevant skills exist. However, even with these skills at an extraordinarily high level, an individual will not produce creative work if creativity-relevant skills are lacking.

This component includes, first, a cognitive-perceptual style characterized by a facility in understanding complexities and an ability to break set during problem-solving. Several specific aspects of cognitive-perceptual style appear to be relevant to creativity (cf. Amabile 1983a): (a) breaking perceptual set; (b) breaking cognitive set, or exploring new cognitive pathways; (c) keeping response options open as long as possible; (d) suspending judgment; (e) using "wide" categories in storing information; (f) remembering accurately; and (g) breaking out of performance "scripts".

The creativity-relevant skills component also includes knowledge of heuristics for generating novel ideas. A heuristic can be defined as "any principle or device that contributes to a reduction in the average search to solution" (Newell et al. 1962, p. 152). Thus, a heuristic may be considered as a general strategy that can be of aid in approaching problems or tasks. Several theorists and philosophers of science have proposed creativity heuristics: (a) "When all else fails, try something counterintuitive" (Newell et al. 1962); (b) "Make the familiar strange" (Gordon 1961); (c) generate hypotheses by analyzing case studies, use analogies, account for exceptions, and investigate paradoxes (McGuire 1973). Clearly, creativity heuristics are best considered as methods of approaching a problem that are most likely to lead to set-breaking and novel ideas, rather than as strict rules applied by rote. Although these heuristics may be stated explicitly by the person using them, they may also be known at a more implicit level and used without direct awareness.

A work style conducive to creativity appears to be a critical element of creativity-relevant skills. For example, an ability to concentrate effort for long periods of time may be an important facet of such a work style (Campbell 1960; Hogarth 1980), along with an ability to use "productive forgetting"—the ability to abandon unproductive search strategies and temporarily put aside stubborn problems (Simon 1966).

In an important way, creativity-relevant skills depend on personality characteristics related to independence, self-discipline, ability to delay gratification, perseverance in the face of frustration, and an absence of conformity in thinking or dependence on social approval (Feldman 1980; Golann 1963; Hogarth 1980; Stein 1974).

In addition, though, creativity-relevant skills also depend on training, through which they may be explicitly taught, or simply on experience with idea generation, through which an individual may devise his or her own strategies for creative thinking. A great deal of previous research has investigated these elements, including work on creativity-training programs, such as brainstorming (Osborn 1963) and synectics (Gordon 1961), and research on the "creative personality" (e.g. Barron 1955; Cattell and Butcher 1968; MacKinnon 1962; Wallach and Kogan 1965).

Task motivation
This is the component of individual creative performance that, perhaps, has been most neglected by creativity researchers and theorists. Yet, in some ways, this may be the most important component. No amount of skill in the domain or in methods of creative thinking can compensate for a lack of appropriate motivation to do an activity. But, to some extent, a high degree of proper motivation *can* make up for a deficiency of domain-relevant skills or creativity-relevant skills. And, because task motivation appears to depend strongly on the work environment, this may be the easiest component to address in attempts to stimulate creativity.

Within my model, task motivation includes two elements: the individual's baseline attitude toward the task and the individual's perceptions of his or her reasons for undertaking the task in a given instance. A baseline attitude toward the task is simply the person's natural inclination toward or away from activities of that sort. Perceptions of one's motivation for undertaking the task in a given instance, however, appear to depend largely on external social and environmental factors—the presence or absence of salient extrinsic constraints in the work environment. Extrinsic constraints are external factors intended to control or seen as controlling the individual's performance on the task in a particular instance. As such, the constraint is extrinsic to the work itself; it is not an essential feature of task performance, but it is introduced by the social environment. A salient extrinsic constraint is one whose controlling implications are clear to the individual during task engagement.

In addition to external constraints, internal factors, such as a person's ability to cognitively minimize the salience of such extrinsic constraints, might also influence the self-perception of motivation. Thus, the final level of task motivation in a particular instance varies from the baseline level of intrinsic motivation as a function of extrinsic constraints that may be present in the situation and the individual's strategies for dealing with these constraints.

This, then, is the intrinsic motivation hypothesis of creativity: Intrinsic motivation is conducive to the idea-generation stage of creativity, but extrinsic motivation is detrimental. People who are motivated primarily by factors extrinsic to the work itself are less likely to produce creative ideas than people who are motivated primarily by their own interest in the intrinsic properties of the work itself.

Over the past few years, a number of studies have shown that extrinsic constraints in the work environment can indeed undermine creative performance. I cited some of these studies earlier. They have demonstrated the negative impact of constraints as varied as evaluation, surveillance, reward, competition, restricted choice, and a general extrinsic motivational orientation toward work (e.g. Amabile 1979, 1982a, 1985; Amabile and Gitomer 1984; Amabile, Goldfarb, and Brackfield 1982; Amabile, Hennessey, and Grossman 1984; Kruglanski, Friedman, and Zeevi 1971; McGraw and McCullers 1979). Thus, this model proposes that any of a wide variety of extrinsic constraints will, by impairing intrinsic motivation, have detrimental effects on creative performance. In other words, task motivation is the most important determinant between what an individual *can* do and what he *will* do, in terms of creative performance. The former depends on the level of domain-relevant skills and creativity-relevant skills. But it is task motivation that determines the extent to which domain-relevant skills and creativity-relevant skills will be fully and appropriately engaged in the service of creative performance.

Stages of the Individual Creative Process

Figure 2 provides a schematic representation of the componential model of the creative process (Amabile 1983a). This model describes the way in which an individual might assemble and use information in attempting to arrive at a solution, response, or product. In information-processing terms, task motivation is responsible for initiating and sustaining the process; it determines whether the search for a solution will begin and whether it will continue, and it also determines some aspects of response generation. Domain-relevant skills are the material drawn on during operation. They determine what pathways will be searched initially and what criteria will be used to assess the response possibilities that are generated. Creativity-relevant skills act as an executive

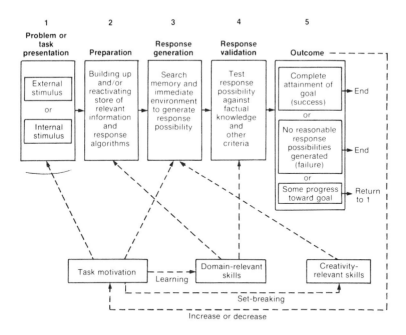

Fig. 2. Componential framework of creativity. Broken lines indicate the influence of particular factors on others. Solid lines indicate the sequence of steps in the process. Only direct and primary influences are depicted here.

controller; they can influence the way in which the search for responses will proceed.

This model resembles previous theories of creativity in the specification of the stages of problem presentation, preparation, response generation, and response validation (e.g. Wallas 1926; Hogarth 1980)—although there are a number of variations on the exact number and naming of stages in the sequence. This model is more detailed than previous ones, however, in its inclusion of the impact of each of the three components of creativity at each stage in the process.

The process outlined in Figure 2 applies to both high and low levels of creativity; the level of creativity of a product or response varies as a function of the levels of each of the three components. Each component is necessary, and no one component is sufficient for creativity in and of itself. Thus, although this framework cannot be considered a detailed mathematical model of the creative process, it is, in a general sense, a multiplicative model. No component may be absent if some recognizable level of creativity is to be produced, and the levels of all the components together determine the final level of creativity achieved.

The initial step in this sequence is the presentation of the task or the problem. Task movtivation has an important influence at this stage. If the individual has a high level of intrinsic interest in the task, this interest will often be sufficient to begin the creative process. Under these circumstances, the individual, in essence, poses the problem to himself. In other situations, however, the problem is presented by someone else. The problem might, of course, be intrinsically interesting under these circumstances, as well. However, it is likely that, in general, an externally posed problem is less intrinsically interesting to the individual.

The second stage is preparatory to the actual generation of responses or solutions. At this point, the individual builds up or reactivates a store of information relevant to the problem or task, including a knowledge of response algorithms for working problems in the domain in question. In the case where domain-relevant skills are rather impoverished at the outset, this stage may be quite a long one during which a great deal of learning takes place. On the other hand, if the domain-relevant skills are already sufficiently rich to afford an ample set of possible pathways to explore during task engagement, the reactivation of this already-stored set of information and algorithms may be almost instantaneous.

The novelty of the product or response is determined in the third stage. Here, the individual generates response possibilities by searching through the available pathways and exploring features of the environment that are relevant to the task at hand. Both creativity-relevant skills and task motivation play an important role at this stage. The existing repertoire of creativity-relevant skills determines the flexibility with which cognitive pathways are explored, the attention given to particular aspects of the task, and the extent to which a particular pathway is followed in pursuit of a solution. In addition, creativity-relevant skills can influence the subgoals of the response-generation stage by determining whether a large number of response possibilities will be generated through a temporary suspension of judgment. If task motivation is intrinsic rather than extrinsic, it can add to the existing repertoire of creativity-relevant skills a willingness to take risks with this particular task and to notice aspects of the task that might not be obviously relevant to attainment of a solution.

Domain-relevant skills again figure prominently in the fourth stage—the validation of the response possibility that has been chosen on a particular trial. Using domain-relevant techniques of analysis, the response possibility is tested for correctness or appropriateness against the knowledge and the relevant criteria included within domain-relevant skills. Thus, it is this stage that determines whether the product or response will be appropriate, useful, correct, or valuable—the second response characteristic that, together with novelty, is essential for the product to be considered creative according to the conceptual definition of creativity.

The fifth stage represents the decision making that must be carried out on the basis of the test performed in stage 4. If the test has been passed perfectly—if there is complete attainment of the original goal—the process terminates. If there is complete failure—if no reasonable response possibility has been generated—the process will also terminate. If there is some progress toward the goal—if at least a reasonable response possibility has been generated (or if, in Simon's (1978) terms, there is some evidence of "getting warmer")—the process returns to the fifth stage, where the problem is once again posed. In any case, information gained from the trial is added to the existing repertoire of domain-relevant skills. If task motivation remains sufficiently high, another trial will be attempted, perhaps with information gained from the previous trial being used to pose the problem in a somewhat different form. If, however, task motivation drops below some critical minimum, the process will terminate.

For complex tasks, the application of this model to the production of creative responses also becomes complex. Work on any given task or problem may involve a long series of loops through the process, until success in a final product is achieved. Indeed, work on what seems to be one task may actually involve a series of rather different subtasks, each with its own separate solution.

The Feedback Cycle

The outcome of one cycle of the individual creative process can directly influence task motivation, thereby setting up a feedback cycle through which future engagement in the same or similar tasks can be affected. If complete success has been achieved, there will be no motivation to undertake exactly the same task again, because that task has truly been completed. However, with success, intrinsic motivation for similar tasks within the domain should increase. If complete failure has occurred—if no reasonable responses were generated—intrinsic motivation for the task should decrease. If partial success has been met, intrinsic motivation will increase when the problem solver has the sense of getting warmer in approaching the goal. However, it will decrease when the outcome of the tests reveals that the problem solver is essentially no closer to the goal than at the outset.

Harter's theory of "effectance motivation" (1978) suggests this influence of process outcome on task motivation. Harter built on White's (1959) definition of the "urge toward competence", a definition proposing a motivational construct "which impels the organism toward competence and is satisfied by a feeling of efficacy" (Harter 1978, p. 34). According to Harter's theory, failure at mastery attempts leads eventually to decreases in intrinsic motivation and striving for competence. However, success (which will be more probable the higher the level of skills) leads to intrinsic gratification, feelings of efficacy, and increases in intrinsic motivation, which, in turn, lead to more mastery attempts. In essential agreement with Harter, a number of social-psychological theorists have proposed that success (confirmation of competence) leads to increased intrinsic motivation.

Through its influence on task motivation, outcome assessment can also indirectly affect domain-relevant and creativity-relevant skills. A higher level of intrinsic task motivation may make set-breaking and cognitive risk-taking more probable and more habitual, thereby increasing the permanent repertoire of creativity skills. Also, a higher level of motivation may motivate learning about the task and related subjects, thereby increasing domain-relevant skills.

Why Does Motivation Make a Difference?

I stated earlier that task motivation, though extremely important, has been the most neglected component of individual creativity. And, over the past few years, there has been a growing body of evidence to suggest just how important motivation can be. We are only beginning to gather information, though, on *how* motivational state has an impact on creative performance. Why should it make a difference? What might be the mechanism whereby intrinsic motivation leads to higher levels of creativity than extrinsic motivation?

Three theoretical works can be useful in developing a model for this mechanism. The first of these comes from McGraw (1978), who suggested that the type of task is important in determining whether intrinsic or extrinsic motivation will be most conducive to performance. After reviewing an extensive literature on the effects of reward (an extrinsic constraint) on performance, McGraw proposed that reward (extrinsic motivation) will enhance performance on *algorithmic* tasks—tasks where there is a clear and straightforward path to solution. By contrast, reward (extrinsic motivation) will undermine performance on *heuristic* tasks—tasks where there is no straightforward path to solution, where some search is required. Clearly, by definition, a creativity task is a heuristic task.

The second theoretical work that is relevant here is that of Campbell (1960). When a task is heuristic, necessitating a search of possible pathways, what determines which pathways are explored? Campbell suggested that possibilities are produced more or less by a blind or random process. Certainly, the search can be narrowed down by various methods. However, Campbell proposed that some amount of blind search is always required with tasks of this nature. The more possibilities there are to be explored, he suggested, and the better the strategies for exploring them rapidly, the greater the likelihood of producing a novel yet appropriate response. Some degree of luck, however, is always an element (Hogarth 1980).

Finally, conceptualizations from the cognitive psychology of Simon, Newell, and their colleagues can be useful in understanding the mechanism by which task motivation influences the response-generation stage of the creative process. Simon (1967) postulated that the most important function of motivation is the control of attention. He proposed that motivation determines which goal hierarchy will be activated at any given time, and suggested that the more intense the motivation to achieve an original goal, the less attention will be paid to aspects of the environment that are irrelevant (or seemingly irrelevant) to achieving that goal. But attention to seemingly irrelevant aspects might be precisely what is required for creativity. For a creative response to be produced, it is often necessary to "step away" temporarily from the perceived goal (Newell, Shaw, and Simon 1962), to direct attention toward seemingly "incidental" aspects of the task and the environment.

It is now possible to postulate a mechanism for the influence of intrinsic or extrinsic motivational state on creativity, if we make one assumption about tasks and combine that assumption with key points from these three theorists. The assumption is that most tasks on which creativity might be shown can be done in either a relatively algorithmic way (by relying on well-worn, familiar methods) or a relatively heuristic way (by exploring new methods). The key points are these: (1) From McGraw, the notion that extrinsic motivation is most appropriate to algorithmic problem-solving; (2) From Campbell, the notion that creative activity always involves at least some heuristic exploration of possibilities; and (3) From the Simon-Newell team, the notion that motivation controls attention, which is a crucial determinant of whether creative discoveries will be made.

All of this leads to the proposition that motivational state affects creativity by influencing the likelihood that alternative—and potentially more creative—response possibilities will be explored during task engagement. The more single-mindedly a goal is pursued, the less likely it will be that creative response possibilities will be explored. An extrinsic motivation is one in which the individual is motivated primarily by the extrinsic goal, and not by the intrinsic aspects of the task itself. It is precisely under these conditions that the goal will be most single-mindedly pursued, and that creativity will be least likely. It is under these conditions that the creativity heuristics of exploration, set-breaking, and risk-taking are least likely to be used.

Perhaps the best way to explain this distinction is by using a metaphor. Imagine that a task or activity is a maze. For the sake of simplicity, assume that there is only one starting point for the maze. From that starting point, there may be a very clear, well-worn, and straight path to the outside; all that must be done to reach an exit is to proceed down that path. That exist *is* an acceptable way out; it is an acceptable solution. At the same time, however, it is not particularly new, exciting, or elegant; it is not particularly creative.

There may well be other exits from the maze, exits that would provide more novel, exciting, and elegant solutions; in other words, there are more creative ways out of the maze. But those exits cannot be reached by following the well-worn pathway. They can only be reached by exploration, and by taking the risk of running into a dead end here and there.

Someone who is extrinsically motivated is motivated primarily by something *outside* of the maze—the extrinsic goal. Since that goal can only be achieved once the maze has been exited, the best strategy for the extrinsically motivated person is to take the safest, surest, and fastest way out of the maze: the well-worn pathway, the uncreative route.

Someone who is intrinsically motivated, on the other hand, is motivated primarily by the interest, challenge, and enjoyment of *being in the maze.* Surely, there is no point in being in the maze if there is no desire to exit—to find a solution. Indeed, there may be strong desire to exit, often caused by external factors (such as the dire need of the organization for a solution to this problem, or strong competition from other organizations who are trying to achieve the same thing). But the important distinction between intrinsic and extrinsic motivation arises from both the individual's basic interest in the activity and the amount of freedom from extrinsic constraint in the immediate work environment. The intrinsically motivated person, because enjoyment of being in the maze is so high and concern about extrinsic pressures is so low, will be more likely to spend the cognitive energy exploring the maze, and will not be overly concerned about the possible dead-end risks involved. Thus, it is the intrinsically motivated person, working in an environment with low extrinsic controls, who is most likely to happen upon one of those elegant exits.

Matching the Motivation to the Job

It might be tempting, upon hearing about the negative impact of extrinsic motivation on creativity, to make the simplistic—and incorrect—assumption that external controls and extrinsic motivation are always harmful. In fact, there are a number of situations in which extrinsic motivation should not undermine performance and might, in fact, enhance it. Recall McGraw's (1978) conclusion that extrinsic motivation stimulates performance on algorithmic tasks; he also concluded that extrinsic motivation is helpful when the task is aversive rather than attractive. Obviously, there are any number of tasks in organizations and in other settings that are either quite algorithmic or quite aversive (or both). On such tasks, extrinsic motivation might be necessary to ensure that the task will be done and that it will be done quickly and effectively. It is only on heuristic tasks (creativity tasks) and on initially attractive tasks that extrinsic motivation will lead to decrements in performance. These, of course, constitute only a subset of the tasks that must be done within an organization; however, they might well be considered among the most important tasks.

Within the creative process itself, task motivation appears to be most crucial at the response generation stage—when the actual exploration of "the maze" is taking place. At other stages, extrinsic motivation might be harmless or might actually help. For example, if domain-relevant skills are initially lacking—as in the case of a new scientist in the lab who must learn the background of the research questions under consideration—extrinsic motivation might be useful at some points of the potentially tedious learning process. As another example, once a creative idea is generated, many of us need the extrinsic pressure of deadlines or evaluation to spur us on to validate the idea, test the idea, and communicate it clearly to others. To take a random example, generating and sketching out ideas on the psychology of creativity is a heuristic task that can be a great deal of fun and, for that reason, it happens best under intrinsic motivation. But the working through of those ideas in correct and readable prose would probably not happen at all were it not for the extrinsic motivation provided by editors and deadlines.

The moral, then, is that motivational state (and the work environment conditions that influence motivational state) should be matched to the type of task and the individual's initial motivational state toward the task. If the task is heuristic and the individual initially finds the task intrinsically motivating, then extrinsic constraints should be used cautiously and sparingly. In other cases, however, extrinsic constraints might be quite appropriate.

TOWARD A THEORY OF ORGANIZATIONAL INNOVATION

There are two major reasons for studying organizational innovation and for developing a theory of how innovation occurs in organizations. First, the problem has enormous practical significance, both socially and economically. Quite simply, organizations cannot continue to survive and thrive and have an impact in the modern world if they are incapable of successful innovation. Second, as I have shown in the earlier sections of this chapter, there is abundant evidence that social factors can have a major influence on an individual's creativity. Certainly, various sorts of social factors can have an impact on anyone's creativity, even a person who works in a relatively solitary mode (such as an independent artist for a freelance writer). But organizations represent a particularly good setting to systematically consider the impact of social factors on individual creativity because, in organizations, many of those factors are so systematized, structured, well-defined, and easy to identify.

In articulating a theory of organizational innovation, it might be reasonable to consider the process of organizational innovation as similar, in a broad sense, to the process of individual creativity. If the two can be considered as roughly analogous, then the theory of individual creativity can be used as a beginning framework for the organizational theory. Operating in this way, organizational innovation could be modeled by asking a series of basic questions: What are the components that contribute to organizational innovation? Is each of those components necessary for innovation? Are any sufficient in and of themselves? What are the stages by which organizational innovation occurs (i.e. what is the process of organizational innovation)? How and at what stages do the components of organizational innovation impact on the process?

Obviously, there are important differences between the process of individual creativity and the process of organizational innovation. The individual process models something that occurs within the mind and activity of a single person or, at most, within the minds and activities of a small number of people working together on the same specific problem. The organizational process occurs at the level of a *system*: a large number of individuals working together in different units on different aspects of the very general problem of implementing a new idea.

But, in a number of ways, the individual process and the organizational process seem to be similar. These similarities become evident when we consider the various aspects of individual creativity and organizational innovation discussed earlier and reviewed extensively in other sources (Amabile 1983b, 1984; Amabile and Gryskiewicz 1985). First, just as the individual needs some component of basic resources to draw on in generating and evaluating creative ideas (Domain-relevant skills), so does an organization need basic resources for all stages of the innovation process. Second, just as the individual needs some component of mechanisms for controlling the generation of creative ideas (Creativity-relevant skills), so the organization needs mechanisms for appropriately managing various stages of the innovation process. Third, just as the individual will be more creative under an intrinsic Task Motivation, it appears that an organization will be more innovative when the highest levels of leadership within the organization communicate a genuine motivation for innovation. Fourth, the organizational innovation process seems to go through stages that are analogous to the stages of individual idea-generation. There appears to be a stage where the general problem or innovational direction is set, a stage where the stage is set for generating the innovation, a stage where the idea is actually produced, a stage where the innovation is tested and implemented, and a stage where the outcome is evaluated.

A theory of organizational innovation must include all major facets of an organization, since all facets could have a potentially important impact on some stage of the innovation process. So, for example, if we are considering a traditionally-structured corporation, the theory should have some way to account for the impact of Corporate Administration, Research and Development, Manufacturing, Marketing, Finance, and Personnel. In the theory, each of these organizational facets should appear as an element within one or more of the components of innovation.

In order that the theory be a dynamic one, it will be important to specify the impact of each of the components on each stage of the innovation process, as well as the feedback-type impact of any innovation outcome on the components, and possible interactive influences between the components. Perhaps the most important specifications, however, are the nature and direction of the influences of organizational factors on individual creativity within the organization, and vice versa. I suspect, from my review of the current literature on creativity and organizational innovation, that individual creativity impacts the organizational innovation process at just one point: the actual generation of ideas. I also suspect that organizational factors impact the individual creative process at a number of different points, both directly and indirectly. The specification of these influences will be perhaps the most difficult but the most important part of building the theory of organizational innovation.

In summary, there is now a great deal of evidence suggesting that the process of individual creativity and the process of organizational innovation have significant influences on each other. While progress has been made in the description of the individual creative process, we now need a general model of organizational innovation that includes individual idea-generation as an important part. Once we have such a theory, we will be able to use it to understand a growing body of research evidence on the factors that influence innovation in organizations. With that, the stimulation of innovative growth in organizations will be made, if not easy, at least considerably clearer.

REFERENCES

Amabile, T. M. 1979: Effects of external evaluation on artistic creativity. *Journal of Personality and Social Psychology*, 37, 221–33.
Amabile, T. M. 1982a: Children's artistic creativity: Detrimental effects of competition in a field setting. *Personality and Social Psychology Bulletin*, 8, 573–78.
Amabile, T. M. 1982b: Social psychology of creativity: A consensual assessment technique. *Journal of Personality and Social Psychology*, 43, 997–1013.
Amabile, T. M. 1983a: Social psychology of creativity: A componential conceptualization. *Journal of Personality and Social Psychology*, 45, 357–77.
Amabile, T. M. 1983b: *The social psychology of creativity*. New York: Springer-Verlag.

Amabile, T. M. 1984 (August): Creativity motivation in research and development. In D. Campbell (Chair), *Creativity in the corporation.* Symposium conducted at the meeting of the American Psychological Association, Toronto.

Amabile, T. M. 1985: Motivation and creativity: Effects of motivational orientation on creative writers. *Journal of Personality and Social Psychology, 48,* 393–9.

Amabile, T. M. & Gitomer, J. 1984: Children's artistic creativity: Effects of choice in task materials. *Personality and Social Psychology Bulletin, 10,* 209–15.

Amabile, T. M., Goldfarb, P. & Brackfield, S. 1982: Effects of social facilitation and evaluation on creativity. Unpublished manuscript, Brandeis University.

Amabile, T. M. & Gryskiewicz, S. S. 1985: *Creativity in research and development.* Technical Report, Center for Creative Leadership, Greensboro, N.C.

Amabile, T. M., Hennessey, B. A. & Grossman, B. G. (in press): Social influences on creativity: The effects of contracted-for reward. *Journal of Personality and Social Psychology.*

Barron, F. 1955: The disposition toward originality. *Journal of Abnormal and Social Psychology, 51,* 478–85.

Campbell, D. 1960: Blind variation and selective retention in creative thought as in other knowledge processes. *Psychological Review, 67,* 380–400.

Cattell, R. & Butcher, H. 1968: *The prediction of achievement and creativity.* Indianapolis, Ind.: Bobbs-Merrill.

Einstein, A. 1949: In P. Schilpp (Ed.), *Albert Einstein: Philosopher-scientist.* Evanston, Ill.: Library of Living Philosophers.

Feldman, D. 1980: *Beyond universals in cognitive development.* Norwood, N.J.: Ablex.

Golann, S. 1963: Psychological study of creativity. *Psychological Bulletin, 60,* 548–65.

Gordon, W. 1961: Synectics. *The development of creative capacity.* New York: Harper & Row.

Harter, S. 1978: Effectance motivation reconsidered: Toward a developmental model. *Human Development, 21,* 34–64.

Hogarth, R. 1980: *Judgement and choice.* Chichester, England: Wiley.

Kruglanski, A., Friedman, E. & Zeevi, G. 1971: The effects of extrinsic incentives of some qualitative aspects of task performance. *Journal of Personality, 39,* 606–17.

MacKinnon, D. 1962: The nature and nurture of creative talent. *American Psychologist, 17,* 484–95.

McGraw, K. 1978: The detrimental effects of reward on performances: A literature review and a prediction model. In M. Lepper & D. Greene (Eds.), *The hidden costs of reward.* Hillsdale, N.J.: Erlbaum.

McGraw, K. & McCullers, J. 1979: Evidence of a detrimental effect of extrinsic incentives on breaking a mental set. *Journal of Experimental Social Psychology, 15,* 285–94.

McGuire, W. 1973: The yin and yang of progress in social psychology: Seven koan. *Journal of Personality and Social Psychology, 26,* 446–56.

Newell, A., Shaw, J. & Simon, H. A. 1962: The process of creative thinking. In H. Gruber, G. Terrell & M. Wertheimer (Eds.), *Contemporary approaches to creative thinking.* New York: Atherton Press.

Newell, A. & Simon, H. A. 1972: *Human problem solving.* Englewood Cliffs, N.J.: Prentice-Hall.

Osborn, A. 1963: *Applied imagination. Principles and procedures of creative thinking.* New York: Schribner's.

Schank, R. & Abelson, R. 1977: *Scripts, plans, goals, and understanding.* Hillsdale, N.J.: Erlbaum.

Simon, H. A. 1966: Scientific discovery and the psychology of problem solving. In R. G. Colodny (Ed.), *Mind and cosmos: Essays in contemporary science and philosophy.* Pittsburgh, Penn.: University of Pittsburgh Press.

Simon, H. A. 1967: Motivational and emotional controls of cognition. *Psychological Review*, *74*, 29–39.

Simon, H. A. 1978: Information-processing theory of human problem solving. In W. K. Estes (Ed.), *Handbook of learning and cognitive processes: Vol. 5. Human information processing*. Hillsdale, N.J.: Erlbaum.

Stein, M. I. 1974: *Stimulating creativity* (Vol. 1). New York: Academic Press.

Stein, M. I. 1975: *Stimulating creativity* (Vol. 2). New York: Academic Press.

Wallach, M. & Kogan, N. 1965: *Modes of thinking in young children*. New York: Holt, Rinehart & Winston.

Wallas, G. 1926: *The art of thought*. New York: Harcourt.

White, R. 1959: Motivation reconsidered: The concept of competence. *Psychological Review*, *66*, 297–323.

Wickelgren, W. 1979: *Cognitive psychology*. Englewood Cliffs, N.J.: Prentice-Hall.

6

Educational Implications of Creativity Research: An Updated Rationale for Creative Learning

SCOTT G. ISAKSEN

Creativity is a subject with a wide variety of implications for educational practices. This paper will present some historical background relating to the educational contexts of creativity as well as a description of the multi-faceted and interdisciplinary aspects of creativity research. Some examination of the application areas in education will be provided along with a research synthesis, implications, and conclusions.

HISTORICAL BACKGROUND

Most people who provide a historical perspective of creativity research usually highlight the year 1950 as a significant starting point. It was during this year that Guilford gave his presidential address to the American Psychological Association (APA) (Guilford 1950). He pointed out the neglect of the study of creativity and backed up his claim by stating that only 186 out of 121,000 titles listed in *Psychological Abstracts* had anything to do with creativity.

Although there were some attempts to study creativity before 1950 (Patrick 1937; Rossman 1931; Spearman 1931; Wallas 1926; and others) the bulk of creativity research has been conducted during the thirty-some years since 1950. There have been some who have asserted that interest in creativity research in education has been on the decline. Torrance (1975) made a case for the popularity and sustained interest in this type of research and concluded that creativity research is still alive and thriving. In fact, as of June, 1984, using roughly the same descriptors Guilford used (and only going back to 1967), we found in excess of 5,628 citations relating to creativity (Stievater 1984).

It appears that creativity research is not only alive for education, but also on an international level. About thirty years after his address to the APA, Guilford (1980) indicated that there was strong international interest in creativity research. In a foreword to a book on the international perspective of this type of research, he indicated:

This volume provided substantial evidence that there is indeed a creativity movement and that it now has nearly world-wide proportions. This is a hopeful situation, for a world population of creative solvers should be more productive and happy as well as more self-confident and more tolerant and, therefore, more peaceful.

Much of the research on creativity during the past thirty years has been produced through the efforts of various centers. One of the earliest was at the University of Southern California where Guilford conducted the Aptitudes Research Project (Guilford 1967). The Institute of Personality Assessment and Research was another early center for creativity research. It was started in 1949 at the Berkeley campus of the University of California (MacKinnon 1975). Another major center has been at the University of Utah, where the National Science Foundation sponsored research conferences on the Identification of Creative Scientific Talent (Taylor 1963).

Of course, anyone doing any reading on creativity in education would come across Torrance's work. He started his work at the University of Minnesota and then moved to the University of Georgia. The Georgia Studies of Creative Behavior are well known for their importance regarding education (Torrance 1980). Perhaps no other researcher has done as much to illustrate the teachability of creativity than Torrance.

Other centers of research in creativity have a more recent history. The Creative Education Foundation (CEF), housed at the State University College at Buffalo, under the direction of Parnes, was responsible for the Creative Studies Project (Parnes and Noller 1972) and the formation of the Interdisciplinary Center for Creative Studies. The founder of the CEF, Alex Osborn, wished to bring a more creative trend to American education. During the 1970s the Center for Creative Leadership at Greensboro, North Carolina was formed by the Smith-Richardson Foundation and has an active research program dealing with both creativity and leadership (Gryskiewicz 1980). One of the more recent centers has been at the Harvard Graduate School of Education where Project Zero has provided a number of important research reports (Perkins 1981; Gardner 1982).

In terms of the educational implications of all this creativity research, there are two major points to form the broad context for our examination. The first is that education can do something about nurturing creativity. Torrance (1981, p. 99) made the following assertation:

A few years ago, it was commonly thought that creativity, scientific discovery, the production of new ideas, inventions, and the like had to be left to chance. Indeed many people still think so. With today's accumulated knowledge, however, I do not see how any reasonable, well-informed person can still hold this view. The amazing record of inventions, scientific discoveries, and other creative achievements amassed through deliberate methods of creative problem solving should convince even the most stubborn skeptic.

The second point is that the educational context is an appropriate one within which to focus creativity research. Guilford (1980, p. viii) provided support for this point:

Of all the environmental influences on development of creativity, education has received special interest. It is the business of education more than any other institution to determine to what extent creativeness and creative production can be improved and how this shall be done: It is apparently no longer doubted that there can be improvement in creative thinking and problem solving. There is increasing realization of education's responsibility in this direction.

CREATIVITY AS A MULTI-FACETED PHENOMENON

Before proceeding with the examination of creativity research, we must first establish what we mean by the term "creativity". To start, creativity must be seen as a multi-faceted phenomenon rather than as a single construct to be precisely defined. About ten years after Guilford's address to the APA, Rhodes (1961) responded to the criticism leveled at those attempting to study creativity due to the loose and varied meanings assigned to the word "creativity". Rhodes set out to find a single definition of the word. He collected in excess of 56 different definitions and despite the profusion, he reported:

> . . . as I inspected my collection I observed that the definitions are not mutually exclusive. They overlap and intertwine. When analyzed, as through a prism, the content of the definitions form four strands. Each strand has unique identity academically, but only in unity do the four strands operate functionally (p. 307).

The four strands Rhodes discussed included information about: the personality, intellect, traits, attitudes, values, and behavior (PERSON); the stages of thinking people go through when overcoming an obstacle or achieving an outcome which is both novel and useful (PROCESS); the relationship between people and their environment, the situation which is conducive to creativity (PRESS); and the characteristics of artifacts of new thoughts and ideas, inventions, designs, or systems (PRODUCT). Each of these four strands (which Rhodes called the four "P's" of creativity) operate as identifiers of some key components of the larger, more complex, concept of creativity.

This classification scheme has been used quite extensively in the creativity literature and helps to provide some frame of reference in studying creativity (Hallman 1981; MacKinnon 1978; and Welsh 1973). This general approach to the definition of creativity appears to be more fruitful than attempting to specify a single definition. This approach does, however, feed the notion that creativity is a complex concept.

Definitions of creativity are varied, to say the least. Rhodes was not alone in finding a multiplicity. Taylor (1959) found in excess of one hundred definitions available for analysis. Welsch (1980) sought to find elements of agreement and disagreement across twenty-two definitions of creativity. She was searching for a definition that would be applicable to a variety of creative activities. She provided a definition having all four strands (the four P's):

> The definitions of creativity are numerous, with variations not only in concept, but in the meaning of subconcepts and of terminology referring to similar ideas. There appears to be, however, a significant level of agreement on key attributes among those persons most closely associated with work in this field. Significantly for this study, the greater disagreements occur in relation to aspects that are less relevant to educational purposes.

On the basis of the survey of the literature, the following definition is proposed: Creativity is the process of generating unique
products by transformation of existing products. These products,
tangible and intangible, must be unique only to the creator, and
must meet the criteria of purpose and value established by the
creator (p. 97).

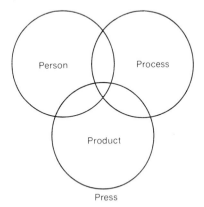

Fig. 1.

One of the factors that contributes to the complexity of the
conceptions of creativity is that it is an interdisciplinary phenomenon. Certainly, no single discipline can claim to have exclusive
rights to creativity. Studies of creativity are found in the arts
(Barron and Welsh 1952; Getzels and Csikszentmihalyi 1976), as
well as in the sciences (Mansfield and Busse 1981; Taylor, Smith,
and Ghiselin 1963).

There are many possible contexts within which to study creativity
(Isaksen, Stein, Hills, and Gryskiewicz 1984). One of the earliest
focal areas was the study of exceptional talent or genius (Alberts
1983; Cox 1926; Galton 1869; Goertzel, Goertzel and Goertzel
1978; Simonton 1984; and Terman 1954). Related to this area was
the study of great individuals and the method of thought which
permitted innovative breakthroughs (Gruber 1981; Kuhn 1970;
Wallas 1926; and Wertheimer 1945).

Creativity has been studied in managerial, business and industrial
areas (Basadur 1981; Ekvall and Parnes 1984; and Johansson 1975);
in disciplines such as engineering (Arnold 1959; Rubinstein 1975);
mathematics (Helson and Crutchfield 1970; Shoenfeld 1982; and
Schoenfeld and Herrmann 1982), philosophy (Hausman 1984; and
Lipman, Sharp and Oscanyan 1980), physics (Larkin 1980), and
English (Elbow 1983; Langer 1982; and Olson 1984); and in teacher

preservice and inservice educational programs (Brooks 1984; Gibney and Meiring 1983; Juntune 1979; Krulik and Rudnick 1982; and Martin 1984) as well as in the general counseling process (Heppner 1978).

The educational interest in creativity stems from a vast collection of writers and extends far beyond those areas already cited. One current label to use for this area of interest is the "teaching of thinking". Two entire recent issues of *Educational Leadership* (Vol. 42, nos. 1 and 3, 1984), the official journal of the Association for Supervision and Curriculum Development, have focused on the teaching of thinking skills! In short, there is continued and extensive interest, writing, research and discussion regarding ways to effectively improve the thinking of learners at all levels.

Aside from being studied within specific disciplines and contexts, the research of Torrance (1974), Biondi and Parnes (1976), Khatena (1982) and Amabile (1983) indicates that creativity can be assessed systematically and scientifically. A related and extremely important finding is that creativity can be enhanced through deliberate instructional procedures (Goor and Rapoport 1977; Heppner, Neal, and Larson 1984; Mansfield, Busse, and Krepelka 1978; Parnes and Noller 1972; Reese, Parnes, Treffinger, and Kaltsounis 1976; Rose and Lin 1984; and Torrance 1972).

Another reason for the complexity of the field of study of creativity is due to its link with a wide array of theoretical perspectives (Treffinger, Isaksen, and Firestien 1983) as well as with the concepts of problem solving and creative learning. Guilford (1977) defined problem solving as facing a situation with which you are not fully prepared to deal. Problem solving occurs when there is a need to go beyond the information given, thus there is a need for new intellectual activity. Guilford (1977) reported that:

> . . . problem solving and creative thinking are closely related. The very definitions of those two activities show logical connections. Creative thinking produces novel outcomes, and problem solving involves producing a new response to a new situation, which is a novel outcome (p. 161).

This description is also very closely related to a framework for describing the process of creative learning described by Torrance and Myers (1970). They described the process as:

> . . . (B)ecoming sensitive to or aware of problems, deficiencies, gaps in knowledge, missing elements, disharmonies, and so on; bringing together available information; defining the difficulty

or identifying the missing elements; searching for solutions, making hypotheses, and modifying and retesting them; perfecting them; and finally communicating the results (p. 22).

Creativity has been referred to in a wide variety of ways ranging from those who hold differing theoretical positions to those who place various labels on essentially the same concept.

VARIETIES OF EDUCATIONAL USE

There has been a general increase in research and writing in the area of creativity despite the many different approaches to defining it; the differences in assumptions, presuppositions, and contexts; and the differences in research methodologies and strategies among and within groups of various orientations. Given this complex array of literature, it is not surprising to find some individuals who feel creativity is something mystical and, as such, too difficult to understand and analyze. There are others who assert that creativity is magical and shouldn't be explained "lest we lose it". Then there are those who suggest that creativity must be linked with madness and, as such, this sickness should be avoided (along with other forms of pathological behavior)!

Those concerned with educational and training implications for creativity hold a different point of view regarding this confusing state of affairs. This view of creativity suggests that *all* people have, to varying levels and in varying styles, the potential to be creative. This view rejects the notions of creativity as magical, mystical or mad, and asserts that it is the legitimate domain for those concerned with optimizing human potential. The assumption that creativity is a natural, human resource leads to the educational practice of dealing with the concept in three basic ways. The applications include weaving creativity into the existing curriculum, teaching creative thinking and problem solving skills directly, and using creativity in the process of planning for learning.

The first, and most ubiquitous method for dealing with creativity appears to be through weaving it into the existing curriculum. This method includes applying what is known about creativity into the subjects and existing instructional programs. An example of this type of activity is the inclusion of questions and suggestions for activities to foster creativity in elementary school reading programs. There are many examples across disciplines and grade levels, where creativity is explicitly planned for within specific subjects and grade levels.

The second approach to using what we know is the direct teaching of the skills, methods, and processes associated with creativity. This approach includes separate units of instruction, courses or programs designed to enhance creativity. Many programs for academically talented students follow this approach.

Another method employed by those that hold the view that creativity is human, is the use of the processes and skills in planning for learning. This approach includes using what is known about creativity to actually plan lessons or units of instruction. This planning occurs independently, with other professionals or with the learners themselves.

All these approaches may be combined and used together in varying degrees. They are not entirely separate, nor should they be! However, what does appear to be rather consistent in all approaches is that the planning must be deliberate and explicit to foster creative thinking skills (Isaksen 1983; Isaksen and Parnes 1985; and Whitman 1983).

Another common thread through all three approaches is the belief that spending time and energy working toward creativity is worth while. If instructional resources and time are focused away from the regular program, what will be lost? As long as the new focus is on creative learning, the loss is non-existent. In fact, gains can be found in reading and mathematics (Brandt 1982), SAT scores (Worsham and Austin 1983) and in overall school achievement (Johnson, Maruyama, Johnson, Nelson, and Skon 1983).

These findings ought not to be very surprising as they are consistent with some very well designed and established research studies. The most comprehensive of these was the Eight-Year Study (Aikin 1942). This study was designed to discover ways the secondary schools of the United States might better "serve all our young" (p. 11). The Commission of the Relation of School and College of the Progressive Education Association was concerned that the "traditional subjects of the curriculum had lost much of their vitality and significance" (p. 6). The Commission was aware of the pervasive attitude within the secondary schools, that creative learning was on the fringe of the educational experience. The creative energies of students were seldom released and developed. Aiken (1942) described the condition in the following manner:

> Students were so busy doing assignments, meeting demands imposed upon them, that they had little time for anything else. When there was time they were seldom challenged or permitted to carry on independent work involving individual initiative,

fresh combination of thought, invention, construction, or special pursuits (p. 6).

The purpose of the Eight-Year Study was to respond to these concerns and demonstrate that the curriculum could be changed without damaging the students' ability to succeed in college. Five volumes contain a more comprehensive study than can be reported here. Thirty schools provided 1,475 matched pairs and a vast amount of data. How did those students do in comparison to those who did not engage in a more creative type of learning?

In the comparison, the College Follow-up Staff found that the graduates of the Thirty Schools:

★ earned a slightly higher total grade average;
★ earned higher grade averages in all subject fields except foreign language;
★ specialized in the same academic fields as did the comparison groups;
★ did not differ from the comparison group in the number of times they were placed on probation;
★ received slightly more academic honors in each year;
★ were often judged to possess a high degree of intellectual curiosity and drive;
★ were more often judged to be precise, systematic, and objective in their thinking;
★ were more often judged to have developed clear or well-formulated ideas concerning the meaning of education—especially in the first two years in college;
★ more often demonstrated a high degree of resourcefulness in meeting new situations;
★ did not differ from the comparison group in ability to plan their time effectively;
★ had about the same problems of adjustment as the comparison group, but approached their solution with greater effectiveness;
★ participated somewhat more frequently, and more often enjoyed appreciative experiences in the arts;
★ participated more in all organized students' groups except religious and "service" activities;
★ earned in each college year a higher percentage of non-academic honors (officership in organizations, election to managerial societies, athletic insignia, leading roles in dramatic and musical presentations);
★ did' not differ from the comparison group in the quality of adjustment to their contemporaries;

★ differed only slightly from the comparison group in the kinds of judgments about their schooling;
★ had a somewhat better orientation toward the choice of a vocation; and
★ demonstrated a more active concern for what was going on in the world. (pp. 111–112)

In general, it was found that the Thirty Schools graduates, as a group, did a better job than the comparison group (judged by college standards, contemporaries, or by students themselves). An interesting follow-up was conducted on the graduates of the six participating schools in which the least change had taken place in the curriculum and the graduates of the six schools in which the most marked departures from conventional college preparatory courses had been made. The findings for this aspect of the study indicated that "the graduates of the most experimental schools were strikingly more successful than their matchees. Differences in their favor were much greater than the differences between the total Thirty Schools and their comparison group" (p. 113). In addition, the differences for those from the least experimental schools were smaller and less consistent.

Not only does there appear to be no loss when providing creative learning, there may actually be some very important gains. These findings provide a striking contrast to what is currently the prevailing method of responding to the "rising tide of mediocrity" in education. In response to the series of national reports, many states within the U.S. are tightening their requirements for the curriculum. This type of response may be politically expedient, but simply ignores what the research says:

If colleges want students of sound scholarship with vital interests, students who have developed effective and objective habits of thinking, and who yet maintain a healthy orientation toward their fellows, then they will encourage the already obvious trend away from restrictions which tend to inhibit departures or deviations from the conventional curriculum patterns (p. 113).

Movement away from such rigid requirements may have been obvious in the 1940s, but making the curriculum more responsive and creative does not appear to be what has happened. Indeed, if what Goodlad (1983) has reported is true, then not much has changed in American education since the early 1900s despite the research findings and their implications. Goodlad reported that schools appear to be spending most time, energy, and resources to provide for lower level thinking and recall through lecture and recitation methodology. This phenomenon has been well documented by those in the fields of mathematics (Carpenter, Lindquist, Matthews, and Silver 1983; NCTM 1980), reading and literature (Berkenkotter 1982; Glatthorn 1980; NAEP 1981) and in teacher questioning practices (Dillon 1984; Gall 1984).

EDUCATIONAL IMPLICATIONS

Although the creativity research is not always conclusive, there does appear to be sufficient evidence to warrant the consideration of its educational implications. The study of creativity, rather than being an exact science, appears to be like a diamond. It is certainly worthwhile, and you can see the entire jewel, or you can focus on one of its many facets. When your attention is directed at only one of the facets, care must be taken to avoid the tendency to forget that you are only looking at one part and not the whole. Real value, operationally, occurs when all facets are taken into consideration. Even then, there remain many critical issues to be investigated to shed further light on the conceptions of creativity and their educational implications.

Investigation and analysis of creativity is facilitated when consideration is given to each of its facets. This section will provide a brief summary of each of the four facets of creativity. It is beyond the scope of this paper to provide a comprehensive overview and summary of creativity research findings, as this has been done elsewhere (see Bloomberg 1973; Guilford 1977; Isaksen 1985; MacKinnon 1978; Parnes and Harding 1962; Rothenberg and Hausman 1976; Stien 1974; Taylor and Getzels 1975; and Torrance 1979). Following the brief summary, the implications for educational planning and practice will be outlined.

The Creative Person

> Creative personality is . . . a matter of those patterns of traits that are characteristic of creative persons. A creative pattern is manifest in creative behavior, which includes such activities as inventing, designing, contriving, composing, and planning. (Guilford 1950)

The questions within this area of study include: Are there traits or characteristics which can differentiate creative persons from their less creative peers? How important are attitudes, habits, and motivations in predicting creative behavior? Can a creative person be identified?

The major response to these questions has been research through biographical, descriptive, and empirical methodologies taking readily identified "creators" and attempting to distill their attributes. The end products of these investigations are lists of characteristics and traits that have something to do with being creative. Torrance (1974) has designed a battery of tests to measure such abilities as fluency, flexibility, originality, and elaboration of thinking. Williams (1980) has been concerned with identifying affective or emotional characteristics including: risk-taking, curiosity, complexity, and imagination. These characteristics provide the elements of the creative personality that are traditionally measured in school settings. I refer to them as the "eight-pack" because of the convenience of their use.

SOME COMPONENTS OF THE CREATIVE
PERSONALITY
("8-PACK")

Cognitive	*Affective*
Fluency	Curiosity
Flexibility	Complexity
Originality	Risk-taking
Elaboration	Imagination

This is, of course, much too simple a picture for identifying something as complex as the creative personality. As MacKinnon (1978) has emphasized, ". . . there are many paths along which persons travel toward the full development and expression of their creative potential, and there is no single mold into which all who are creative will fit. The full and complete picturing of the creative person will require many images" (p. 186).

Other characteristics revealed through this line of research are: a high level of effective intelligence, openness to experience, freedom from crippling restraints and impoverishing inhibitions, aesthetic sensitivity, cognitive flexibility, independence in thought and action, unquestioning commitment to creative endeavor, and an unceasing striving for solutions to the ever more difficult problems constantly being set by him or herself.

Psychological theorists (Fromm 1959; Maslow 1962; and Rogers 1959) have identified other characteristics of the creative person including:

★ An ability to accept conflict by being able to tolerate bipolarity and to integrate opposites;

★ The capacity to be puzzled, able to accept tentativeness and uncertainty and being unfrightened by the unknown and ambiguous;

★ Having an internal locus of evaluation, a high degree of self-discipline, an ability to concentrate, and a belief in one's ability to succeed;

★ Is uninhibited in expressions of opinion, sometimes radical and spirited in disagreement, tenacious;

★ Is unusually aware of his impulses and more open to the irrational (freer expression of feminine interest for boys, greater than usual amount of independence for girls), shows emotional sensitivity and constructive discontent; and

★ Non-conforming, accepting disorder, not interested in details, individualistic, does not fear being different, spontaneous.

Torrance (1979) has provided a model for thinking about the search for creativity. The model below has been useful to many people in understanding, predicting, and developing creative behavior. "It takes into consideration, in addition to creative abilities, creative skills and creative motivations . . . (a) high level of creative achievement can be expected consistently only from those who have creative motivations (commitment) and the skills necessary to accompany the creative abilities" (p. 12).

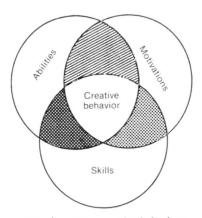

Fig. 2. A model for studying/predicting creative behaviour.

This multi-faceted conception of the creative personality has been well documented in the literature (Amabile 1983b; Renzulli 1978).

The pervasive question for examining these and other educational implications for the creative personality is: In what ways might the curriculum be designed to enhance the characteristics of the creative personality?

One of the major educational implications of the study of the creative person relates to knowing more about the personal orientation toward problem solving and creative thinking. Some of the current research within this domain focuses on studying the different styles of creativity (Kirton 1976; Gryskiewicz 1985; and Myers and Myers 1980) and how these styles may affect different elements of creativity. For example, certain personality characteristics will influence preferences regarding how information is collected and utilized. Knowledge of these characteristics and preferences and how they might effect creativity ought to be a part of any curriculum designed to nurture creative thinking and problem-solving skills.

Much of the current literature on the creative personality emphasizes intrinsic motivation as a key variable (Amabile 1983). Much of the current focus of the curriculum is on recitation and recall. A major difference exists in what "ought to be" and "what is" regarding curricula for creativity. Much more attention needs to be focused on providing educational programs which are designed with consideration of the characteristics of the creative personality.

A major implication of this approach is the consideration of readiness factors. In short, our energies currently and previously directed at identifying and measuring aspects of the creative personality ought to be fused with the work of others who seek to understand better cognitive developmental stages and how to more adequately match student learning styles and levels of readiness to various learning opportunites (see Grennon [1984] and Toepfer [1982] for more information).

One aspect of the creativity styles research (Kirton 1976) examines a possible continuum of creativity style ranging from innovative creativity (that type of creativity focused on doing things differently) to adaptive creativity (focused on doing things better). Much of the current emphasis of the curriculum which deals with the development of inventions appears to stress only the innovative type of creativity. Perhaps, we need to see how creativity, as a natural human characteristic, is in those who may choose to use it adaptively as well as innovatively. This may shorten the "distance" between students and those who are noted innovators. There is already some evidence that type of change is occurring. Students across the U.S. are being provided opportunities to become involved in "Invention Conventions" and "Odysseys of the Mind" which provide challenges to develop inventive solutions to problems (see Gourley and Micklus 1981).

These implications have strong historical precedent for curriculum planning. There have been many spokespersons for making the curriculum more responsive to individuals' differences and characteristics. One of the earliest comprehensive statements regarding this approach to education was provided by Rugg and Shumaker (1928). They pointed out that the child-centered school was guided by "new articles of faith".

Rugg and Shumaker outlined these articles of faith in a chapter of their book entitled *The Child-Centered School*. The first article was freedom to develop naturally, to be spontaneous and unself-conscious, which revealed itself in an easier, more natural group life. The next article of faith for the new schools was that they be reoriented around the child. This means that children should participate in governing, program planning, and conducting the life of the school. The third article pointed to "activity" which grew toward something more mature, a changing for the better, involving prolonged attention and concentrated effort. The fourth article was that child interest forms the orienting center of the new school's program. In the new school the creative spirit from within ought to be encouraged rather than conformity to a pattern imposed from without. It was this new emphasis; not upon finished work, skill, and technical perfection, but upon the release of the child's creative capacities, upon growth in his power to express his own unique ideas naturally and freely, which formed the fifth article of faith for the "new" school. The final article of faith related to the problem of providing an environment in which children can learn to live with others, and retain their personal identities. The old school ignored this area, in practice, while the new school encouraged activities in which the child can make personal contributions to group enterprises while feeling accepted and respected.

The Creative Product

> Creations are products which are both new and valuable and creativity is the capacity or state which brings forth creations. (Rothenberg 1971)

Although many creativity researchers point out the importance of studying the creative product there is a paucity of empirical investigation on this topic. The centrality and importance of this line of investigation has been pointed out by MacKinnon (1978, p. 187):

In a very real sense, then, the study of creative products is the basis upon which all research on creativity rests and, until this foundation is more solidly built than it is at present, all creativity research will leave something to be desired.

One of the possible reasons for the lack of empirical research in this area is the opinion that the problem is too easy. In other words, the identification of creative products is "obvious". Everyone knows a creative product when they see one. MacKinnon (1975) pointed out that this view might account for the derth of scientific investigation of creative products:

> In short, it would appear that the explicit determination of the qualities which identify creative products has been largely neglected just because we implicitly know—or feel we know—a creative product when we see it.

There are some who have conducted investigations of creative products (Ward and Cox 1974; Taylor and Sandler 1972; Amabile 1982; Besemer and Treffinger 1981; Pearlman 1983). Much of this work has dealt with creative products in specific contexts. Very little has been done beyond individual disciplines and contexts to gain a more general picture of the characteristics of creative products. Besemer and O'Quinn (1985) have reviewed the use of a Creative Product Analysis Matrix (CPAM) which was developed from an earlier survey of the literature and model development (Besemer and Treffinger 1981). The CPAM has three clusters of characteristics used to identify creativity in products.

The first cluster is labeled novelty. This includes examining the degree of originality illustrated in the product (originality), the likelihood of products being created which might result as spin-offs or off-shoots (germinal), and the degree of influence a product has in terms of impact upon society or culture (transformational).

Novelty is perhaps the most "obvious" of the dimensions. Certainly some element of newness or originality is necessary for a product to be considered creative. This does not provide a complete picture of what makes a product creative. The degree of relevance and appropriateness of the solution offered by the product appears to be necessary as well. Briskman (1980) pointed this out in a theoretical piece which put forth another model for analyzing creative products:

> . . . the novelty of a product is clearly only a necessary condition of its creativity, not a sufficient condition: for the madman who,

in Russell's apt phrase, believes himself to be a poached egg may very well be uttering a novel thought, but few of us, I imagine, would want to say that he was producing a creative one. (Briskman 1980, p. 95)

This concern of relevance and appropriateness is labeled resolution on the CPAM. This dimension considers the adequacy of the product, its appropriateness, logical qualities, usefulness and its valuable characteristics.

The final dimension of the CPAM examines the stylistic attributes of the product; how it is presented or the way it is manifested so that it can be used or interacted with. This dimension is called elaboration and synthesis. This would include how elegant, attractive, or expressive the product is.

Besemer and O'Quinn (1985) report four studies to determine whether or not subjects would evaluate creative products in a manner consistent with the model. Responses on the CPAM were subjected to tests of interobject reliability, factor analysis of resulting subscales, and analysis of variance of the subscales. In general, the results offered partial support for the theoretical model.

A related, and more researched area of study dealing with creative products, involves diffusion of innovations. Many people have been interested in how new ideas or products are communicated or accepted by others. In fact, with an increased interest in the process of innovation for organizations and individuals, has come increased concern for studying communication to promote acceptance of new ideas; this area of study is called diffusion of innovations.

When the book *Diffusion of Innovations* was first published in 1962, there were 405 publications about this topic available in the literature (Rogers 1983). By the end of 1983, there were more than 3,000 publications about diffusion, many of which were scientific investigations of the diffusion process. Rogers (1983) described diffusion as an information exchange occurring as a convergence process involving interpersonal networks. He asserted that the diffusion of innovations is a social process for communicating information about new ideas.

Although diffusion is a process, the study of this process has yielded the identification of specific attributes of innovations. This area of study has examined how these effect acceptance. Rogers (1983, pp. 210–40) has described these characteristics: relative advantage, compatibility, complexity, trialability, and observability.

Relative advantage is the measure of how much better an innovation is than the idea or object it replaces. This can be expressed in terms of cost, profitability, improved status, etc. People who will be adopting a new idea or invention want to know how much better it is than previous ideas.

Diffusion scholars have found that relative advantage is one of the best predictors of an innovations' acceptance and use by others. The specific dimensions of relative advantage include degree of profitability, low cost, decrease in discomfort, savings in time and effort, and how soon the reward of use occurs. There is a strong, positive relationship between an innovation's relative advantage and its success rate for adoption.

Compatibility is the consistency of a new idea or object with current values, past experiences, and needs of those who will be potential adopters. Generally, the more compatible the innovation is, the more it is adopted.

Complexity relates to the difficulty of understanding and using an innovation. The more complex a new idea or object, the less likely it will be accepted and used.

Diffusion researchers have found this characeristic to be more strongly related to how ideas are adopted than all the others except relative advantage. The general advice to keep things simple appears to be warranted.

Trialability is a characteristic involving the degree of experimentation that can be conducted with new ideas or objects. Some new ideas are limited in the degree to which experimentation is possible, and others must be an "all or nothing" arrangement. Some innovations are more difficult than others to subdivide, try out, modify, or field test.

Diffusion researchers have found that the easier it is to try out or experiment with an innovation, the better the rate of adoption. This may be the case because an innovation that can be modified or used experimentally offers more flexibility and control for the potential adopter.

Observability is the characteristic of new ideas and objects that deal with the innovation's visibility to others. Some ideas provide results which are easily observed and communicated to others; others may be extremely complex and difficult for others to understand. Rogers suggested that the level of observability of an innovation, as seen by others, is positively related to its diffusion. The more readily an innovation can be observed and understood, the greater the likelihood of its adoption.

Although these five categories account for many of the reasons why individuals choose to accept innovation, there are also other variables which have an impact on the diffusion of new ideas and inventions. The relative speed with which a new idea is adopted by others is referred to as the rate of adoption. This is generally measured as the number of people who adopt a new idea in a specified period. Other variables include (among others): the number of people involved in making a decision; the type of communication used; the environment or culture; and who is supporting (or selling) the new idea or object.

These variables have not been researched extensively by those interested in diffusion, but they will undoubtedly have an effect on acceptance of creative products.

The educational implications of the study of creative products and their diffusion address the following question: In what ways might we foster the development and study of creative products and their diffusion in educational settings?

One of the most easily seen implications is that students and those who share responsibility for their learning ought to be engaged in the development of creative products. This implies that learning ought to be made more active, culminating in something tangible or intangible of real meaning for the students and others.

This more active type of curriculum can be used to enhance the critical and creative thinking of students by involving them in the evaluation of their products. They could use the characteristics mentioned above to determine the degree of creativity manifested in a particular product and provide rationale for the evaluation. Teachers could use diffusion criteria to assist students in planning for acceptance on their new products. Students could be informed about these characteristics for diffusion and could consider them in developing their own approach to communicating the creative product.

This approach to education is certainly not new. Kilpatrick (1918, p. 18) described it as the project method. In his paper he put forth the following statement:

> The contention of this paper is that whole hearted purposeful activity in a social situation as the typical unit of school procedure is the best guarantee of the utilization of the child's native capacities now too frequently wasted.

The Creative Process

We can . . . take a single achievement of thought—the making of a new

> generalization or invention, or the
> potential expressions of a new idea—
> and ask how it was brought about. We
> can then roughly dissect out a
> continuous process, with a beginning
> and a middle and an end of its own.
> (Wallas 1926)

One of the earliest descriptions of the creative process was provided by Wallas (1926). He described four stages for this process: Preparation, incubation, illumination, and verification.

WALLAS' STAGES IN THE CREATIVE PROCESS

Stage One: Preparation	Problem is investigated in all directions
Stage Two: Incubation	Thinking about the problem in a "not-conscious manner"
Stage Three: Illumination	Appearance of the "happy idea"
Stage Four: Verification	Validity-testing of the idea; idea reduced to exact form

Research regarding the creative process has relied upon retrospective reports, observation of performance on a time-limited creative task, factor analysis of the components of creative thinking, experimental manipulation and study of variables presumably relevant to creative thinking and simulation of creative processes on computers.

Some questions relating to the creative process include: What are the stages of the thinking process? Are the processes identical for problem solving and for creative thinking? What are the best ways to teach the creative process? How can the creative process be encouraged? Is the creative process similar in different contexts?

The area of creative process also provides the framework for describing the creative learning process. As was previously mentioned the description of the process of creative learning (Torrance and Myers 1970) is sometimes equated with what is meant by creativity, in general.

Creative learning combines a variety of types of thinking into a series of stages. Current thinking about the process of creative problem solving describes the process as having two mutually important types of thinking. Creative thinking involves making and communicating meaningful new connections to: think of many possibilities; think and experience in various ways and use different points of view; think of new and unusual possibilities; and guide in generating and selecting alternatives (Isaksen and Treffinger 1985). Critical thinking involves analyzing and developing possibilities to: compare and contrast many ideas; improve and refine promising alternatives, screen, select, and support ideas; make effective decisions and judgments; and provide a sound foundation for effective action (Treffinger 1984). These two types of thinking are seen as mutually important components of effective problem solving. Much of the emphasis within programs designed to teach for creativity has been on divergent thinking. Current developments regarding the direct instruction on the Creative Problem Solving Process attempt to provide a balanced approach, including methods and techniques of divergent and convergent thinking (Isaksen and Treffinger 1985).

The figure of the six-stage model of Creative Problem Solving (overleaf) visually demonstrates the concern for both types of thinking along the process.

Much of the educational emphasis regarding the creative process involves the teaching of explicit methods and techniques in order to help the learner solve problems and think more effectively. The major question underlying this emphasis is whether or not energy directed upon learning and practicing these strategies and skills will be of assistance across disciplinary boundaries. In short, are there generic problem-solving skills that cut across fields?

A review of research with this exact question was supported by the Program Planning Research Council of the Educational Testing Service (Baird 1983). Despite the difficulties inherent in the problem-solving literature (research based on highly artificial problems, a wide variety of tasks and studies, and others), several lines of inquiry appeared fruitful:

> First, there is some evidence that various heuristics are used by effective problem solvers in many areas of activity when confronted by new types of problems and that these heuristics can be identified. Second, there are converging lines of evidence that a major role is or can be, played by a managerial function that selects strategies and plans attacks on problems. Finally, the study of how problem solvers within specific fields learn to solve the field-specific problems they face sugests several generic skills that cut across fields. (Baird 1983, p. 19)

These findings need to be qualified. Baird is careful to point out that the actual field within which the problem solving occurs provides the requisite knowledge as well as the procedures and outlets necessary to implement the generic skills.

These findings relate strongly to the current writing on meta-cognition (Costa 1984). The abilities associated with the managerial

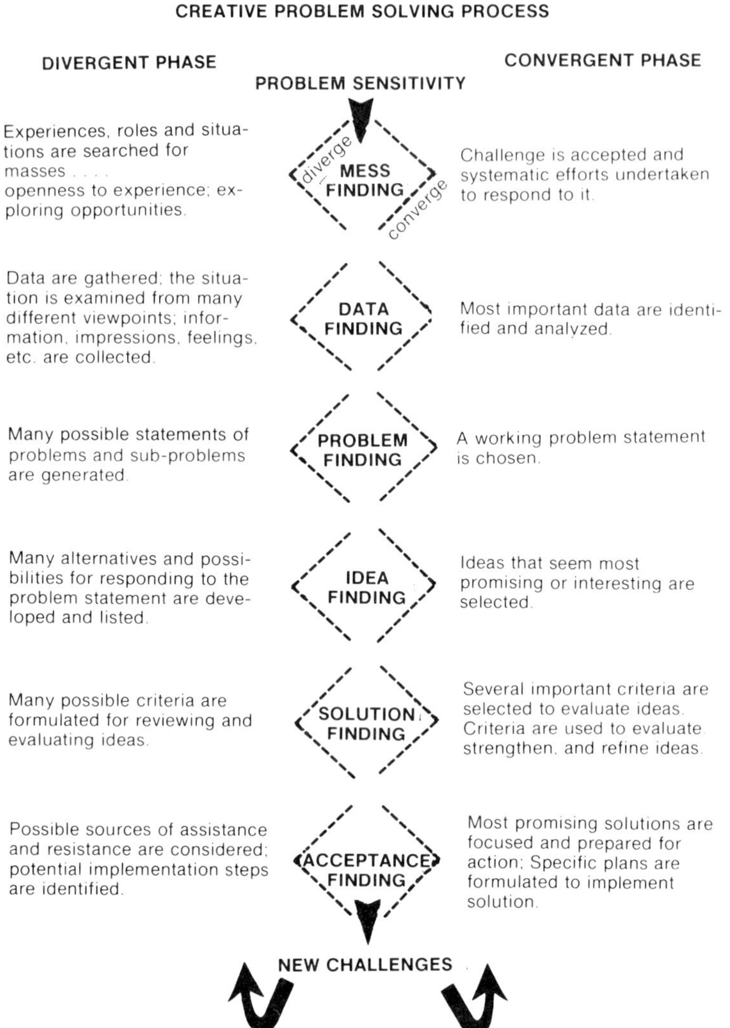

CREATIVE PROBLEM SOLVING PROCESS

DIVERGENT PHASE   **CONVERGENT PHASE**

PROBLEM SENSITIVITY

Experiences, roles and situations are searched for masses . . . . openness to experience; exploring opportunities.

MESS FINDING

Challenge is accepted and systematic efforts undertaken to respond to it.

Data are gathered; the situation is examined from many different viewpoints; information, impressions, feelings, etc. are collected.

DATA FINDING

Most important data are identified and analyzed.

Many possible statements of problems and sub-problems are generated.

PROBLEM FINDING

A working problem statement is chosen.

Many alternatives and possibilities for responding to the problem statement are developed and listed.

IDEA FINDING

Ideas that seem most promising or interesting are selected.

Many possible criteria are formulated for reviewing and evaluating ideas.

SOLUTION FINDING

Several important criteria are selected to evaluate ideas. Criteria are used to evaluate strengthen, and refine ideas.

Possible sources of assistance and resistance are considered; potential implementation steps are identified.

ACCEPTANCE FINDING

Most promising solutions are focused and prepared for action; Specific plans are formulated to implement solution.

NEW CHALLENGES

Fig. 3.

function are remarkably similar to those of metacognition. Both seem related to an individual's ability to make the thought process more explicit and deliberate.

There are many implications of the creative process for educational practices. The general question for these implications is: In what ways might we engage learners in the creative process?

There are many methods and techniques and other resources to aid in teaching for the creative process. In addition, there are many models to guide this teaching process (Williams 1970; Treffinger 1980). These resources and materials can be used in the three approaches described earlier in this chapter (directly, indirectly, and in planning).

Another implication, and one especially easy to see when considering the facet of the creative process, is that it is impossible to separate any of the four facets of creativity in practice. For example, a teacher cannot teach the creative process the same way as a lesson on some spelling words. To teach for the creative process, fruitfully, the teacher must also consider the characteristics of the students, what the students are engaged in producing and the appropriate climate within which to facilitate the occurrence of creative behavior. In short, it doesn't appear easy or simple to provide creative learning!

These implications are extremely well-documented in a vast collection of educational literature (see: Treffinger, Isaksen, and Firestien 1982, p. 18), but one writer who ought to be acknowledged is Dewey. His work on reflective thinking provided great influence for many who have an interest in students' thinking. Dewey (1933) charged teachers with the responsibility to know the process of reflective thought and facilitate its development, indirectly, in students by providing appropriate conditions to stimulate and guide thinking. This approach was described in even more detail by Hullfish and Smith (1961).

They reported:

> Where there is no problem, where no snarl appears in the normal flow of experience, there is no occasion to engage in thought . . . it is important that teachers understand the intimate relationship between problem solving and thought. (p. 212)

The Creative Press

> To speak of a creative situation is to imply that creativity is not a fixed trait of personality but something that changes over time . . . being facilitated by some conditions and situations, and inhibited by others. (MacKinnon 1978)

The term press refers to the relationships between individuals and their environments. This facet of creativity includes the study of social climates conducive or inhibitive to the manifestation of creativity, differences in perception and sensory inputs from varying environments, and the various reactions to certain types of situations. The questions guiding study within this area are: What are the environmental conditions that have an effect on creative behavior? How do these conditions effect creativity? How can an environmental atmosphere be established to facilitate creativity?

Torrance (1962) synthesized the findings of various investigators (including Kris, Maslow, Rogers, Stein, Barron, Kubie, MacKinnon and others) and listed the following as necessary conditions for the healthy functioning of the preconscious mental processes which produce creativity:

★ The absence of serious threat to the self, the willingness to risk.
★ Self awareness . . . in touch with one's feelings.
★ Self-differentiation . . . sees self as being different from others.
★ Both openness to the ideas of others and confidence in one's own perceptions of reality or ideas.
★ Mutuality in interpersonal relations . . . balance between excessive quest for social relations and pathological reflections of them. (p. 143)

Aside from the work on the creative environment for education, there has been much recent attention to the climate conducive for creativity and innovation from the business and industrial community (Ekvall 1983; Amabile 1984). The emphasis of this research has been to identify those factors, in certain organizations, that account for creative behavior. The findings from business and education are somewhat similar in that the climates in both types of organizations appear to be supportive of the intrinsic motivation hypothesis put forth by Amabile (1983a).

VanGundy (1984) identified three categories of factors that determine a group's creative climate. They are: the external environment, the internal climate of the individuals within the group, and the quality of the interpersonal relationships among group members. He acknowledged that there would be considerable overlap among these categories and that each category would include suggestions that deal with both task and people oriented issues.

The following list of twenty suggestions provides a representative synthesis of the work done by Torrance (1962), Torrance and Myers (1970), MacKinnon (1978), Amabile (1984) and VanGundy (1984). The list is not totally comprehensive or conclusive. In short, the suggestions constitute recommendations to shape an atmosphere conductive to creativity and innovation. The list provides some necessary conditions (but not sufficient) for creativity:

1. Provide freedom to try new ways of performing tasks; allow and encourage individuals to achieve success in an area and in a way possible for him/her.
2. Permit the activities, tasks, or curriculum to be different for various individuals; point out the value of individual differences, styles, and points of view.
3. Support and reinforce unusual ideas and responses of individuals when engaged in both critical and creative types of thinking; establish an open atmosphere.
4. Encourage individuals to have choices and be a part of goal setting and the decision-making process; build a feeling of individual control over what is to be done and how it might best be done.
5. Let everyone get involved and demonstrate the value of involvement by supporting and helping to develop individual ideas and solutions to problems and projects; encourage the use of the Creative Problem-Solving Process where appropriate.
6. Provide an appropriate amount of time for the accomplishment of tasks; the right amount of work in a realistic time-frame.
7. Communicate that you are confident in the individuals you work with rather than against them; provide a non-punitive environment.
8. Recognize some previously unrecognized and unused potential; challenge individuals to solve problems and work on new tasks.
9. Respect an individual's need to work alone; encourage self-initiated projects.
10. Tolerate complexity and disorder, at least for a period; even the best organization and planning using clear goals requires some degree of flexibility.
11. Use mistakes as positives to help individuals realize errors and meet acceptable standards in a supportive atmosphere; provide constructive feedback and use appropriate evaluation procedures.
12. Criticism is killing . . . use it carefully and in small doses; use encouragement and reduce concern over failure.
13. Adapt to individual interests and ideas whenever possible.

14. Allow time for individuals to think about and develop their creative ideas; not all creativity and innovation occurs immediately and spontaneously.
15. Create a climate of mutual respect and acceptance among individuals so that they will share, develop, and learn cooperatively; encourage a feeling of interpersonal trust.
16. Be aware that creativity is a multi-faceted phenomenon; it enters wide variety of contexts . . . not just arts and crafts!
17. Encourage divergent activities by providing resources and room rather than controlling every element of the tasks to be accomplished.
18. Listen to and laugh with individuals; a warm supportive atmosphere provides freedom and security in exploratory and developmental thinking.
19. Encourage and use provocative questions; move away from the sole use of convergent, one-answer questions.
20. Encourage a high quality of interpersonal relationships and be aware of factors like: a spirit of cooperation, open confrontation of conflicts and the encouragement for expression of ideas.

There are many contingencies, qualifications and factors to consider when applying these guidelines. For example, a common thread running through many of these suggestions is the encouragement of group involvement and ownership. It is important to point out that there are plenty of times when you wouldn't care to use group resources when making a decision. Situational variables such as: the needed quality of decision, the amount of information available, the needed level of commitment to the decision, the amount of conflict in existence, and many other factors could have an impact on deciding when and where to use group resources.

Generally, it is important to keep the concept of balance in mind when using the suggestions for establishing a creative climate. Taking many factors under consideration when using these guidelines will help to moderate the many variables effecting their appropriate application.

The educational implications of the findings and writing in the area of the creative environment are numerous. The kind of environment which is supportive of creativity will allow individuals to be aware of their own blocks to creative thinking and provide a climate where these can be minimized. Some of these blocks may be personal (such as the inability to take risks), problem solving (such as working only within a fixed "set") or situational (like a great deal of emphasis on negative criticism). Following the suggestions provided earlier may reduce the likelihood of the manifestation of blocks. Taking time to deliberately develop an orientation to these inhibitors may provide reinforcement of the "ground rules" for the creative environment.

Another implication for education is that the leadership role for creative learning needs to be examined. There are different kinds of leadership appropriate for different kinds of situations (Hersey and Blanchard 1982). After studying the kind of situation where creativity flourishes, it becomes apparent that a different role is called for in educational settings. Instead of being mainly a provider of factual information and recall-type questions, the teacher needs to become more facilitative of inquiry, problem finding and solving, and guide student's independent and creative learning. This role is more fully described in Wittman and Myrick (1974) and Isaksen (1983b).

Support for this implication is provided by much of the educational literature. Judd (1963, p. 17) clearly indicated the need to modify existing circumstances if education is to nurture more than factual recall. He concluded:

> Memorization of facts frequently fails to result in the development of higher mental processes. If the higher mental processes of application and inference are really to be cultivated, learning conditions appropriate for their cultivation are necessary.

CONCLUSIONS

There appears to be a wealth of research, methods, techniques, and programs related to creativity in education settings. There is, however, a shortage of usable information regarding how to apply these complex and multi-faceted issues and findings. Isaksen (1984) provided the following guidelines for those interested in planning for creative learning:

1. *Provide a responsive environment*—Capitalize on situations that pique students' interests and enthusiasm. This may involve using teacher-pupil planning and working on real and relevant problems. A responsive environment is sensitive to the individual needs of learners. The climate should promote a sense of freedom for thinking. The conditions of the

situation at hand need to promote a desirable physical, emotional, social, and intellectual environment.

2. *Develop a commitment toward facilitation*—Creative learning requires a different role for the teacher (or trainer). This role involves effective listening and skill in interpersonal communications as well a awareness and expertise in the process of facilitation. Since the interaction between facilitators and learners is a key component, the distance between learners and teachers (in the traditional sense) is lessened through genuine respect and seeking to understand the frames of reference of learners. The teachers or trainer needs a deep-seated belief in the necessity and usefulness of creativity.

3. *Use an experiential approach*—Creative learning involves the learner actively so that he/she can understand and feel the benefits of this type of activity. The teacher or trainer may have a specific goal or objective in mind, but the activity or content of the learning comes from the learner. This aspect of creative learning provides linkages to the real world of the learner and makes these situations more intrinsically meaningful.

4. *Work from a knowledge base*—The facilitator or planner seeks to know the psychological and philosophical background of creative learning. Since this type of learning is process-oriented, it is important for the facilitator to be knowledgeable about related processes and skills. It is just as important for the facilitator to recognize that this type of learning occurs in some meaningful context. Content knowledge surrounding the domain of inquiry (subject matter) is important as well.

5. *See learning as continuous*—Learning is viewed as a dynamic process rather than as static information. Facilitators involve themselves in an ongoing process of inservice training to gain up to date and new information and resources. This bolsters their skills, but also demonstrates to the learners that the teacher or trainer is an active learner as well.

6. *Be aware of politics*—Effective planners work with various spheres of influence to share what creative learning is and why it is important. Build on the successes of small groups of supportive practitioners. Also, recognize the need to work with various levels of leadership to help gain acceptance and support. Work toward creating a broad base of support. Top level leaders can provide a safe climate and organizational

goals. Networks can provide shared-experiences, new resources, and other forms of support.

7. *Have specific purposes*—Whether the context is regular subject matter or separate training, the facilitator needs to have explicit goals and objectives to guide the formation and direction of activities. By focusing on specific skills to be developed, it is easier to provide linkages so that learners can "take the skills home". Knowing and communicating your purpose can help in providing this transfer of training.

8. *Work toward integration*—Creative learning needs to be woven into various aspects of the curriculum. Although the initial attempts may be as separate distinct courses or units of instruction, the activities, skills, and situations appropriate for creative learning can gradually permeate the entire curriculum.

9. *Use a planning system*—One of the ways to accomplish some of these guidelines is through the use of some explicit preparation. You may choose an interdisciplinary team approach, cooperative group planning, teacher-pupil planning, or some other method. The important thing is that some actual process is used in the planning of the curriculum. No single process is most appropriate, but the planning framework needs to allow for "preplanned flexibility".

10. *Use an eclectic approach*—There are no shortcuts to creative learning. The overall approach should provide enough flexibility to provide for a variety of learning and personality styles. Use a range of instructional methods such as: discussion-questioning, individualized instruction, group investigation, simulation-role play, or inquiry. The emphasis should not be in the material or package provided by publishers or "program pushers". Rather, the key is the environment and the quality of interaction of people. Successful curriculum planners develop their own creative learning programs tailored to specific situational constraints.

These guidelines were drawn from a variety of sources and would be applicable to a wide range of situations. Much of the support for these guidelines comes from a descriptive survey and interviews of planners who were engaged in planning and designing programs to enhance creative thinking and problem solving (Isaksen 1983a). Planning, as a practical matter, fuses all four facets of creativity into a more meaningful whole.

AN UPDATED RATIONALE

On the basis of all the literature cited earlier in this paper, as well as the work of Combs (1962), Berman (1967), and Cole (1972), the following rationale is offered for those examining the question: Why is it important to consider the implications of creativity in education?

★ *The nature of knowledge*—The accumulation of factual information is growing to the point that total comprehensive awareness is not feasible. More comprehensive states of awareness are possible within selected specific disciplines. This may lead to isolated learning of static information. Data can be "looked up", skills of creative problem solving cannot.

★ *The importance of creative thinking skills*—Since the world is changing so rapidly and it is impossible to accurately predict what knowledge or information will be needed, it is important to focus on the development of skills which help individuals become more adaptable to new and changing circumstances. This focus can help shape alternative images of future circumstances.

★ *Skills more transferable than knowledge*—The ability and facility of using knowledge are more generalizable and more widely applicable than memorization of data. Skills and abilities are more permanent and related to the process of solving problems.

★ *Situations may demand creativity*—There are many situations where there is no immediate or single right answer. These frequent, real-life conditions clearly call for a creative type of thinking.

★ *Creative thinking can be enjoyable*—Learning that calls for the student to actively produce, rather than passively recall, is more motivating. These situations encourage commitment by providing opportunities for learners to follow through on intrinsically-motivated tasks. This increases motivation and relevance for learning.

★ *Creativity is natural*—All students benefit from involvement in creative learning. There may be varying levels and styles in the responses, but everyone can use the level or style they have when provided with the appropriate opportunity.

★ *Builds on knowledge*—Creative learning is not an "either/or" situation. You cannot focus purely on creativity. All creativity has a context; and data surrounding that context. Creative learning uses traditional content as raw material to be used when there is some relevance and need. The focus on process is not entirely independent or exclusive of content, and may actually increase the retention and transfer of learned data.

Finally, lest the reader is left with the notion that providing for creativity in educational contexts is an easy matter, it needs to be recognized that there are more questions than answers when it comes to acting on these implications. To a large degree, it takes a certain amount of risk-taking ability and tolerance for ambiguity to deal effectively with creative learning. There is some data, a wealth of materials and strategies, and a legitimate need for considering creativity and its educational implications. Combs (1962) stated the case for creative learning this way:

> At one time the cultural heritage comprised the sum total of curriculum content. Mastery of the traditional academic subject matter learning", who would legislate what ought to be taught the various disciplines into which man's knowledge has been organized was to be fully and finally educated. There are watch-dogs of education today who urge the return to "solid subject matter learning," who would legislate what ought to be taught at every grade level, and who equate skills and factual infor-mation with substantial knowledge. They overlook the fact, long apparent to responsible educators, that such learning is little more than indoctrination and that what one does with infor-mation and skills is dependent on factors other than the degree of mastery. In short, education is more than the acquisition of facts, and skills are merely means to the end; the development of effective, thinking, creative people demands more than a pouring in of information. (p. 153)

It has been the thesis of this paper that Combs' statement is as applicable today as it was over twenty years ago.

REFERENCES

Aikin, W. M. 1942: *The story of the eight-year study*. New York & London: Harper & Brothers.

Albert, R. S. (Ed.) 1983: *Genius and eminence: The social psychology of creativity and exceptional achievement*. New York: Pergamon Press.

Amabile, T. M. 1982: Social psychology of creativity: A consensual assessment technique. *Journal of Social Psychology*, *43* (5), 997–1013.

Amabile, T. M. 1983a: *The social psychology of creativity*. New York: Springer Verlag.

Amabile, T. M. 1983b: The social psychology of creativity: A componential con-ceptualization. *Journal of Personality and Social Psychology*, *45* (2), 357–76.

Amabile, T. M. 1984: Creativity motivation in research and development. Paper presented as a part of a Division 14 symposium, Creativity in the Corporation, American Psychological Association, Toronto.

Arnold, J. E. 1959: Creativity in engineering. In P. Smith (Ed.), *Creativity: An examination of the creative process.* New York: Hastings House, 33–46.

Baird, L. L. 1983: *Research report: Review of problem solving skills.* Princeton, N.J. Educational Testing Service (March).

Barron, F. & Welsh, G. S. 1952: Artistic perception as a possible factor in personality style: Its measurement by a figure preference test. *Journal of Psychology*, 199–203.

Basadur, M. S. 1981: Research in creative problem solving training in business and industry. *Creativity Week IV: 1981 Proceedings.* Greensboro, NC: Center for Creative Leadership, 40–59.

Berkenkotter, C. 1980: Writing and problem solving. In T. Fulwiler & A. Young (Eds.), *Language connections: Writing and reading across the curriculum.* Urbana, IL: National Council of Teachers of English.

Berman, L. M. 1967: *From thinking to behaving: Assignments reconsidered.* New York: Teachers College Press.

Besemer, S. P. & O'Quinn, K. 1985: Creative product analysis: Beyond the basics. In S. G. Isaksen (Ed.), *Frontiers of creativity research: Beyond the basics.* Buffalo, NY: Bearly Limited.

Besemer, S. P. & Treffinger, D. J. 1981: Analysis of creative products: Review and synthesis. *Journal of Creative Behavior, 15* (3), 158–78.

Biondi, A. M. & Parnes, S. J. 1976: *Assessing creative growth.* Buffalo, N.Y.: Creative Education Foundation.

Bloomberg, M. (Ed.) 1973: *Creativity: Theory and research.* New Haven, CT: College and University Press.

Brandt, A. 1982: Teaching kids to think. *Ladies Home Journal* (September), 104–6.

Briskman, L. 1980: Creative product and creative process in science and art. *Inquiry, 23* (1), 83–106.

Brooks, M. 1984: A constructivist approach to staff development. *Educational Leadership*, (November), *42* (5), 23–8.

Carpenter, T. P., Lindquist, M. M., Matthews, W. & Silver, E. 1983: Results of the third NAEP mathematics assessment: Secondary school. *Mathematics Teacher*, (December), 652–9.

Cole, H. P. 1972: *Process education: The new direction for elementary secondary schools.* Englewood Cliffs, N.J. Educational Technology Publications.

Combs, A. W. 1969: *Perceiving, behaving, becoming: A new focus for education.* Washington, D.C.: Association for Supervision and Curriculum Development, NEA, Yearbook.

Costa, A. L. 1984: Mediating the metacognitive. *Educational Leadership, 42* (5), 57–8.

Cox, C. M. 1926: *Genetic studies of genius.* Stanford: Stanford University Press.

Dewey, J. 1933: *How we think: A restatement of the relation of reflective thinking to the educative process.* Lexington, MA: Heath & Company.

Dillon, J. T. 1984: Research on questioning and discussion. *Educational Leadership* (November), *42* (5), 50–6.

Ekvall, G. 1983: *Climate, structure and innovativeness of organizations: A theoretical framework and an experiment.* The Swedish Council for Management and Organizational Behavior, Stockholm, Sweden.

Ekvall, G. & Parnes, S. J. 1984: *Creative problem solving methods in product development—a second experiment.* The Swedish Council for Management and Work Life Issues. Stockholm: FA Radet.

Elbow, P. 1983: Teaching thinking by teaching writing. *Change,* (September), *15*, 37–40.

Fromm, E. 1959: The creative attitude. In H. H. Anderson (Ed.), *Creativity and its cultivation*. New York: Harper & Brothers, 44–54.

Gall, M. 1984: Synthesis of research on teacher's questioning. *Educational Leadership*, (November), *42* (5) 40–7.

Galton, F. 1968: *Hereditary genius*. London: Macmillan.

Gardner, H. 1983: *Frames of mind: The theory of multiple intelligences*. New York: Basic Books.

Getzels, J. W. & Csikszentmihalyi, M. 1976: *The creative vision: A longitudinal study of problem finding in art*. New York, Wiley.

Gibney, T. C. & Meiring, S. P. 1983: Problem solving: A success story. *School Science and Mathematics*, (March), *83* (3), 194–203.

Glatthorn, A. A. 1982: *A guide for developing an English curriculum for the eighties*. Urbana, IL: National Council of Teachers of English.

Goertzel, M. G., Goertzel, V. & Goertzel, T. G. 1978: *300 eminent personalities*. San Francisco: Jossey-Bass.

Goodlad, J. I. 1983: A study of schooling: Some findings and hypotheses. *Phi Delta Kappa*. (March), *64* (7), 465–70.

Goor, A. &. Rapoport, T. 1977: Enhancing creativity in an informal educational framework. *Journal of Educational Psychology*, *69* (5), 636–43.

Gourley, T. &. Micklus, C. S. 1981: Creative competitions: Now that's creativity! *Gifted/Creative/Talented*, (November/December), 35–7.

Gowan, J. C., Khatena, J. & Torrance, E. P. (Eds.). 1981: *Creativity: Its educational implications*. Dubuque, IA: Kendall/Hunt.

Grennon, J. 1984: Making sense of student thinking. *Educational Leadership*, *43* (5), 11–16.

Gruber, H. E. 1981: *Darwin on man: A psychological study of scientific creativity*. Chicago: The University of Chicago Press.

Gryskiewicz, S. G. 1985: Predictable creativity. In S. G. Isaksen (Ed.), *Frontiers of creativity research: Beyond the basics*. Buffalo, NY: Bearly Limited.

Gryskiewicz, S. G. 1980: Targeted innovation: A situational approach. *Creativity Week III: 1980 Proceedings*. Greensboro, NC: Center for Creative Leadership, 1–27.

Guilford, J. P. 1950: Creativity. *American Psychologist*, *5*, 444–54.

Guilford, J. P. 1967: *The nature of human intelligence*. New York: McGraw-Hill.

Guilford, J. P. 1977: *Way Beyond the IQ*. Buffalo, NY: Bearly Limited.

Guilford, J. P. 1980: Foreword. In M. K. Raina (Ed.), *Creativity research: International perspective*. New Delhi, India: National Council of Educational Research and Training.

Hadamard, J. 1954: *The psychology of invention in the mathematical field*. New York: Dover.

Hallman, R. J. 1981: The necessary and sufficient conditions of creativity. In Gowan, J. C., Khatena, J. & Torrance, E.P. *Creativity: Its educational implications*. Dubuque, IA: Kendall/Hunt, 19–30.

Hausman, C. S. 1984: *A discourse on novelty and creation*. Albany, N.Y.: State University of New York Press.

Helson, R. & Crutchfield, R. S. 1970: Creative types in mathematics. *Journal of Personality*, *38*, 177–97.

Heppner, P. P. 1978: A review of the problem solving literature and its relationship to the counseling process. *Journal of Counseling Psychology*, *25* (5), 366–75.

Heppner, P. P., Neal, G. W. & Larson, L. M. 1984: Problem-solving training as prevention with college students. *The Personnel and Guidance Journal*, (May), 514–19.

Hersey, P. & Blanchard, K. 1982: *Management of organizational behavior: Utilizing human resources* (4th ed.). Englewood Cliffs, N.J.: Prentice Hall.

Hullfish, H. G. & Smith, P. G. 1961: *Reflective thinking: The method of education*. New York: Dodd, Mead.

Isaksen, S. G. 1983a: A curriculum planning schema for the facilitation of creative thinking and problem-solving skills. Unpublished doctoral dissertation. State University of New York at Buffalo.

Isaksen, S. G. 1983b: Toward a model for the facilitation of creative problem solving. *Journal of Creative Behavior*, *17* (1), 18–31.

Isaksen, S. G. 1984: Implications of creativity for the middle school "education". *Transescence: The Journal on Emerging Adolescent Education*, *12* (2).

Isaksen, S G. 1985: *Frontiers of creativity research: Beyond the basics.* Buffalo, N.Y.: Bearly Limited.

Isaksen, S. G. & Parnes, S. J. 1985: Curriculum planning for creative thinking and problem solving. *Journal of Creative Behavior*, *19* (1).

Isaksen, S. G., Stein, M. I., Hills, D. A. & Gryskiewicz, S. S. 1984: A proposed model for the formulation of creativity research. *Journal of Creative Behavior*, *18* (1), 67–75.

Isaksen, S. G. & Treffinger, D. J. 1985: *Creative problem solving: The basic course.* Buffalo, N.Y.: Bearly Limited.

Johansson, B. 1975: Creativity and creative problem-solving courses in United States industry. Special project funded by the Center for Creative Leadership, Greensboro, NC.

Johnson, D. W., Maruyama, G. Johnson, R., Nelson, D. & Skon, L. 1981: The effects of cooperative, competitive, and individualistic goal structures on achievement: A meta-analysis. *Psychological Bulletin*, *89* (1), 47–62.

Judd, C. H. 1936: *Education as cultivation of the higher mental processes.* New York: Macmillan.

Juntune, J. 1979: Project REACH: A teacher's training program for developing creative thinking skills in students. *Gifted Child Quarterly*, *23* (3), 461–71.

Khatena, J. 1982: Myth: Creativity is too difficult to measure. *Gifted Child Quarterly*, *26* (1), 21–3.

Kilpatrick, W. H. 1918: *The project method: The use of the purposeful act in the educative process.* New York: Teacher's College Bulletin.

Kirton, M. J. 1976: Adaptors and innovators: A description and measure. *Journal of Applied Psychology*, *61* (5), 622–9.

Krulik, S. & Rudnick, J. A. 1982: Teaching problem solving to preservice teachers. *Arithmetic Teacher*, (February), 42–5.

Kuhn, T. S. 1970: *The structure of scientific revolutions.* Chicago: The University of Chicago Press.

Langer, J. A. 1982: Reading, thinking, writing . . . and teaching. *Language Arts*, (April), *59* (4), 336–41.

Larkin, J. H. 1980: Teaching problem solving in physics: The psychological laboratory and the practical classroom. In D. T. Tuma & F. Reif (Eds.), *Problem solving and education: Issues in teaching and research.* Hilldale, N.J.: Lawrence Erlbaum Associates, 111–26.

Lipman, M., Sharp, A. M. & Oscanyan, F. S. 1980: *Philosophy in the classroom.* Philadelphia: Temple University Press.

MacKinnon, D. W. 1975: IPAR's contribution to the conceptualization and study of creativity. In Taylor, I. A. & Getzels, J. W. (Eds.), *Perspectives in creativity.* Chicago: Aldine, 60–89.

MacKinnon, D. W. 1978: *In search of human effectiveness: Identifying and developing creativity.* Buffalo, N.Y.: Creative Education Foundation.

Mansfield, R. S. & Busse, T. V. 1981: *The psychology of creativity and discovery: Scientists and their work.* Chicago: Nelson-Hall.

Mansfield, R. S., Busse, T. V. & Krepelka, E. J. 1978: The effectiveness of creativity training. *Review of Educational Research*, *48* (4), 517–36.

Martin, D. S. 1984: Infusing cognitive strategies into teacher preparation programs. *Educational Leadership*, (November), *42* (5), 68–72.

Maslow, A. 1959: Creativity in self-actualizing people. In H. H. Anderson (Ed.), *Creativity and its cultivation.* New York: Harper & Brothers, 83–95.

Myers, I. B. & Myers, P. B. 1980: *Gifts differing*. Palo Alto, CA: Consulting Psychologists Press.

National Assessment of Educational Progress 1981: *Reading, thinking, and writing*. Denver, CO: NAEP.

National Council of Teachers of Mathematics 1980: *An agenda for action: Recommendations for school mathematics of the 1980's*. Reston, VA: The Council.

Olson, C. B. 1984: Fostering critical thinking skills through writing. *Educational Leadership*, (November), *42* (5), 28–39.

Parnes, S. J. & Harding, H. 1962: *A sourcebook for creative thinking*. New York: Scribners.

Parnes, S. J. & Noller, R. B. 1972: Applied creativity: The creative studies project (part II—results of the two-year program). *Journal of Creative Behavior*, *6* (3), 164–86.

Patrick, C. 1937: Creative thought in artists. *Journal of Psychology*, (January), *4*, 35–73.

Pearlman, C. 1983: Teachers as an informational resource in identifying and rating student creativity. *Education*, *103* (3), 215–22.

Perkins, D. N. 1981: *The mind's best work*. Cambridge, MA: Harvard University Press.

Raina, M. K. 1980: *Creativity research: International perspective*. New Delhi, India: National Council of Educational Research and Training.

Reese, H. W., Parnes, S. J., Treffinger, D. J. & Kaltsounis, G. 1976: Effects of a creative studies program on structure-of-intellect factors. *Journal of Educational Psychology*, (August), *68* (4), 401–10.

Renzulli, J. 1978: What makes giftedness? *Phi Delta Kappan*, (November), 180–251.

Rhodes, M. 1961: An analysis of creativity. *Phi Delta Kappan*, (April), 305–10.

Rogers, C. 1959: Toward a theory of creativity. In H. H. Anderson (Ed.). *Creativity and its cultivation*. New York: Harper & Brothers, 69–82.

Rogers, E. M. 1983: *Diffusion of innovations* (3rd ed.). New York: The Free Press.

Rose, L. H. & Lin, H. T. 1984: A meta-analysis of long-term creativity training programs. *Journal of Creative Behavior*, *18* (1), 11–22.

Rossman, J. 1931: *The psychology of the inventor: A study of the patentee*. Washington, D.C.: The Inventors Publishing Company.

Rothenberg, A. & Hausman, C. R. 1976: *The creativity question*. Durham, N.C.: Duke University Press.

Rubinstein, M. F. 1975: *Patterns of problem solving*. Englewood Cliffs, N.J.: Prentice-Hall.

Rugg, H. & Shumaker, A. 1928: *The child-centered school*. New York: World Book.

Schoenfeld, A. H. & Herrmann, D. J. 1982: Problem perception and knowledge structure in expert and novice mathematical problem solvers. *Journal of Experimental Psychology: Learning, Memory and Cognition*, 8 (5), 484–94.

Simonton, D. K. 1984: *Genius, creativity and leadership: Historiometric inquiries*. Cambridge, MA: Harvard University Press.

Spearman, C. E. 1931: *The creative mind*. New York: Appleton-Century.

Stein, M. I. 1974: *Stimulating creativity: Individual procedures* (vol. I). New York: Academic Press.

Stein, M. I. 1975: *Stimulating creativity: Group procedures* (vol. II). New York: Academic Press.

Stievater, S. 1984: Personal correspondence in response to a request to update Guilford's initial search. State University College at Buffalo, Memorandum dated August 7, 1984.

Taylor, I. A. 1959: The nature of the creative process. In P. Smith (Ed.), *Creativity: An examination of the creative process*. New York: Hastings House, 51–82.

Taylor, I. A. & Getzels, J. W. (Eds.) 1975: *Perspectives in creativity*. Chicago: Aldine.

Taylor, I. A. & Sandler, B. J. 1972: Use of a creative product inventory for evaluating products of chemists. *Proceedings of the 80th Annual Convention of the American Psychological Association*, 7, 311–12.

Taylor, C. W. (Ed.) 1963: *Widening horizons in creativity*. New York: Wiley.

Taylor, C. W., Smith, W. R. & Ghiselin, B. 1963: The creative and other contributions of one sample of research scientists. In C. W. Taylor & F. Barron (Eds.), *Scientific creativity: Its recognition and development*. New York: Wiley, 53–76.

Terman, L. M. 1954: The discovery and encouragement of exceptional talent. *American Psychologist*, 9, 221–30.

Toepfer, C. F. 1982: Curriculum design and neuropsychological development. *Journal of Research and Development in Education*, (Spring), 15 (2), 1–11.

Torrance, E. P. 1962: *Guiding creative talent*. Englewood Cliffs, N.J.: Prentice-Hall.

Torrance, E. P. 1972: Can we teach children to think creatively? *Journal of Creative Behavior*, 6 (2), 114–43.

Torrance, E. P. 1974: *Torrance tests of creative thinking: Norms and technical manual*. Lexington, MA: Personnel Press/Ginn Xerox.

Torrance, E. P. 1975 Creativity research in education: Still alive. In I. A. Taylor & J. W. Getzels (Eds.), *Perspectives in creativity*. Chicago: Aldine, 278–96.

Torrance, E. P. 1979: *Search for satori and creativity*. Buffalo, N.Y.: Creative Education Foundation 12.

Torrance, E. P. 1980: Georgia studies of creative behavior: A brief summary of activities and results. In M. K. Raina (Ed.), *Creativity research: International perspective*. New Delhi, India: National Council of Educational Research and Training, 253–71.

Torrance, E. P. 1981: Can creativity be increased by practice? In J. C. Gowan, J. Khatena & E. P. Torrance (Eds.), *Creativity: Its educational implications*. Dubuque, IA: Kendall/Hunt.

Torrance, E. P. & Myers, R. E. 1970: *Creative learning and teaching*. New York: Dodd, Mead.

Treffinger, D. J. 1980: *Encouraging creative learning for the gift and talented*. Ventura, CA: Ventura County Schools/LTI.

Treffinger, D. J. 1984: Critical and creative thinking: Mutually important components of effective problem solving. Unpublished paper prepared as a part of a series of papers on Gifted Education for the Language and Learning Improvement Branch of the Division on Instruction of the Maryland State Department of Education.

Treffinger, D. J., Isaksen, S. G. & Firestien, R. L. 1982: *The handbook of creative learning, vol. I*. Honeoye, N.Y.: Center for Creative Learning.

Treffinger, D. J., Isaksen, S. G. & Firestien, R. L. 1983: Theoretical perspectives on creative learning and its facilitation: An overview. *Journal of Creative Behavior*, 17 (1), 9–17.

Van Gundy, A. B. 1984: *Managing group creativity: A modular approach to problem solving*. New York: American Management Association.

Wallas, G. 1926: *The art of thought*. New York: Franklin Watts.

Ward, W. C. & Cox, P. W. 1974: A field study of nonverbal creativity. *Journal of Personality*, 42, 202–19.

Welsch, P. K. 1980: The nurturance of creative behavior in educational environments: A comprehensive curriculum approach. Unpublished doctoral dissertation, University of Michigan, 96.

Welsch, G. S. 1973: Perspectives in the study of creativity. *Journal of Creative Behavior*, 7 (4), 231–46.

Wertheimer, M. 1945: *Productive thinking*. New York: Harper & Brothers.

Whitman, N. 1983: Teaching problem solving and creativity in college courses. *AAHE Bulletin: Research Currents*, (February), 9–13.

Williams, F. E. 1970: *Classroom ideas for encouraging thinking and feeling.* Buffalo, N.Y.: D.O.K. Publishers.

Williams, F. E. 1980: *Creativity assessment packet (CAP).* Buffalo, N.Y.: D.O.K. Publishers.

Wittmer, J. & Myrick, R. D. 1974: *Facilitative teaching: Theory and practice.* Pacific Palisades, CA: Goodyear Publishing.

Worsham, A. W. & Austin, G. R. 1983: Effects of teaching thinking skills on SAT scores. *Educational Leadership*, (November), 50–7.

7

Trial by Fire in an Industrial Setting: A Practical Evaluation of Three Creative Problem-Solving Techniques

STANLEY S. GRYSKIEWICZ

In previous research, when asked to evaluate conditions which enhance the generation of ideas in group settings, Dunnette (1964), along with several others (Taylor, Berry, and Block 1958; Rotter and Portugal 1969; Hill 1982), concluded that groups are indeed mediators of uniform thinking. Furthermore, the general conclusion was that more efficient creative problem solving would occur if time were provided for individuals to generate ideas alone and avoid the uniformity pressures spawned within interacting groups (Lamm and Trommsdorff 1973; Graham and Dillon 1974). Interactive group creative problem solving may be an inefficient way of generating ideas.

Such a conclusion is, however, not uncontested in the literature. Studies by Maier (1967) and Hoffman (1965, 1979) suggest that groups have the potential to produce higher quality ideas than individuals because (1) a group contains more information than any one of its members, and (2) group discussion can lead to conflict, its resolution, and hence, to a novel solution. Hoffman (1979), and in the following chapter in this book, argues that group problem solving may enhance the level of creative solutions and facilitate their implementation. Glover (1979) and Glover and Chambers (1978) suggest that the group setting impacts group performance. They have demonstrated that a laissez-faire small group structure results in higher levels of creative behaviors on a group product than did either a democratic or authoritarian form of group structure. The completion of creative thinking tests in dyads (Torrance 1971) has not only led to higher levels of creativity but subjects report stronger feelings of stimulation, enjoyment and originality of expression than individuals who worked alone. Again, inconsistency reigns with a study that indicated "production blocking" in dyads that worked jointly (Barkowski, Lamm, and Schwinger 1982).

In line with these inconsistent findings, it is my experience that the use of group problem solving in organizations will continue and its occurrence is predicated upon three assumptions: (1) the necessary information for solving problems is distributed among group members, (2) the temporal association of two or more possible ideas can lead to yet another, and (3) the shared ownership of ideas generated in a group setting will lead to idea acceptance at the implementation stage. It would seem that the refusal to use group problem solving in favor of high performing individuals may result in a decrement in idea quality and idea acceptance. Such trade-offs should be considered, however, the information necessary for making that decision has not been thoroughly explored. Hopefully, this research will provide informed alternatives.

Before we accept the findings of Dunnette (1964) and others (Lamm and Trommsdorff 1973; Jablin and Siebold 1978) that individuals are superior to groups, let us trace the genesis of this research with a literature review focused on the relevant paradigm. The authors, Taylor, Berry, and Block (1958) took up the challenge to evaluate individual versus group problem solving. Specifically, their aim was to evaluate the effectiveness of the brainstorming technique. Alex Osborn, in his now classic work, *Applied Imagination* (1957), stated that group brainstorming produced 44 percent more ideas than individuals working alone. Osborn's guidelines for effective brainstorming in groups are that: (1) criticism is ruled out, (2) unusual ideas are welcome, (3) quantity is encouraged, and (4) combination and improvement of ideas are sought. Taylor set the stage for his effort in an earlier review of problem-solving research (Taylor and McNemar 1955). He ended the review by suggesting a research paradigm designed to evaluate Osborn's claims. Subsequently, this real group/nominal group paradigm suggested by Taylor influenced the thirty years of research which followed.

Real group was defined as a given number of individuals working together for a stated period of time, usually 30 minutes. Nominal group (group by name only) was defined as a given number of individuals working alone for the same period of time. As a baseline to compare the performance with real groups, the individuals' products would be summed as if people had actually interacted together in a group, generating ideas.

A series of research studies flowed from this paradigm, and the consistency in the central findings of these experiments is quite striking as to the quantity of ideas produced. The equivalent number of individuals produced more ideas (even when duplicate ideas were removed) than individuals working in a real group setting. Clearly the social dynamics of a real group accounted for the decrement in the number of ideas generated, since the four brainstorming rules, especially cross fertilization, did not increase the idea count in the real group performance. Qualitative ratings of the products using dimensions such as feasibility, effectiveness, generality, significance and probability of solving the problem produced mixed results. Some categories rated higher in the nominal condition, and when adjustments were made for quantity differences, some real condition categories were rated higher. While consistent quantity findings resulted in favor of individuals (nominal groups), the inconsistency of qualitative ratings persisted.

Dunnette, Campbell, and Jaastad (1963) suggested that the inhibitory effects of group brainstorming may result in a tendency for groups to "fall in a rut" (uniformity pressures) and for members to pursue similar thought trains. The pressure towards uniform thought could preclude the generation of diverse themes and ideas to the problem.

Kelly and Thibaut (1969) suggested that brainstorming might be useful to overcome uniformity pressures that occur during the discussion phase of group problem solving (Thomas and Fink 1961). Vroom, Grant, and Cotton (1969) reported that interaction during the generation phase of problem solving is, in fact, dysfunctional. These groups produced a smaller number of varied solutions and fewer high quality solutions than groups in which members interaction was moderated during idea generation. This is consistent with Taylor et al. (1958) findings and reinforces the power of group interaction to limit ideas, but from a less dramatic paradigm.

Several researchers have tried to reduce the uniformity pressures found in real groups. Meadow, Parnes, and Reese (1959) varied instructions for brainstorming. Those groups told to formulate solutions without immediate evaluation were superior to groups told to formulate and evaluate. Evaluation scenarios (immediate or delayed) showed no difference in real group conditions nor when comparing their corresponding nominal group control condition (Maginn and Harris 1980). Attempts to improve group problem solving extended to the selection of group members by using personality test scores, through the training of the group members and in the variation of problem content (Mann 1959; Heslin 1964).

Cohen, Whitmyre, and Funk (1960) tried to influence group performance by using problems of interest to the participants. These problems, called "ego-involving", were taken from the work settings of the participants. The results suggest that groups using ego-involving problems produced more unique ideas.

Thomas Bouchard has published more than any other researcher to demonstrate the impact of the person, the process, and the situation upon the products of group problem solving. These frameworks provided Bouchard with the variables needed to investigate the effect of personality upon creative problem-solving performance (Bouchard 1969). Group membership was selected using an individual differences construct measured by five scales on the California Psychological Inventory (Gough 1957). Bouchard and Hare (1970) varied group size and sex of the experimenter. In both studies, real and nominal groups were compared on the number of unique ideas. In all cases, nominal groups outperformed real groups.

Bouchard (1972b) manipulated process and person variables in two studies. The process changes included training and the use of a modified brainstorming technique called "sequencing". This technique was designed to increase participation by requiring each member to put forward an idea in "round-robin" sequence. Group membership was assigned on high versus low interpersonal effectiveness measured by the California Psychological Inventory scales. The sequencing procedure appeared to have an impact upon real group performance when compared with nominal group. Bouchard optimistically concluded that the motivation of forced participation resulted in a level of real group performance that was nearly identical to nominal group performance. Prior training on the use of the brainstorming technique also increased quantity. In all these manipulations, while the real/nominal differences was reduced, nominal groups still did out perform real groups significantly. Group assignment based on high/low interpersonal effectiveness had no effect.

Additional research in process variation has resulted in mixed findings. Bouchard (1972a) demonstrated that a process called "personal analogy" produced ideas judged more innovative than brainstorming. Bouchard, Barsaloux, and Drauden (1974) evaluated yet another process (brainstorming versus environment concordance) and situation (group size and real versus nominal) to observe the impact upon product (number of different ideas). Environment concordance involves placing subjects in a setting which closely replicates the environmental theme of the problem, i.e., if the problem related to blind people, subjects sat in a darkened room. Process variation had no effect. The dependent variable was responsive to a variety of factors such as group size, and real group or nominal group settings. Increasing real group size added little to performance. Bray, Kerr, and Atkins (1978) advanced the concept "functional size" to explain the phenomenon they reported of non-participators increasing as group size increases resulting in a functional group smaller than the actual large group's size. The larger group's performance suffered from the non-participation of some members and as these numbers increased, performance decreased when compared to equal numbered nominal groups. Increasing nominal group size led to drastic performance increases and further support for the thesis that increasing real group size mediates uniformity pressure. Another situational variation included written feedback to real group and nominal group brainstorming members. The experiment resulted in a higher production of ideas for the nominal group (Madsen and Finger 1978). Others (Klimoski and Karol 1976; Glover and Chambers 1978; Glover 1979) found that by varying process conditions they could impact the quality and quantity of the ideas produced.

The dynamic to be avoided in creative problem solving is uniformity of thinking. Social pressures occurring within groups have been observed and result in uniformity pressures which in turn lead to fewer themes being examined by group members and fewer ideas produced within those themes. What follows is a listing of some of the factors which contribute to the uniformity pressures found in real creative problem-solving groups.

CONSTRUCTS MEDIATING UNIFORMITY IN REAL PROBLEM-SOLVING GROUPS

While there are many constructs which mediate uniformity pressure in groups (Lamm and Trommsdorff 1973; Stein 1975; Jablin and Siebold 1978; Hill 1982), I would like to review those addressed directly by this research.

★ *Communication Patterns.* Research suggests that communication between group members should be open, allowing members to share ideas directly. Direct sharing increased the occurrence of novel associations (Meyer 1957; Shaw 1964; Van de Ven and Delbecq 1971; Shaw 1981). Vroom, Grant, and Cotton (1969) show, however, that direct interaction "tends to result in group members developing a common set in their approach to the problem" and a total idea production that is not very diverse.

★ *Group Climate.* Generally it is believed that a problem solving group milieu which permits the evaluation of members' ideas will inhibit idea flow (Gordon 1961; Maginn and Harris 1980). Rogers (1983) states that non-evaluative settings are most conducive to the creative performance of individuals in groups and this is confirmed by non-competitive laissez-faire, small groups which result in more creative behaviors (Glover 1979). Deferred judgment is important for overcoming uniformity pressures in groups (Brilhart and Jochem 1964; Basadur and Finkbeiner 1985).

★ *Highly Vocal Group Members.* The person who talks the most in the group often controls and directs the idea flow of the group. Uniformity of thinking in 28 of 44 groups, whether correct or incorrect, was apparently caused by the actions of a highly vocal person (Thomas and Fink 1961). The frequency of oral communication leads to interpersonal judgments by group members which, in turn, inhibits the brainstorming productivity of groups (Milton 1965; Jablin and Siebold 1978). Delbecq, Van de Ven, and Gustafson (1975) report that the highly expressive individual dominates the group and thereby influences group uniformity.

★ *Perceived Expertise of Group Members.* In brainstorming groups, members evaluate their own ideas according to the perceived competency of fellow members. Inhibition of responses was greater, and members reported a reluctance to contribute in groups perceived to include experts (Collaros and Anderson 1969). The research confirms Hoffman's (1965) findings that status differentials among group members affects the products of group problem solving.

★ *Process.* Some techniques are more effective in reducing uniformity pressures than others (Warren and Davis 1969; Bouchard 1972a; Bouchard, Barsaloux, and Drauden 1974). Idea generation variations designed to respond to particular aspects of group process which limits idea generation were tried. In all cases the real/nominal relationship stayed the same; however, the magnitude of difference was reduced. Instructions to "be creative" (Datta 1963) or instructions designed to foster originality in problem solving (Colgrove 1968) have facilitated divergent thinking.

★ *Problem Topic.* The few researchers who used "ego-involving" problems were more successful in reducing the uniformity pressures inherent in groups (Lorge and Solomon 1959; Cohen, Whitmyre, and Funk 1960; Dillon, Graham, and Aidells 1972). In these instances, the problems reflected instances from the participants' work setting or the current political debate taking place on the college campus at local and national levels.

We turn now to an empirical examination of the importance of these factors. Communication pattern (Item 1) will be addressed by the real/nominal comparison. Items 2–5 will be addressed by the use of three creative problem-solving techniques and Item 6 relates to the type of problem used for this research.

FIVE RESEARCH CONCERNS

Five issues emerged from the past research. Our greatest overall concern and self imposed guidelines for the research was the transfer of findings to applied settings.

Issue One—A Question of Methodology: Are Nominal Groups, Groups?

To what degree is a disservice done to the understanding of creative problem-solving technology by referring to a control condition for baseline comparison data whose membership is determined randomly, after the fact, as a nominal group?

For a collection of people to be considered a group, there must be some interaction (Hare 1976) there must be roles for members to play along with developed norms (Sherif 1954). This label, "nominal group", has became confused with the construct "real group".

MacKinnon (1975) defined the creative situation in terms of the social/work milieus which "facilitate or inhibit the appearance of creative thought and action". Therefore, separate techniques unique to each situation may suggest an alternative interpretation of the research results.

Support for this notion was reported by Street (1974), who implemented a third condition called co-acting groups. This condition required subjects to sit together in the same room performing a task with instructions not to interact. There was no difference between nominal and co-acting groups; however, both were superior to real groups. Most of the nominal versus real group research would suggest that only when one's actions become public does uniformity pressure increase, resulting in a decrease in the number of ideas generated. Both public, real group problem-solving technology and private, nominal group problem-solving technology must be considered as separate entities to be evaluated and better understood. Only then can they be effectively used by practitioners.

Confusion increased further when Delbecq, Van de Ven, and Gustafson (1975) used the name nominal group technique (NGT) to describe an idea generation technique expressly designed to overcome the uniformity pressures found in small groups. NGT combines both nonverbal and verbal stages, but does not capitalize on the potential of cross fertilization.

Issue Two—Research Populations

Of all the published studies, from 1957 through 1982, there were only three (Dunnette, Campbell, and Jaastad 1963; Campbell 1968; Basadur, Green, and Green 1982) that used managerial populations. The remainder used college students and younger students in laboratory settings (Smith 1968). While the results were the same (nominal superiority), the student populations remain a major factor inhibiting the generalization of findings to organizational settings.

This chapter reports the findings of two studies conducted in the United Kingdom and the United States using first-level and middle-level managers. The United Kingdom study used 120 managers while the United States study answered an additional question and required a total of 160 managers.

Issue Three—Type Problem Solved

It was clear that researchers had used "unreal" problems. Harari and Graham (1975) concluded that the tasks in most studies represent little more than innocuous puzzles or games. Two examples are:

★ You woke up this morning and you had an extra thumb on your hand. What are the practical benefits or difficulties of having an extra thumb?

★ What are the implications of having a large population of ten-inch tall people?

In adherence to sound application of methodology, these problems and others like them were used in most reported research. This author, however, was influenced by a need to be applied and by related research which suggested that "ego-involving" problems were more successful in reducing group uniformity than puzzles or games (Dillon et al. 1972; Harari and Graham 1975). As an alternative to unreal problems, industry was consulted for problems amenable to the research project. The problem finally used had been of primary importance to the packaging industry.

Issue Four—Additional Dependent Measures

After initial attempts to measure both quantity and quality as distinct variables in his early research (Bouchard 1969), quality measures were dropped (Bouchard and Hare 1970; Bouchard 1972b) due to their consistently high correlation with quantity measures. "Previous work (Bouchard 1969) has shown that more sophisticated scoring is unnecessary, since all other scores that can be derived from this type of data correlate highly with total quantity" (Bouchard and Hare 1970). This position was also adopted by subsequent researchers (Dillon, Graham, and Aidells 1972; Graham and Dillon 1974; Klimoski and Karol 1976; Graham 1977) who cited the same references.

The author believes that the inclusion of qualitative measures distinct from, yet along with, quantity measures are necessary to clearly identify and evaluate the assets and liabilities of any idea generation technique.

Included were measures generated by trained raters and by experts from the industry that provided the problem. The trained raters generated ratings similar to those reported by earlier researchers such as effectiveness, novelty, and a combination score of these two called creativity (Meadow, Parnes, and Reese 1959). An additional dependent measure of four distinct categories was derived from the adaption-innovation construct developed by Kirton (1976). The use of expert raters in this context reflect a novel departure from the literature. These multiple dependent measures add to the strength of the data analysis and to the generalization of the findings.

Issue Five—Alternative Creative Problem-Solving Techniques?

From the considerations mentioned above, it seems reasonable to wonder what effect a broader spectrum of procedures hypothesized to reduce uniformity pressure will have. Among the most promising is brainwriting (Geschka, Schaude, and Schlicksupp 1973), a non-oral technique which follows the same rules as brainstorming, but idea generation is written.

A second addition is excursion. The excursion (Gordon 1961; Davis 1973) creates a novel stimulus, the fantasy scenario, whose content is removed from the content of the original problem statement. The scenario, in turn, becomes a stimulus and is juxtaposed with the problem and used to trigger some novel responses to the problem.

The Independent Variables

It was hypothesized that three techniques would facilitate, to different degrees, the reduction of uniformity pressure present in real group problem solving.

Brainstorming
Designed by Osborn (1957) to encourage the verbal exchange of divergent ideas in a non-evaluative climate, its continued use is an indicator of its perceived value for producing a climate which permits the production of novel ideas (Rickards 1974; VanGundy 1980). Its rules were listed earlier.

Excursion
This is a modification of fantasy analogy devised by Gordon (1961) in which users generate a fantasy scenario having no apparent relation to the problem. The novel fantasy is in turn juxtaposed with the problem as a means to stimulate ideas.

Schaude (1979) lists excursion under the generic topic "creative confrontation". This technique brings fantasy into the group process in a structured way, and is designed to disrupt idea fixation via the confrontation of problem statement with the fantasy introduced into the idea generation session.

There was great variability (5 to 16 minutes) in the amount of time subjects took to complete the imaging and fantasy generation phase of the excursion technology. This meant that idea generation time varied from 25 to 14 minutes. A more accurate comparison of the 30 minutes in brainwriting and brainstorming required that excursion in Study 2 have two data notations; one after 30 minutes total and one after 30 minutes of idea generation.

Brainwriting
This method was originally developed by Bernd Rohrback (Johanson 1978) and is described in detail in Geschka, Schaude, and Schlicksupp (1973). Participants are asked to write their ideas in stages on a form and then asked to exchange forms with other members. Reading the written ideas of others (this process is non-oral) acts as a stimulus for even more ideas. Use of written rather than oral statements is meant to stimulate idea generation without one member orally disturbing, dominating, or demonstratively influencing the group. By comparing this procedure with traditional brainstorming, a direct test is possible of the thesis that oral dominance by members contributes to uniformity pressures.

A non-technique control condition was deemed non-feasible in this case since earlier research had already demonstrated the value of group brainstorming over no-special instruction groups (Bouchard 1969). While this omission is a possible criticism of the study, in this case, the earlier findings in the case of brainstorming were accepted. It was decided to implement this design addition for direct comparison with brainwriting in study two. It is called 7-3-4 (BW-2) and participants were presented with the problem and the brainwriting form but were not given the brainstorming rules.

Development of the Measures

Dependent measures were collected from the research subjects (Level One), trained raters (Level Two), and from industry experts (Level Three). Inclusion of the Level Three measure is a novel feature in the present study.

Level One
Along with biographical data subjects were asked to complete the Barron-Welsh Art Scale and to evaluate the CPS technique they experienced. The Art Scale was used to measure preference for perceiving and dealing with complexity of the participating managers as a check on random assignment to condition. While the scale was normed on various artist populations, we had available, for comparison purposes, scores for related managerial populations.

User evaluation was based on three 50-point scales ranging from very unsuccessful or unenjoyable to very successful or enjoyable and a rating of value which ranged from no value at all to extremely valuable.

The quantity indicators used in Level One (number of ideas generated, number of repeats, and number of novel ideas) reflect some of the measures used in earlier research.

Level Two

Trained and selected raters were asked to categorize and rate each idea. Every idea was seen by two raters. The raters reliability ranged between 0.83 and 0.60.

Each idea was rated along a dimension of effectiveness, its ability to contribute to solving the problem. Novelty scores, an idea appearing only once within the total population of ideas in a condition, were determined. Finally, a creativity score was established by combining the first two. To be rated as creative an idea had to be both rare in occurrence (novel) and given a high effectiveness rating.

Four product categories were developed using Kirton's (1976) Adaption–Innovation construct which hypothesizes individual differences in problem defining behavior. Adaption, operating within the confines of the problem definition, and innovation, seeing the confines of the definition as the problem, represent the construct's extremes. The four categories are defined below:

Category 1—Solution answers question, "What else can be packed in a tea bag?"
Category 2—Solution involves a new use for the bag.
Category 3—Solution involves a structural change (size, shape, color and materials used) in the bag.
Category 4—Solution involves different functions for the materials used in the construction of the bag.

Level Three

Three development managers from the world's largest supplier of tea and tea packaging used industry-wide criteria described in detail elsewhere (Gryskiewicz and Shields 1985). Ultimately, the experts' ratings were used to evaluate the techniques for their ability to produce ideas of industrial value.

Subjects

All subjects were practicing managers who volunteered. Subjects represented banking, sales, manufacturing, and personnel. Probability sampling (Jahoda, Sellitz, Deutsch, and Cook 1965) was used where each element of the population had an equal chance of being selected. To ensure that all classifications were represented equally, stratified random sampling was pursued attempting to include subjects from every organization in every condition.

Subjects were made available to the writer in multiples of four. Assignment to condition was made randomly prior to data collection.

RESULTS

Level One

There were no significant differences found between our condition populations and their scores on the Barron-Welsh Art Scale (T = 0.82 and T = 1.00; U.K. and U.S. respectively) and we assumed comparability with regard to this measure of creativity and verified the random assignment to conditions.

The evaluation of the process by the users is summarized. We are limited with regard to generalization since the data reflect perceived value. The reports of the managers suggest to practitioners the continued use of real group techniques and provides

Table 1. Process evaluation.

U.K.: a. The most successful technique was perceived to be real brainstorming.
b. Real and nominal brainstorming was most valued.
c. Real and nominal brainwriting was enjoyed least.
U.S.: a. Regardless of technique, managers preferred real groups over working alone.
b. Real brainwriting was seen as having most value.

even more support to our attempt to better understand the manipulation of group uniformity pressures.

What impact does technology have on quantity indicators?

Hypothesis One
Due to the absence of brainstorming rules, excursion will produce less ideas than the other two in both real and nominal conditions.

This hypothesis was confirmed at $P < 0.05$ in both studies using Duncan's multiple range test comparison.

Hypothesis Two
Based upon previous findings, nominal brainstorming should produce more ideas than real brainstorming.

Hypothesis two was confirmed at $P < 0.05$ in both studies using
Duncan's multiple range test comparison with mean magnitude
difference of nearly three to one.

Attempts to see if uniformity pressure can be manipulated by
using a non-verbal technique was addressed in the next two
hypotheses.

Hypothesis Three
Real brainwriting will produce more ideas than nominal brainwriting since the oral pressures of a real group are monitored and
synergism is permitted.

Hypothesis three was confirmed in Study Two, but not Study
One using the same statistical comparison of treatment means
mentioned in the above hypothesis testing and same level of
significance.

Hypothesis Four
Real brainwriting will produce more ideas than real brainstorming
by overcoming oral condition uniformity.

Hypothesis four was confirmed in Study Two, but not Study One
(same analysis as above).

Table 2. Hypothesis evaluation

Level 1 quantity		
Number of ideas and Number of novel ideas	U.K.	U.S.
H1 − RE < Others	+	+
H2 − NBS > RBS	+	+
H3 − RBW > NBW	−	+
H4 − RBW > RBS	−	+

H = Hypothesis, RE = Real Excursion, NBS = Nominal
Brainstorming, RBS = Real Brainstorming, RBW = Real
Brainwriting, NBW = Nominal Brainwriting.

These first four hypotheses are summarized in Table 2.

In Study Two, real brainwriting produced more ideas than real
7-3-4. This confirms earlier oral brainstorming rules comparisons,
but is the first time the brainstorming rules have been tested using

MEANS FOR NUMBER OF IDEAS BY TREATMENT

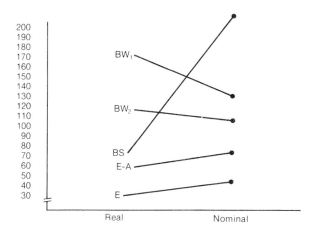

Fig. 1. Means for number of ideas by treatment.

a non-oral setting. There was no significant difference between real 7-3-4 (BW2) and nominal 7-3-4 (BW2). See Figure 1.

Level Two

Four dependent measures of product quality were assessed, and the following hypotheses were tested:

Hypothesis Five

Nominal group conditions would produce more novel ideas than real group conditions.

As seen in Table 3, the hypothesis was confirmed. More novel ideas (appearing only once in the sample) occurred in conditions that limited cognitive uniformity, that is, the nominal condition. This finding was also established by Maginn and Harris (1980) using brainstorming and, unfortunately, led them to conclude that individual brainstorming remained the method of choice. One exception of note in our data is brainwriting. Almost equal numbers of novel ideas occurred in real brainwriting as in nominal brainwriting, and real brainwriting almost doubled the number of novel ideas of real brainstorming. These findings further support the non-oral brainwriting's ability to reduce uniformity pressure.

Hypothesis Six

Real groups will produce more effective ideas than nominal groups.

The author believes that real group discussion would lead to more evaluation and idea perfection. This hypothesis was not confirmed. Higher percentages of top effectiveness ratings (top $\frac{1}{3}$) were given to ideas generated in the nominal conditions (see Table 3).

Table 3. Summary data, novel ideas and effectiveness ratings.

	U.K.		U.S.			
	Real	Nominal	Real	Nominal		
Hypothesis 5	46%	54%	38%	62%	+,	+
Hypothesis 6	15%	19%	17%	20%	−,	−

Hypothesis Seven

Real groups would produce more creative ideas (ideas that were novel and effective) than nominal groups.

This hypothesis was not confirmed. Twenty percent of the U.K. sample and eleven percent of the U.S. sample met the creative criteria; however, more creative ideas appeared in the nominal conditions in both samples (U.K. = 57 percent, U.S. = 64 percent).

Hypothesis Eight

Across conditions, brainwriting will produce more Category 1 ideas.

This was confirmed in both studies. Brainwriting produced more Category 1 products between all techniques and also a higher percentage of category one ideas in all the ideas produced by brainwriting.

Hypothesis Nine

Across conditions, excursion will produce more Category 4 ideas.

This hypothesis was not confirmed. Excursion was the second greatest producer of Category 4 ideas in the U.K. and the third greatest produced in the U.S. All we can point to is a trend. Brainstorming resulted in a greater number of Category 4 products.

Hypothesis Ten

Across conditions, brainstorming will produce more Category 2 and 3 ideas.

This was confirmed in both samples and a summary of the above three hypotheses can be seen in Table 4.

Table 4. Category responses (%).

a. Percent Category 1 products

	U.K.		U.S.	
	Real	Nominal	Real	Nominal
Brainstorming	61	57	67	64
Brainwriting	78	86	83	80
Excursion	69	72	75	67

b. Percent Category 4 products

	U.K.		U.S.	
	Real	Nominal	Real	Nominal
Brainstorming	18	14	14	12
Brainwriting	04	03	05	08
Excursion	13	15	07	15

c. Percent Category 2 and 3 products

	U.K.		U.S.	
	Real	Nominal	Real	Nominal
Brainstorming	20	29	19	24
Brainwriting	18	10	12	12
Excursion	18	13	18	18

We were able to demonstrate that certain techniques are more likely to produce a certain product quality variation along the four categories.

Level 3

This represents our attempt to use industrial criteria as an alternative validation. Thirty ideas were randomly selected from each of the fourteen conditions in both studies. These ideas were rated for potential industrial use. Table 5 summarizes the experts' ratings.

Nominal brainwriting produced the highest numbers of ideas eliminated as worthless by the experts in both samples. While this suggests the ineffectiveness of nominal brainwriting, it also can be interpreted to suggest the efficiency of real brainwriting as a filtering mechanism to eliminate bad ideas before they move forward for expert evaluation.

The most useful ideas seen by the experts were produced by the excursion technique in the U.K. while the excursion technique

Table 5. Level three analysis.

	Ideas eliminated		$\overline{X}$ Usefulness	Total usefulness score	Sum
	N	%			
U.K. data					
Brainstorming					
Nominal	16	53	4.5	49	
Real	18	60	3.1	37	
					86
Brainwriting					
Nominal	21	70	2.3	21	
Real	17	57	3.9	43	
					64
Excursion					
Nominal	13	43	3.5	59	
Real	16	53	3.8	53	
					112
U.S. data					
Brainstorming					
Nominal	14	47	5.0	82	
Real	10	33	4.0	80	
					162
Brainwriting					
Nominal	16	53	2.8	39	
Real	14	47	4.5	70	
					109
7-3-4					
Nominal	14	47	2.2	35	
Real	15	50	3.1	41	
					76
Excursion					
Nominal	15	50	2.5	37	
Real	12	40	4.1	76	
					113

ranked second in the U.S., behind brainstorming. The results obtained provide an intriguing perspective.

Experts were not aware of the techniques which produced the ideas and they evaluated the products uncontaminated by knowledge of the procedures. Though Level One findings would suggest that brainstorming and brainwriting techniques may be more acceptable in the perception of the users, and they each produced significantly more ideas than the excursion technique, experts selected as most useful ideas produced by the same excursion technique. Both samples support the conclusion that even though excursion techniques do not generate a large quantity of ideas, they are capable of producing ideas that experts find particularly useful.

This suggests that quantity, at least from the experts' viewpoint, is not the way to produce quality ideas. This conclusion contradicts earlier laboratory research where experimenters concluded that high correlations between quantity and quality lead to the recommendation of using techniques capable of producing a high number of ideas (e.g. nominal group techniques or individual idea generators to solve problems). It should also be noted that this indication of high quality by experts contradicts the brainstorming dictum that quantity breeds quality. Understanding the dependent measures in a more applied context has, in fact, resulted in an important challenge to traditional thinking about brainstorming.

RESEARCH IMPLICATIONS

Issue 1

The real group/nominal group paradigm has constricted the conclusions drawn from past research. Most of the research suggested that only when ideas are made public in groups does a decrease in the number of ideas occur. To test this more comprehensibly, brainwriting variations were used. Past findings were not unconditionally corroborated. Real condition brainwriting produced more ideas than the nominal condition. The use of these non-oral techniques helps us gain a more differentiated view of the impact of verbal behavior, and also increase the options available to practitioners beyond the recommendation of individual ideation.

In both studies users evaluated the real condition techniques more highly in most cases on all three criteria of success, value and enjoyment. One important exception did occur, real brainwriting, which produced more ideas than the nominal condition and reversed earlier real/nominal technique comparisons was not viewed as a successful or valuable technique by the users.

There will be worthwhile trade-offs between real and nominal techniques which go beyond uniformity pressure manipulation.

Issue 2

The findings regarding uniformity pressure manipulations and user reaction were obtained with managers as subjects. It is now easier for practitioners to use the findings to generalize to an industrial setting.

Issue 3

The industry problem provided us with sound examples of the techniques ability to solve real problems and with some valuable qualitative dependent measures. Level Two analysis provided us with four categories of quality ratings which suggest a goal reference model of creative problem solving which links product quality occurrence with a particular technique. More detailed implications follow below in an explanation of the Targeted Innovation model.

Issue 4

The problem used provided an opportunity for industry experts to rate the ideas along industrially important dimensions. The finding challenged earlier indications which suggested that dependent measures of quantity and quality were so highly correlated that no independent operational definition of quality was needed. The use of the excursion technique, by far the lowest quantity producing technique, resulted in ideas experts found of use. We are provided with a new dimension for the evaluation of creative problem solving techniques. Quantity (one of the four basic brainstorming rules) did *not* in this case lead to higher quality. Rather, the results suggest the stimulation provided by the excursion technique was, in fact, more powerful as a genesis for useful ideas than the brainstorming rule of quantity.

Issue 5

Two new techniques were used to compare real/nominal differences, to assess the manipulation of uniformity pressure upon product quantity, to determine their impact upon product quality, and to assess the impact of the brainstorming rules upon a non-oral technique called brainwriting. New information was forthcoming.

In each case, the findings offer understandable and workable alternatives to practitioners. As in the issue above, some of the past findings and implications with regard to the sanctity of brainstorming will have to be re-evaluated.

THE TARGETED INNOVATION MODEL

Issue 3 provides the basis for developing a more general model to explain the complexities found in a more complete articulation of the use of creative problem solving techniques (see Figure 2). The model provides a framework for reducing the idea variability, which often accompanies creative problem solving techniques.

Goal reference models, which suggest the value of knowing the approximate product outcome desired, demand techniques for reducing the variance in pursuit of that goal. Such a tool would offer to the practitioner a means for predicting, and controlling, the products of creative problem solving groups. While earlier research suggest that brainstorming and its variations are superior to just asking groups to generate ideas, Bouchard and others have indicated the value of these techniques in group and individual settings. What has been missing, for reasons mentioned, has been 1) a model which takes into account a variety of creative problem solving techniques and 2) a framework for understanding the techniques' impact upon idea quality. Qualitative manipulation affords practitioners more important options than the manipulation of product quantity. Given the findings addressed by Issue 4, the Targeted Innovation model offers an especially useful alternative framework since quantity does not seem to lead to ideas of high

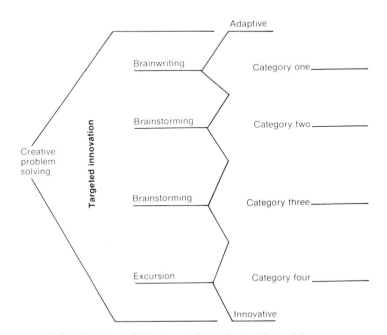

Fig. 2. Goal reference model for targeted creative problem solving.

quality as determined by experts. The Targeted Innovation model is given more credence in light of these Level Three findings. This finding should not be taken lightly.

We have replicated here the impact of a known technique (brainstorming) in an attempt to improve upon the dependent measures used to assess its value. The implication runs counter to the long held dictum that quantity breeds quality. Such findings disturb our "realities", test our theories, and stretch our tight models. As research in this vein continues, we should have a proliferation of models and their variations to better understand the effective and efficient use of idea generation techniques beyond brainstorming.

Van Gundy (1980, 1981, 1983) and others (Souder and Ziegler 1977; Barrett 1978) have attempted to list and describe the known creative problem solving techniques and they have accounted for more than 100 in number. The next step will be to evaluate how to best utilize them to achieve some specific end need defined by practitioners.

The climate is right for growth in the use of creativity techniques. However, it has been my experience that practicing managers prefer to use techniques which have some demonstrated value as a way to solve specific problems. Managers can articulate their problems, but what seems needed are techniques which provide targeted and efficient solutions.

The model in Figure 2 addresses the bottom line which constantly confronts the practitioner. Calling together individuals to participate in problem solving groups can cost money; therefore, the requirements are for techniques that are efficient, generate user satisfaction, and are accurate or at the very least, directional in goal attainment. Once we can do this effectively and consistently then, and only then, will we influence legitimately the demand for these techniques. Fads and fashions will be replaced by legitimate alternatives.

The Targeted Innovation model represents a move away from the traditional shotgun approach found in all creative problem solving techniques to a more targeted model, which takes its direction from the needed output. The model offers more predictability of outcome with problems that require multiple alternatives.

IMPLICATIONS FOR MANAGERS

While managers may not know exactly what specific creative solution they would like to claim as a hit, they do know the constraints that will be imposed upon the solution, and can therefore identify the ballpark in which a solution must be found. For example, solutions frequently have time and cost restraints. ("It must be solved in the next six months", or "I want an answer tomorrow", or "It must cost less than $25,000 amount".) If a manager has an understanding of the terrain of these environmental constraints and knows that the solution must be in the vicinity of, say, $30,000 and five months, the manager can then strike a course in the appropriate direction. The Targeted Innovation Model may be a useful road map of problem solving; by way of providing guidance to the best routes to reach a specific goal.

To select an appropriate, efficient route, one must look for several landmarks that indicate the vicinity in which one may find the appropriate solution. The following is a list—not necessarily a complete one—of factors that impinge upon solution appropriateness. The list was compiled by problem solvers in Research and Development centers and is based on their experience. These factors are situational and it is suggested that they be consulted when considering the style of creative idea needed. The style considered reflects the four categories this author established using Kirton's Adaption–Innovation construct (Kirton 1976).

★ *Time and Money.* The more time and money there is available, the more opportunity there is to try something in the innovative mode (towards category four). But because this mode is less tested and less polished, it is often less efficient in the short run. In case of a tight schedule and budget, an adaptive approach (towards category one) might be better. In the case of being *impossibly* short on resources and it is clear that the old ways of doing things will not work, there may be no other alternative but to try an innovative solution and hope that it will be quicker or cheaper.

★ *Degree of Risk.* Innovative solutions are riskier than adaptive ones. If some sort of insurance is available which allows a buffer in case a solution should fall through, then it may be worth risking money, time, and even credibility with innovative approaches. There is also the possibility that one will lose everything if something more innovative is not found.

★ *Past Solutions Not Working.* If the problem continues to recur, it may be that the "band-aid" solutions developed may have been the answers to the wrong problems. A willingness to try a different, innovative problem approach may be what is needed.

★ *Solution User.* Also to be considered is the character and demeanor of bosses, clients, and solution implementors, and chose a mode that is most likely to generate approval and cooperation. When it is important to stay on the side of credibility, an adaptive approach is advised. When the task is to do something really different, an innovative approach is helpful.

★ *Project Phase.* Early in the project phase is the time to be open to the innovative exploration of all the different ways that a project can be done. It is, however, bad form to decide after many plans have been drawn and construction started that there is a better way to do this. Adaptiveness is needed to accomplish closure.

★ *Crisis.* When things are bad, people may be pushed toward the extreme of either the adaptive or the innovative direction. Some say that when the ship is sinking, they will try anything to survive—innovation at all costs. Others say that when their back is to the wall, they will dearly hold on to what they know best—adaptive approaches.

★ *Success.* When things are going well, there is little need for dramatic change. Change for the sake of change is foolhardy. One must, of course, beware of excess comfort. Success may lead to such a complacency with adaptiveness that one may fall prey to the successful innovative ideas of competitors.

Each of these factors may have an impact (and admittedly sometimes conflicting impacts) on the style of solution that is chosen. These considerations, while not tested empirically, can be used as a checklist of factors to keep in mind, while realizing that additional judgment will be needed to determine the way each of these factors will affect the choice of problem-solving style. There is no simple, automatic formula. But, by considering factors like these in order to assess the situation and determine which technique is appropriate, the use of creative problem-solving techniques within the Targeted Innovation model at least becomes more explicit.

Conclusion

The paramount value influencing this research has been its practical orientation for managers. This research has demonstrated that despite strong opinion and conventional wisdom to the contrary, the creativity of idea generation may be manipulated in a predictable and controllable way. It is this author's belief and experience that such a practical orientation represents the kind of creativity that makes sense to managers.

We challenge others who produce models in the creative problem solving field to include the innovative, cutting edge perspective along with a realistic view which comes from a model that withstands a trial by fire in the industrial setting. The creative tension inherent in such an approach should spark its own creativity.

REFERENCES

Ackoff, R. L. & Vergara, E. 1981: Creativity in problem solving and planning: A review. *European Journal of Operational Research*, 7, 1–12.

Barkowski, D., Lamm, H. & Schwinger, T. 1982: Brainstorming in group (dyadic) and individual conditions. *Psychologische Beitrage*, 24 (1), 39–46.

Barrett, F. D. 1978: Creativity techniques: Yesterday, today, and tomorrow. *Advanced Management Journal*, 43 (1), 25–35.

Basadur, M., Graen, G. B. & Green, S. G. 1982: Training in creative problem solving: Effects on ideation and problem finding and solving in an industrial research organization. *Organizational Behavior and Human Performance*, 30, 41–70.

Basadur, M. & Finkbeiner, C. T. 1985: Measuring preferences for ideation in creative problem-solving training. *Journal of Applied Behavioral Science*, 21 (1), 37–49.

Bouchard, T. J., Jr. 1969: Personality, problem-solving procedure, and performance in small groups. *Journal of Applied Psychology Monograph*, 53 (1), Part 2.

Bouchard, T. J., Jr. 1972a: A comparison of two group brainstorming procedures. *Journal of Applied Psychology*, 54, 418–21.

Bouchard, T. J., Jr. 1972b: Training, motivation, and personality as determinants of the effectiveness of brainstorming groups and individuals. *Journal of Applied Psychology*, 56, 324–31.

Bouchard, T. J., Jr., Barsaloux, H. J. & Drauden, G. 1974: Brainstorming procedure, group size, and sex as determinants of the problem-solving effectiveness of groups and individuals. *Journal of Applied Psychology*, 59, 135–8.

Bouchard, T. J., Jr. & Hare, M. 1970: Size, performance, and potential in brainstorming groups. *Journal of Applied Psychology*, 54, 51–5.

Bray, R. M., Kerr, N. L. & Atkin, R. S. 1978: Effects of group size, problem difficulty, and sex on group performance and member reactions. *Journal of Personality and Social Psychology*, 36, 1224–40.

Brilhart, J. K. & Jochem, L. M. 1964: Effects of different patterns on outcomes of problem-solving discussions. *Journal of Applied Psychology*, 48, 175–9.

Campbell, J. 1968: Individual versus group problem solving in an industrial sample. *Journal of Applied Psychology*, 52, 205–10.

Cohen, D., Whitmyre, J. W. & Funk, W. H. 1960: Effect of group cohesiveness and training upon creative thinking. *Journal of Applied Psychology*, 44, 319–22.

Colgrove, M. A. 1968: Stimulating creative problem solving: Innovative set. *Psychological Reports*, 22, 1205–11.

Collaros, P. A. & Anderson, L. R. 1969: Effect of perceived expertness upon creativity of members of brainstorming groups. *Journal of Applied Psychology*, 53, 159–63.

Datta, L. E. 1963: Test instructions and identification of creative scientific talent. *Psychological Reports*, 13, 495–500.

Davis, G. A. 1973: *Psychology of problem solving: Theory and practice*. New York: Basic Books, Inc.

Delbecq, A. L., Van de Ven, A. H. & Gustafson, D. H. 1975: *Group techniques for program planning*. Glenview, Ill: Scott, Foreman and Company.

Dillon, P. C., Graham, W. K. & Aidells, A. L. 1972: Brainstorming on a "hot" problem: Effects of training and practice on individual and group performance. *Journal of Applied Psychology*, 56, 487–90.

Dunnette, M. D. 1964: Are meetings any good for solving problems? *Personnel Administration*, March–April, pp. 12–29.

Dunnette, M. D., Campbell, J. & Jaastad, K. 1963: The effects of group participation on brainstorming effectiveness for two industrial samples. *Journal of Applied Psychology*, 47, 30–37.

Geschka, H. 1980: Perspectives on using various creativity techniques. In S. S. Gryskiewicz (Ed.), *Creativity week II proceedings*, pp. 49–61. Greensboro, North Carolina: Center for Creative Leadership.

Geschka, H., Schaude, G. R. & Schlicksupp, H. 1973, August 6: Modern techniques for solving problems. *Chemical Engineering*, pp. 91–7.

Geschka, H., Schaude, G. R. & Schlicksupp, H. 1976–1977: Modern techniques for solving problems. *International Studies of Management & Organization*, 6, 45–63.

Glover, J. A. 1979: Group structure and creative responding. *Small Group Behavior*, 10, 62–72.

Glover, J. A. & Chambers, T. 1978: The creative production of the group effects of small group structure. *Small Group Behavior*, 9, 387–92.

Gordon, W. J. J. 1961: *Synectics*. New York: Harper.

Gough, H. G. 1957: *Manual for the California psychological inventory*. Palo Alto, California: Consulting Psychologists Press.

Graham, W. K. 1977: Acceptance of ideas through individual and group brainstorming. *Journal of Social Psychology*, 101, 231–4.

Graham, W. K. & Dillon, P. C. 1974: Creative supergroups: Group performance as a function of individual performance on brainstorming tasks. *Journal of Social Psychology*, 93, 101–5.

Gryskiewicz, S. S. & Shields, J. T. in press: *Targeted innovation*. Greensboro, North Carolina: Center for Creative Leadership.

Harari, O. & Graham, W. K. 1975: Tasks and task consequences as factors in individual and group brainstorming. *The Journal of Social Psychology*, 95, 61–5.

Hare, A. P. 1976: *Handbook of small group research (3rd ed.)*. New York: The Free Press.

Heslin, R. 1964: Predicting group task effectiveness from member characteristics. *Psychological Bulletin*, 62, 249–56.

Hill, G. W. 1982: Group versus individual performance: Are N + 1 heads better than one? *Psychological Bulletin*, 91(3), 517–39.

Hoffman, L. R. 1965: Group problem solving. In L. Berkowitz (Ed.), *Advances in experimental social psychology* (Vol. 2). New York: Academic Press.

Hoffman, L. R. 1979: *The group problem-solving process: Studies of a valence model*. New York: Praeger

Jablin, F. M. & Seibold, D. R. 1978: Implications for problem-solving groups of empirical research on "brainstorming": A critical review of the literature. *The Southern Speech Communication Journal*, 43, 327–56.

Jahoda, M., Sellitz, C., Deutsch, M. & Cook, W. S. 1965: *Research methods in social relations*. London: Methuen.

Johanson, B. 1978: *Kreativitet & marketing*. Hikern A. G., Buchund Offsetdruckerer, Switzerland.

Kelly, H. H. & Thibaut, J. W. 1969: Group problem solving. In G. Lindzey and E. Aronson (Eds.), *Handbook of social psychology (2nd ed.)*. Reading, Mass: Addison-Wesley.

Kirton, M. 1976: Adaptors and innovators: A description and measure. *Journal of Applied Psychology*, *61*, 622–9.

Klimoski, R. J. & Karol, B. L. 1976: The impact of trust on CPS Groups. *Journal of Applied Psychology*, *61*, 630–3.

Lamm, H. & Trommsdorff, G. 1973: Group versus individual performance on tasks requiring ideational proficiency (brainstorming): A review. *European Journal of Social Psychology*, *3*(4), 361–88.

Lorge, I. & Solomon, H. 1959: Individual performance and group performance in problem solving related to group size and previous exposure to the problem. *Journal of Psychology*, *48*, 107–14.

MacKinnon, D. W. 1975: Paris contribution to the conceptualization and study of creativity. In I. A. Taylor & J. W. Getzels (Eds.), *Perspectives in creativity*. Chicago: Aldine Publications Company.

Madsen, D. B. & Finger, J. R. 1978: Comparison of a written feedback procedure. Group brainstorming and individual brainstorming. *Journal of Applied Psychology*, *63*, 120–3.

Maginn, B. K. & Harris, R. J. 1980: Effects of anticipated evaluation on individual brainstorming performance. *Journal of Applied Psychology*, *65*, 219–25.

Maier, N. R. F. 1967: Assets and liabilities in group problem solving. *Psychological Review*, *74*, 239–49.

Mann, R. D. 1959: A review of the relationship between personality and performance in small groups. *Psychological Bulletin*, *56*, 241–70.

Meadow, A., Parnes, S. J. & Reese, H. 1959: Influence of brainstorming instructions and problem sequence on a creative problem-solving test. *Journal of Applied Psychology*, *43*, 413–16.

Meyer, H. H. 1957, June: *Measuring and stimulating employee creativity: Some research findings and implications* (Public and Employee Relations Research Service). General Electric. [city].

Milton, G. A. 1965: Enthusiasm vs. effectiveness in group and individual problem solving. *Psychological Reports 16*, 1197–1201.

Osborn, A. 1957: *Applied imagination*. New York: Scribners.

Restle, F. & Davis, J. H. 1962: Success and speed of problem solving by individuals and groups. *Psychological Review*, *69*, 520–36.

Rickards, T. 1974: *Problem solving through creative analysis*. Epping, England: Gower Press.

Rogers, C. 1983: *Freedom to learn for the 80's*. Columbus, OH: Charles E. Merrill Publishing Company.

Rotter, G. S. & Portugal, S. M. 1969: Group and individual effects in problem solving. *Journal of Applied Psychology*, *53*, 338–41.

Schaude, G. R. 1979: Methods of idea generation. In S. S. Gryskiewicz (Ed.), *Creativity week I proceedings*. Greensboro, North Carolina: Center for Creative Leadership.

Shaw, M. E. 1964: Communications networks. In L. Berkowitz (Ed.), *Advances in experimental social psychology* (Vol. I). New York: Academic Press.

Shaw, M. E. 1981: *Group dynamics: The psychology of small group behavior*. New York: McGraw-Hill.

Sherif, M. 1954: Sociocultural influences in small group research. *Sociology and Social Research*, *39*, 1–10.

Smith, R. M. 1968: Characteristics of creativity research. *Perceptual & Motor Skills*, *26*, 698.

Souder, W. E. & Ziegler, R. W. 1977: A review of creativity and problem solving techniques. *Research Management*, *20*(4), 34–42.

Stein, M. I. 1975: *Stimulating creativity* (Vol. 2). New York: Academic Press.

Street, W. K. 1974: Brainstorming by individuals, co-acting, and interacting groups. *Journal of Applied Psychology*, *59*, 433–6.

Taylor, D. W. & McNemar, O. W. 1955: Problem solving and thinking. In C. P. Stone (Ed.), *Annual Review of Psychology*, *6*, 455–82.

Taylor, D. W., Berry, P. C. & Block, C. H. 1958: Does group participation when using brainstorming facilitate or inhibit creative thinking? *Administrative Sciences Quarterly*, *3*, 23–47.

Thomas, E. J. & Fink, C. F. 1961: Models of group problem solving. *Journal of Abnormal and Social Psychology*, *63*, 53–63.

Torrance, E. P. 1971: Stimulation, enjoyment and originality in dyadic creativity. *Journal of Educational Psychology*, *62*, 45–8.

Van de Ven, A. & Delbecq, A. L. 1971: Nominal versus interacting group processes for committee decision-making effectiveness. *Academy of Management Journal*, *14*(2), 203–12.

VanGundy, A. 1980: Comparing "little-known" creative problem solving techniques. In S. S. Gryskiewicz (Ed.), *Creativity week III proceedings* (pp. 1–25). Greensboro, North Carolina: Center for Creative Leadership.

VanGundy, A. 1981: *Techniques of structured problem solving*. New York: Van Nostrand Reinhold Company.

VanGundy, A. 1983: *108 ways to get a bright idea*. Englewood Cliffs, N.J.: Prentice-Hall.

Vroom, V. H., Grant, L. D. & Cotton, T. S. 1969: The consequences of social interaction in group problem solving. *Organizational Behavior and Human Performance*, *4*, 77–95.

Warren, T. F. & Davis, G. A. 1969: Techniques for creative thinking: An empirical comparison of three methods. *Psychological Reports*, *25*, 207–14.

Zagona, S. U., Willis, J. E. & MacKinnon, W. J. 1966: Group effectiveness in creative problem solving tasks: An examination of relevant variables. *Journal of Psychology*, *62*, 11–137.

8

Applying Experimental Research on Group Problem Solving to Organizations

L. RICHARD HOFFMAN

The size and complexity of modern organizations and the resulting necessary interdependence of technical and functional specialists have created a need for some mechanism by which information from these various sources can be brought to bear on general organizational problems. Recognition of the need for such information exchange has resulted in increasing numbers of meetings; but, even at the very highest management levels, such meetings are rarely classifiable as problem-solving sessions (Argyris and Schön 1974). They are more often designed for reporting, consulting, evaluation, and persuasion, depending on the group leader's purpose in holding the meeting.

However, the concept that subordinates should be consulted and be permitted to object to higher management's directives has become fairly well accepted in modern corporations. It has led to consultative and pseudo-participative conferences that sometimes gain these ends, but often have quite negative effects on subordinates' motivation.

With the increasingly favorable climate for the use of group problem solving in organizations, it is appropriate to ask what research on group problem solving can tell us about the variables that promote effectiveness. Groups *qua* groups neither necessarily produce the best decisions of which their members are capable (Thomas and Fink 1961; Maier and Solem 1952), nor are the members necessarily highly motivated to carry out the group's decision (Hoffman and Maier 1961; Maier and Hoffman 1964).

Although Hare's *Handbook of Small Group Research* (1978) and Berkowitz's *Group Processes* (1978) represent formidable collections of information about small groups, Stogdill's (1974) comment, "the endless accumulation of empirical data has not produced an integrated understanding of leadership", applies as well to the literature on small groups. Besides the absence of a general theory of group problem solving, there are several major factors that limit the applicability of results from the laboratory to organizational practice. With these concerns in mind, I have divided this paper into two parts: (1) a listing and discussion of a number of conclusions from experimental problem-solving groups that are applicable to the administration of organizations, and (2) a discussion of factors that limit the utility of much of this research.

THE GOALS OF PROBLEM-SOLVING GROUPS

Immediately upon considering the effectiveness of problem-solving groups in organizations, we face a problem in generalizing from the laboratory evidence. In laboratory experiments the experimenter tells the group it has a problem and gives them some guides for its solution—e.g., "What action should be taken to improve productivity?"

Problems often enter real organizational groups in very ill-formed ways. "What should we do about the parking situation?" may mean that workers are unhappy about how long they have to wait to get a space or about how long they are delayed in leaving the job. Or complaints about the parking situation may only be symptoms of an entirely different problem concerning working conditions in the plant. Since different problem characteristics require different combinations of abilities and information (Hackman and Morris 1975; Hoffman, Friend, and Bond 1979), conclusions about group performance derived from laboratory studies may or may not be generalizable to all organizational groups.

I view organizational problem-solving groups as having two principal objectives: (1) the maximum utilization of the resources brought by each individual member, including any added group potential; and (2) the generation of a high level of motivation for carrying out the group's decision in each and every member. Operationally, these two objectives are equivalent to Maier's (1963) distinction between the quality and acceptance of group solutions. Presumably, in real-life groups, the highest quality solutions will emerge from the maximum utilization of the group members' resources, since the correct answer is unknown. Members' acceptance of the group's decision, however, shows only a modest positive correlation or no correlation at all with the objective quality of the solution (Hoffman and Maier 1961). Fortunately, those factors which combine to create effective group decisions in organizations are conducive to both quality and acceptance. On the other hand, in untrained groups, especially where the formal leader lacks skills in group leadership, quality is often sacrificed to gain acceptance and vice versa.

MEMBER CHARACTERISTICS

Besides problem-related resources, each individual brings to the group a propensity for thinking and behaving toward the other group members which may have positive or negative impact. These propensities may cause a member's resources to be over- or undervalued and may become barriers to attaining quality or acceptance.

Personality Characteristics

In free and open discussions, certain personality characteristics will influence the rate of participation and the relative influence of members. The extroverted, dominant, socially aggressive individual consistently emerges as the highest participator, with disproportionate influence over the solution (Blake and Mouton 1961). Such personality traits, showing no correlation with measures of cognitive ability (Guilford 1952), are unlikely to be associated with the resources necessary for any particular problem. Since these types of people are also likely to be accepted as leaders by other group members (Blake and Mouton 1961), they tend to arouse the dependent, unthinking acceptance of their ideas (Pepinsky, Hemphill, and Shevitz 1958). Alternatively, their superior social skills and ability to dominate others tend to prevent opposing ideas from being considered (Hoffman and Clark 1979).

Similar to the extroverts in their tendency to dominate group discussion are the self-confident members (Carlston 1977; Hochbaum 1954). Such self-confidence may derive from past successes on similar problems, the feeling of security in the group, or, especially, a sense of being favored by higher management.

Our research on the solution valence-adoption process (Hoffman 1979) has revealed an even more subtle aspect of participation bias. We have found that people's influence over the decision is only partly indicated by their rate of participation (Hoffman and Clark 1979). By consistently supporting (adding valence to) a particular solution, they influence the group's implicit process of adopting that solution.

Particularly susceptible to these types of emergent leaders are people with strong affiliative needs, especially those who are unsure of acceptance by the group. Such people are likely to withhold resources which appear to contradict the prevailing conception of the problem or its solution (e.g. Jackson and Saltzstein 1958).

Also destructive to the group process are counterdependent types, who may be so resistant to group pressure that they block the group from reaching consensus (Stock and Thelen 1958). Both the dependent and counter-dependent types represent a class of people whose participations in the group are more often expressions of "self-oriented needs"—aggression, power, status, insecurity—than they are of attempts at problem solving.

Prior Commitment

Members with prior commitment to a particular solution tend to argue early and forcibly for their favored solution, to be closed to alternative solutions, to make the job of arriving at true consensus difficult, and to be highly dissatisfied if the group adopts an alternative solution (Hoffman and O'Day 1979; Maier 1970). Commitment can occur because members are trying to build an empire, because they have committed themselves to people outside the group, or because they feel the decision will have an important bearing on their own lives (Hoffman, Harburg and Maier 1962).

Although the research evidence emphasizes the negative consequences of these individual characteristics, experience dictates that each has its positive aspects as well: the energizing effects of the extrovert, the enthusiasm for the problem generated by the committed member, and the assistance towards needed consensus given by the dependent member. When our measures of the group problem-solving process has become more precise, we shall certainly find curvilinear relationships between these individual characteristics and problem-solving effectiveness. Since organizations have little control over the personality composition of all their groups, it is important to be aware of the propensities among the group members.

GROUP STRUCTURE

The characteristics of group members have so far been discussed in isolation. However, the relationships among members, both in terms of their individual characteristics and their position in other

social systems, also provide prior conditions for the relative effectiveness of problem-solving groups.

Group Composition

The greater the variety of perspectives on a problem the more
likely a high quality solution is to emerge. Groups with different
personalities (Hoffman and Maier 1961), leadership abilities
(Ghiselli and Lodahl 1958), types of training (Pelz 1956), and
points of view (Hoffman, Harburg and Maier 1962) have been
shown to be more creative and innovative than groups with more
similar member characteristics. Furthermore, groups with varied
personality composition are no less cohesive than groups with
similar personalities. Perceived task success after the resolution of
differences is a source of member satisfaction with the group
(Hoffman and Maier 1966a).

Differences in viewpoint do create strains in groups, however.
Groups whose members are outright antagonistic are unable to
capitalize on their diversity (Fiedler and Meuwese 1963). The
different perspectives brought to meetings by representatives of
organizational departments—e.g. sales, production, accounting—
often reflect extra-group identifications that inhibit creativity
(Dearborn and Simon 1958).

Groups have difficulty maintaining diversity as they age.
Although the members of newly formed groups must learn initially
how to work with each other, continued interaction develops
common perspectives, group norms, and, often, stereotyped and
structured relations among the members (Hoffman and Maier
1961; Wells 1962).

Group Size

Related to the question of diversity among the members is the
size of the problem-solving group. Unfortunately, while potential
resources increase in larger groups, actual resources available to
the group do not increase at the same rate (Steiner 1966). This
effect is due largely to the increased difficulty of entering and
influencing the discussion in larger groups. Hoffman and Clark
(1979) showed that the proportion of influence exercised by the
most influential member remained the same in 3-, 4-, and 5-
person groups, leaving increasingly less opportunity for others to
contribute to the decision. A balance must be struck between the
diversity produced by including a wide variety of people and the
decreased likelihood of a profitable discussion. For purposes of

acceptance, however, everyone who is instrumental in implementing the group's decision should be included, if possible, to permit everyone to influence the outcome (Block 1974). The need for more effective group process, therefore, becomes greater in larger groups.

Relationships Among Group Members

Interpersonal Relationships

Positive feelings among the group members tend to promote the security for members to risk suggesting unusual solutions to problems, as in successful brainstorming sessions (Berkowitz and Levy 1956). Past problem-solving success can create a group self-confidence, however, that causes them to accept an early suggested solution on the assumption that it "must be correct". Initial unanimity about the solution to a problem creates such euphoria that the group fails to recognize when the solution is wrong (Davis 1973; Maier and Solem 1952).

The dangers from positive group feelings are substantially less, however, than the restrictive consequences of negative feelings. Lack of confidence in the other members (Hamblin 1958), fear and mistrust (Golembiewski and McConkie 1975) tend to block or distort communications among group members. Those members who most aspire to higher organizational positions in response to individual reward systems are the ones who are also most likely to distort their communications to others, especially as such information might reveal problems that they face (Read 1962). Kahn, Wolfe, Quinn, Snoek, and Rosenthal (1964) have also shown that members in boundary roles between departments tend to block and distort communications as a function of the conflict they perceive in their roles.

Social System Relationships

The social systems in which the group is embedded impinge on its functioning. The authority relationships deriving from the formal organization structure seem most powerful. People with higher organizational ranks tend to participate more actively and to exercise undue influence in the group (Bass and Wurster 1953). These tendencies are, unfortunately, unrelated to the likelihood of their having the appropriate resources to solve the problem under consideration (Torrance 1955). The traditional stereotype that leaders control the decision process (Argyris and Schön 1974) is held even by college students, with harmful effects on the problem-solving effectiveness of groups in our laboratory studies (Hoffman and Maier 1967; Falk 1978). The typical organizational reward system promotes this dependence on the leader's influence (Maier and Hoffman 1961).

While rank in the formal authority structure promotes influence in problem-solving discussions, rank in other social systems may also confer similar rights (Berger, Cohen, and Zelditch 1972). The relative influence of various departments in a business, e.g. sales over manufacturing (Lawrence and Lorsch 1967), seniority (Maier and Hoffman 1963), "ethnic" or religious group membership (Dalton 1959), the "right" school background, or even friendship relationships may produce distortions in the communication and influence process unrelated to the utilization of resources and divisive in their effect on members' acceptance of the solution (Humpal and Hoffman 1979).

This listing of the individual and group-composition characteristics and the group structural properties has been done to illustrate the preconditions which affect the development of effective decisions. Whether they serve as "assets or liabilities" for the group (Maier 1970) will depend on the quality of the process.

GROUP PROCESS

Despite the title of Berkowitz's book, *Group Processes* (1978), relatively little of the research reported there is truly about the *process* by which groups operate. Most of the systematic study of the ongoing activity of problem-solving groups has been limited to the study of participation (e.g. Fisek and Ofshe 1970), and these results are often aggregated across the entire discussion to produce only a final distribution of participation rates (Kadane and Lewis 1969).

The process models proposed recently tend to be abstractly mathematical and difficult to tie to observable behavior (Steiner 1966; Davis 1973; Shiflett 1979). Or they are concerned primarily with "process loss" (Steiner 1966; Shiflett 1979), whereas groups sometimes "gain" new and creative solutions—as well as the energy to implement them—through discussion (Hoffman 1965). The next section will identify those aspects of the process which inhibit and those which enhance the outcomes of group problem solving.

The discussion will be organized around my hierarchical model of group problem solving, which has evolved from studies of the process by which groups adopt solutions to problems (Hoffman 1979). Although the model cannot be described in detail here,

Figure 1 sketches the essential features. The three aspects—Task–Maintenance, Normative–Localized, and Explicit–Implicit—are assumed to be activated simultaneously whenever a group solves a problem. The phases of problem solving are considered to be the implicit procedures of most problem-solving tasks.

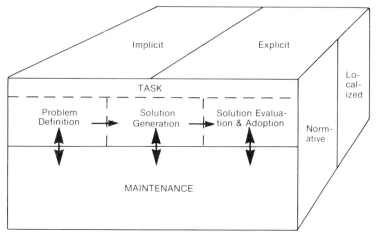

Fig. 1. Sketch of hierarchical model.

Obstacles to Effective Process

Two classes of obstacles to groups' effective utilization of resources may be understood in the model. The first is the pressure to achieve closure by adopting a solution and minimizing the time spent in defining the problem and generating alternative solutions (shown by arrows between phases). The second obstacle is the intrusion of maintenance issues on task functions. (These forces are shown as two-headed arrows, since task activities influence the group's maintenance as well.)

Problem-Related Tensions

Like individuals, groups appear to experience discomfort when faced with a situation which has no obvious answer, and they immediately search for a solution. Solutions which fit habitual directions of thinking then tend to be adopted quickly, though often implicitly. Valence for such a solution builds up as members affirm its merits and it passes the implicit adoption threshold (Hoffman 1979). Conversely, solutions which fit unusual directions or are unusual in themselves tend to die early deaths.

These processes of implicit adoption and rejection are illustrated in Figure 2 (taken from Hoffman, Friend and Bond 1979). Shown there for one group are the cumulative valences for each of five solutions as they changed values during the course of the discussion. Solution G accumulated valence rapidly for the first 50 acts, more slowly during the next 100 acts, and rapidly again in the last half of the meeting to its adoption. Consistent with most of the solution-valence results (Hoffman 1979), the group had adopted this solution implicitly after its valence exceeded the adoption threshold of 15. A later challenge by Solution W then merely stimulated further support for Solution G. Conversely, Solution T, which dropped below the rejection threshold of -8, was never completely resuscitated, even though it subsequently received some favorable comments.

Solution Valence and Participation

The tendency to add valence to habitual solutions illustrates one effect of task activities on the maintenance level. The early acceptance or rejection of members' suggestions affects their later participation, respectively encouraging or discouraging additional comments (Oakes 1962). Leaders are a particularly powerful source of reinforcement. When the leader says, "That's a good idea, Joe", it stimulates Joe to more expression. But the subtle negative response, "That's interesting, Charlie. Does anyone have any other ideas?" tends to depress Charlie's further contributions.

Pressures Towards Uniformity

The tendency for groups to impose the majority will on minority opinion illustrates the intrusion of an implicit maintenance norm on the group's task effectiveness. Members who hold opinions deviant from the majority are subject to extreme pressure, even when they hold the correct answer, thus apparently sacrificing the quality of the solution for group unity (Maier and Solem 1952; Thomas and Fink 1961). Such conformity often reflects compliance rather than personal commitment, so neither quality nor acceptance may be achieved when the deviant is coerced into line (Block 1974; Maier and Hoffman 1964). In contrast, the advantages of attending to minority opinion and of exploiting conflicting opinions to enhance creative solutions to problems have also been well documented (Maier and Solem 1952; Hoffman, Harburg, and Maier 1962).

In the organizational setting, the leaders' attitudes towards conflict are critical. If leaders consider members who disagree as

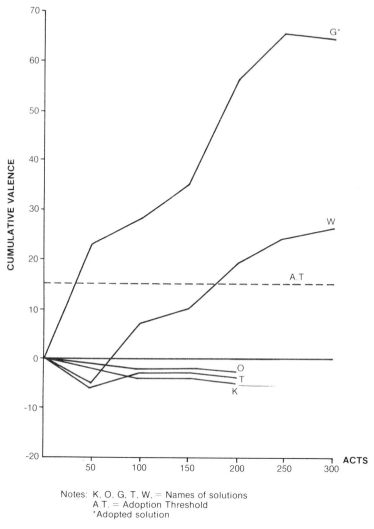

Notes: K, O, G, T, W, = Names of solutions
A.T. = Adoption Threshold
*Adopted solution

Fig. 2. Patterns of solution-valence accumulations.

troublesome and unreasonable, they will tend to force compliance. However, if they see disagreement as an opportunity for innovation, they may be rewarded with a creative solution with higher member acceptance (Maier and Hoffman 1965). Unfortunately, both leaders and members in organizations tend to adopt an implicit norm not to argue too strongly against the leader's favored position (Maier and Hoffman 1961; Janis 1972). Few leaders actively encourage disagreement, since conflict may be disruptive to the group.

Toward Effective Process

Attempts to improve the group problem-solving process have been directed either toward overcoming obstacles or toward facilitating creativity. Methods for improving the problem-solving process may be described according to their effects on the dimensions of the hierarchical model (Figure 1).

Task Methods

Groups tend to move from phase to phase according to an implicit valence process by which problems are defined and solutions adopted (Hoffman, Bond, and Falk 1979; Hoffman, Friend, and Bond 1979). Explicit task methods focus the members' efforts in a single phase and slow the group's tendency to move too quickly toward adopting a solution.

Problem Identification
Conscious search for the character of the problem by surveying the factors that impinge upon it can prevent a group from solving the wrong problem and often leads to innovative solutions (Maier and Solem 1962). The Synectics device of having the members feel what radiating is helped a group define its problem and create a solar energy device.

Separating Solution Generation and Evaluation
Osborn's (1952) brainstorm that the generation of ideas and their evaluation are antithetical processes encouraged the invention of new solutions by delaying their evaluation. While some evidence (Taylor, Berry, and Block 1958; Dunnette, Campbell, and Jaastad 1963) has been misinterpreted as a rejection of the brainstorming method, groups may benefit if the members brainstorm individually before they solve the problem together (Bouchard 1972).

Solution Evaluation and Choice
The separation of idea production and evaluation benefits the idea-production phase, but has little effect on improving evaluation. Maier's Screening Principles (1960) are a set of rules designed to provide an impersonal mechanism by which groups should adopt the solution which is best supported by the available facts. They are intended to reduce the tendency of personal bias, prejudice,

motivation, rivalries, and personal antagonisms to undo the positive efforts in problem identification and solution generation.

The Nominal Group Technique (Delbecq, Van de Ven, and Gustafson 1975), in which each group member contributes in turn, serves two functions. It concentrates the group's activities in the solution-generation phase and it assures that each member has a chance to be heard. In this way the implicit maintenance norm that regulates members' participation does not intrude on the task.

Formal decision rules, like rules of majority or unanimous vote, perform a similar dual function. Falk (1978) showed that a majority decision rule balanced the formal power of a foreman better than a unanimous rule to promote more effective problem solving.

By having members make judgments anonymously, the Delphi Method (Dalkey 1969) prevents status differences from affecting the quality of group outcomes.

Maintenance Methods

Problem-solving groups in organizations rarely discuss their maintenance issues explicitly and almost universally follow a norm that they are not to be discussed. Yet the hierarchical model indicates clearly that problems at the maintenance level can affect the quality of problem solving at any phase.

Hackman, Brousseau, and Weiss (reported in Hackman and Morris 1975) provide almost the sole instance of experimentally investigating the effects of normative changes on task performance. By discussing their performance strategies, when control groups did not, the experimental groups improved performance when coordination among the members was needed but not easily attainable. Since the adoption of any such procedure requires the members to violate their implicit, habitual norms, the members must trust the intervener's intentions to help them and not manipulate them to undesirable ends.

If true consensus and high acceptance are to be achieved, a group atmosphere must be maintained in which the members are free to express reservations about the decision right to the end of the discussion. If the members' reservations are factually based, their expression may force the group to redefine the problem, at the earliest phase of the problem-solving process. Rather than converting such conflict from the task level to the interpersonal level and producing antagonism between the members (Maier and Hoffman 1965), or relying on the leader to use power to move the group immediately to a solution, the group can learn to identify its implicit conflict before the valence for the alternatives becomes too great. Sherwood and Glidewell (1972) called this process "planned renegotiation" and emphasized "[that] [i]t is important that people learn to detect pinches before disruptions develop".

Leaders

In this discussion, I have outlined those aspects of the group process which appear to be potentially inhibiting and enhancing for the quality and acceptance of group solutions. Only occasionally have I distinguished between the leader's and the members' responsibilities for the process. In this way I have implied that the functions to be performed may be carried out by either leaders or members as they are needed and should be equally beneficial. I recognize, however, that the autocratic structure of modern organizations puts the major responsibility for the quality of the group's problem-solving process primarily in the leader's hands (Maier 1963).

CAUTIONS ABOUT THE LIMITS OF OUR KNOWLEDGE AND SOME GUIDES TO RESEARCH

This paper has extrapolated the results of experimental laboratory research to group problem solving in organizational settings. The exercise is fraught with danger and the conclusions should be accepted with utmost caution. Some of the limits to application inhere in the research itself and others lie in the neglect of important variables.

The most striking neglect in experimental research is the contrived nature of groups. Because experimenters bring a number of people together, call them a group, and ask them to solve a problem, they interpret the results as if the group as a whole solved the problem (Hoffman and Maier 1967). Groups in organizations are often problematic. Although people may report to a common superior, they may never be brought together to function as a group. Often the members are more loyal to their personal interests or to those outside than they are to the group itself. The maintenance aspects of groups have received little attention from experimenters, since the *ad hoc* groups of the laboratory need to maintain themselves only for the length of the experimental session.

Related to the maintenance issue and even more difficult to create in the laboratory are the pathological aspects of group functioning (Bion 1959). Even when laboratory groups solved value problems, most agreed to a group solution (Hoffman and O'Day

1979), whereas value conflicts in organizations often result in the disintegration of the group. The level of emotional arousal found in organizations—both loyalty and hatred—could never be approached in a single-session laboratory study. Nor can the effects of laboratory rewards and penalties approach the promise of promotion or the threat of being fired. Even though we may feel safe in suggesting that the stifling effects of power shown in experimental studies will be exaggerated in organizations (Maier and Hoffman 1961), other extrapolations of emotionality in groups should be made cautiously.

The context of the organizational group is also a subject missing from the study of experimental groups (Stern 1970). As all practitioners of organizational development well know, attempts to make a group more effective by introducing norms counter to the general organizational norms usually end in failure (Bennis 1966). An invitation to be "open and honest" in an organization in which power plays and deceit are the rule is often fatal to those who accept it. Merely the concept of a problem-solving group implies an organizational climate different from the traditional bureaucratic, mechanistic model of organization (Burns and Stalker 1961).

Several significant aspects of laboratory research limit the direct applicability of even the results found consistently in such studies. One-shot studies with *ad hoc* groups produce results relevant to newly forming groups (Tuckman 1965), not to groups which have developed some modest social structure, much less those with a highly developed set of norms. Groups which have been trained in the use of procedures have generally been more effective than naive groups with the same instructions (Parnes and Meadow 1959; Bouchard 1972).

The variables studied often lack anchors in the real world. While diverse viewpoints in a group can facilitate creative problem solving (Hoffman and Maier 1961), too much diversity can destroy a group (Fiedler and Meuwese 1963). How can an organization know where the optimum range of diversity lies without some method for measuring it?

A related gap is seen in the measurement of leadership styles without behavioral referents. What behaviors are linked to high consideration scores on Fleishman's Leadership Behavior Description Questionnaire (Fleishman and Harris 1962)?

Finally, as is obvious from the last part of the discussion of research findings, the meager investment in studies of the problem-solving process itself is an extremely limiting factor. The solution valence-adoption relationships are highly stable and replicable and reveal the importance of the timing of events in problem-solving groups (Hoffman 1979). Yet they are only the tip of the iceberg, as suggested by the hierarchical model (Hoffman 1979). Norms, values, pathology all intrude on the task to determine the group's outcomes. Lacking a framework for understanding or a methodology for measuring these phenomena leaves us to our artistic, intuitive resources in attempting to effect improvements.

Despite these recognized limitations, the 25 years I have spent in research and consulting to improve the effectiveness of problem-solving groups have convinced me that groups are potentially powerful mechanisms for processing information, solving problems, and ensuring their members' commitment to the group's decision. They can also enhance the members' self-esteem by providing opportunities for them to express their individuality and exercise their creativity. Unfortunately, groups most often do not perform these functions in most organizations. Rather than abandon this area of research as many have done, we need to press harder to overcome the difficulties outlined above.

REFERENCES

Argyris, C. & Schön, D. A. 1974: *Theory in practice.* San Francisco: Jossey-Bass.

Bass, B. M. & Wurster, C. R. 1953: Effects of company rank on LGD performance of oil refinery supervisors. *Journal of Applied Psychology, 37,* 100–4.

Bennis, W. 1966: *Changing organizations.* New York: McGraw-Hill.

Berger, J., Cohen, B. P. & Zelditch, M., Jr. 1972: Status characteristics and social interaction. *American Sociological Review, 37,* 241–55.

Berkowitz, L. (Ed.) 1978: *Group processes.* New York: Academic Press.

Berkowitz, L. & Levy, B. I. 1956: Pride in group performance and group task motivation. *Journal of Abnormal and Social Psychology, 53,* 300–6.

Bion, W. R. 1959: *Experiences in groups.* New York: Basic Books.

Blake, R. R. & Mouton, J. S. 1961: *Group dynamics—key to decision making.* Houston, Tex.: Gulf Publishing.

Block, M. W. 1974: Member commitment to group decisions: A study of individual and group determinants. Unpublished doctoral dissertation, University of Chicago.

Bouchard, T. J. 1972: Training, motivation, and personality as determinants of the effectiveness of brainstorming groups and individuals. *Journal of Applied Psychology, 56,* 324–31.

Burns, T. & Stalker, G. M. 1961: *The management of innovation.* London: Tavistock.

Carlston, D. E. 1977: Effects of polling order on social influence in decision-making groups. *Sociometry, 40,* 115–23.

Dalkey, N. C. 1969: *The delphi method: An experimental study of group opinion.* Santa Monica, Calif.: The Rand Corporation.

Dalton, M. 1959: *Men who manage.* New York: John Wiley & Sons.

Davis, J. H. 1973: Group decision and social interaction: A theory of social decision schemes. *Psychological Review, 80,* 97–125.

Dearborn, D. C. & Simon, H. A. 1958: Selective perception: A note on the departmental identifications of executives. *Sociometry, 21,* 140–4.

Delbecq, A., Van de Ven, A. H. & Gustafson, D. H. 1975: *Group techniques for program planning.* Glenview, Ill.: Scott, Foresman.

Dunnette, M. D., Campbell, J. & Jaastad, K. 1963: The effect of group participation on brainstorming effectiveness for two industrial samples. *Journal of Applied Psychology, 47,* 30–7.

Falk, G. 1978: An examination of some normative effects of unanimity and majority rules on the quality of solutions in problem solving groups with unequal power. Unpublished doctoral dissertation, University of Chicago.

Fiedler, F. E. & Meuwese, W. A. T. 1963: Leader's contribution to task performance in cohesive and uncohesive groups. *Journal of Abnormal and Social Psychology, 67,* 83–7.

Fisek, M. & Ofshe, R. 1970: The process of status evolution. *Sociometry, 33,* 327–46.

Fleishman, E. A. & Harris, E. F. 1962: Patterns of leadership behavior related to employee grievances and turnover. *Personnel Psychology, 15,* 43–56.

Ghiselli, E. E. & Lodahl, T. M. 1958: Patterns of managerial traits and group effectiveness. *Journal of Abnormal and Social Psychology, 57,* 61–6.

Golembiewski, R. T. & McConkie, M. 1975: The centrality of interpersonal trust in group processes. In C. L. Cooper (Ed.), *Theories of group processes.* New York: John Wiley & Sons.

Guilford, J. S. 1952: Temperament traits of executives and supervisors measured by the Guilford personality inventories. *Journal of Applied Psychology, 36,* 228–33.

Hackman, J. R. & Morris, C. G. 1975: Group tasks, group interaction process, and group performance effectiveness: A review and proposed integration. In L. Berkowitz (Ed.), *Advances in experimental social psychology* (Vol. 8). New York: Academic Press.

Hamblin, R. L. 1958: Leadership and crisis. *Sociometry, 21,* 322–35.

Hare, A. P. 1978: *Handbook of small group research* (2nd ed.). New York: Free Press.

Hochbaum, G. M. 1954: The relation between group members' selfconfidence and their reactions to group pressures to uniformity. *American Sociological Review, 79,* 678–87.

Hoffman, L. R. 1965: Group problem solving. In L. Berkowitz (Ed.), *Advances in Experimental Social Psychology* (Vol. 2). New York: Academic Press.

Hoffman, L. R. (Ed.) 1979: *The group problem-solving process: Studies of a valence model.* New York: Praeger.

Hoffman, L. R., Bond, G. R. & Falk, G. 1979: Valence for criteria: A preliminary exploration. In L. R. Hoffman (Ed.), *The group problem-solving process: Studies of a valence model.* New York: Praeger.

Hoffman, L. R. & Clark, M. M. 1979: Participation and influence in problem-solving groups. In L. R. Hoffman (Ed.), *The group problem-solving process: Studies of a valence model.* New York: Praeger.

Hoffman, L. R., Friend, K. E. & Bond, G. R. 1979: Problem differences and the process of adopting group solutions. In L. R. Hoffman (Ed.), *The group problem-solving process: Studies of a valence model.* New York: Praeger.

Hoffman, L. R., Harburg, E. & Maier, N. R. F. 1962: Differences and disagreement as factors in creative group problem solving. *Journal of Abnormal and Social Psychology, 64,* 206–14.

Hoffman, L. R. & Maier, N. R. F. 1961: Quality and acceptance of problem solutions by members of homogeneous and heterogeneous groups. *Journal of Abnormal and Social Psychology, 62,* 401–7.

Hoffman, L. R. & Maier, N. R. F. 1966a: An experimental reexamination of the similarity-attraction hypothesis. *Journal of Personality and Social Psychology, 3,* 134–52.

Hoffman, L. R. & Maier, N. R. F. 1966b: Social factors influencing problem solving in women. *Journal of Personality and Social Psychology, 4,* 382–91.

Hoffman, L. R. & Maier, N. R. F. 1967: Valence in the adoption of solutions by problem-solving groups. II. Quality and acceptance as goals of leaders and members. *Journal of Personality and Social Psychology*, *6*, 175–82.

Hoffman, L. R. & O'Day, R. 1979: The process of solving reasoning and value problems. In L. R. Hoffman (Ed.), *The group problem-solving process: Studies of a valence model*. New York: Praeger.

Humpal, J. J. & Hoffman, L. R. 1979: Normative conflict in a business setting. Paper delivered at Eastern Psychological Association meeting, April.

Jackson, J. M. & Saltzstein, H. D. 1958: The effect of person-group relationships on conformity processes. *Journal of Abnormal and Social Psychology*, *57*, 17–24.

Janis, I. L. 1972: *Victims of groupthink*. Boston, Houghton Mifflin.

Kadane, J. & Lewis, G. 1969: The distribution of participation in group discussions: An empirical and theoretical reappraisal. *American Sociological Review*, *34*, 710–22.

Kahn, R. L., Wolfe, D. W., Quinn, R. P., Snoek, J. D. & Rosenthal, R. A. 1964: *Organizational stress*. New York: John Wiley & Sons.

Lawrence, P. & Lorsch, J. 1967: *Organization and environment*. Boston: Harvard Business School, Division of Research.

Maier, N. R. F. 1960: Screening solutions to upgrade quality: A new approach to problem solving under conditions of uncertainty. *Journal of Psychology*, *49*, 217–31.

Maier, N. R. F. 1963: *Problem-solving discussions and conferences: Leadership methods and skills*. New York: McGraw-Hill.

Maier, N. R. F. 1970: *Problem-solving and creativity*. Belmont, Calif.: Wadsworth.

Maier, N. R. F. & Hoffman, L. R. 1961: Organization and creative problem-solving. *Journal of Applied Psychology*, *45*, 277–80.

Maier, N. R. F. & Hoffman, L. R. 1963: Seniority in work groups: A right or an honor? *Journal of Applied Psychology*, *47*, 173–6.

Maier, N. R. F. & Hoffman, L. R. 1964: Financial incentives and group decision in motivating change. *Journal of Social Psychology*, *64*, 369–78.

Maier, N. R. F. & Hoffman, L. R. 1965: Acceptance and quality of solutions as related to leaders' attitudes toward disagreement in group problem-solving. *Journal of Applied Behavioral Science*, *1*, 373–86.

Maier, N. R. F. & Solem, A. R. 1952: The contribution of a discussion leader to the quality of group thinking: The effective use of minority opinions. *Human Relations*, *5*, 277–88.

Maier, N. R. F. & Solem, A. R. 1962: Improving solutions by turning choice situations into problems. *Personnel Psychology*, *15*, 151–7.

Oakes, W. F. 1962: Effectiveness of signal light reinforcers given various meanings on participation in group discussion. *Psychological Reports*, *11*, 469–70.

Osborn, A. F. 1957: *Applied imagination*. (Rev. ed.). New York: Scribner.

Parnes, S. F. & Meadow, A. 1959: Effects of "brainstorming" instructions on creative problem solving by trained and untrained subjects. *Journal of Educational Psychology*, *50*, 171–6.

Pelz, D. C. 1956: Some social factors related to performance in a research organization. *Administrative Science Quarterly*, *1*, 310–25.

Pepinsky, P., Hemphill, J. K. & Shevitz, R. N. 1958: Attempts to lead, group productivity, and morale under conditions of acceptance and rejection. *Journal of Abnormal and Social Psychology*, *57*, 47–54.

Read, W. H. 1962: Upward communication in industrial hierarchies. *Human Relations*, *15*, 3–15.

Sherwood, J. J. & Glidewell, J. C. 1972: Planned renegotiation: A norm-setting OD intervention. In W. Burke (Ed.), *Contemporary organization development: Approaches and interventions*. Washington, D.C.: NTL Learning Resources Corporation.

Shiflett, S. 1979: Toward a general model of small group productivity. *Psychological Bulletin*, *86*, 67–79.

Steiner, I. D. 1966: Models for inferring relationships between group size and potential group productivity. *Behavioral Science*, *11*, 273–83.

Stern, G. G. 1970: *People in context.* New York: John Wiley & Sons.

Stock, D. & Thelen, H. A. 1958: *Emotional dynamics and group culture.* Washington, D.C.: National Training Laboratory.

Stogdill, R. M. 1974: *Handbook of leadership.* Glencoe, Ill.: Free Press.

Taylor, D. W., Berry, P. C. & Block, C. H. 1958: Does group participation when using brainstorming facilitate or inhibit creative thinking? *Administrative Science Quarterly*, *3*, 23–47.

Thomas, E. J. & Fink, C. F. 1961: Models of group problem solving. *Journal of Abnormal and Social Psychology*, *68*, 53–63.

Torrance, E. P. 1955: Some consequences of power differences on decision making in permanent and temporary three-man groups. In A. P. Hare, E. F. Borgatta and R. F. Bales (Eds.), *Small groups: Studies in social interaction.* New York: Knopf.

Tuckman, B. W. 1965: Developmental sequence in small groups. *Psychological Bulletin*, *63*, 384–99.

Wells, W. P. 1962: Group age and scientific performance. Unpublished doctoral dissertation, University of Michigan.

Part II

Management, Organization, and Innovation

9

The Innovation Design Dilemma: Some Notes on its Relevance and Solutions

JONNY HOLBEK

A curious ambiguity exists in the organization literature with respect to designs or structures promoting innovation. One reason for this ambiguity is the frequent failure to acknowledge—at least in empirical studies—that innovation and organization(s) meet in a process, extending over the total life-cycle of the innovation. It is this extended "meeting" over time that creates the innovation design dilemma.

This article starts with a presentation of the Dilemma and the reasons for its existence. Some recent studies have put a question-mark to the general relevance of the Dilemma, and this issue is considered next. In the following two sections on the innovation process, and on the *basic* types of Dilemma solutions, a necessary background is provided for a presentation of more *specific* design solutions.

Throughout this article, the following definition of innovation is used: any idea, practice, or material artifact perceived to be new by the relevant unit of adoption (Zaltman et al. 1973).

THE DILEMMA: A ROUTINE PRESENTATION

During the last decade, it has become a reasonably well established proposition of organization theory that the non-routine innovation process in organizations may fruitfully be viewed as consisting of two major subprocesses or stages (Zaltman et al. 1973; Duncan 1976; Rogers 1983; Souder 1983; Huber 1984).

Zaltman et al. (1973) have named the two subprocesses, initiation and implementation. *Initiation* is concerned with how the organization becomes aware of innovation, with the formation of attitudes toward the innovation, and with the innovation's development towards the making of a decision about its implementation; here, gathering and processing of information is required.

Implementation is concerned with the process by which the innovation is put into effect, eventually becoming integrated into the ongoing operations of the same or a different organization; here, development of rules and procedures is required. Together, the two subprocesses form the iterative innovation process. Each of the subprocesses contains a set of phases, linked together by a varied mixture of pooled, sequential and reciprocal dependencies (Thompson 1967).

As the brief descriptions of the two subprocesses suggest, very different tasks must be performed in the initiation and implementation stages of the innovation process. For the tasks to be performed well, they must be performed within the contexts of different organization designs—different types of organization structures. In fact, the existing research indicates that the very structural characteristics that facilitate initiation may impede the implementation of innovation, and vice versa (cf. Sapolsky 1967; Zaltman and Duncan 1977). This contrast between effective structural features of the initiation and implementation subprocesses constitutes the *innovation design dilemma.*

In 1966, James Q. Wilson pointed out that organizational forces which tend to generate innovation ideas and proposals are in conflict with forces which ultimately secure adoption and implementation. Five years earlier, Burns and Stalker (1961) had found that different types of organizational forms or systems are effective in different situations: An *organic* form is best characterized under unstable conditions by high uncertainty, while a *mechanistic* form under stable conditions is best with low uncertainty. These findings suggested that the loose and flexible features of an organic structure might be advantageous for effective performance during the initiation stage, while the tighter and more bureaucratic features of a mechanistic structure might be preferable for efficient performance during the implementation stage of the innovation process.

The ideas of Wilson (1966) and of Burns and Stalker (1961) were subsequently elaborated by Zaltman et al. (1973), who identified more precisely the structural variables which cause the innovation design dilemma. The variables in focus were organization complexity, formalization and centralization. Reviewing the existing literature, Zaltman et al. (1973) found each of the three variables to have opposing influence on the initiation and implementation subprocesses. More specifically, the authors found that a *high* degree of complexity, *low* formalization and *low* centralization (all organic features) stimulate or facilitate the gathering and processing of information that is crucial to the initiation stage. On the other hand, a *low* degree of complexity, *high* formalization and *high* centralization (all mechanistic features) tend to reduce role conflict and ambiguity that will easily impair implementation. In conclusion, *different* configurations of organizational structure appeared to facilitate changes in different parts of the total innovation process.

Faced with conflicting requirements for organization structuring, designers or organizations were forced to answer an old question: how should the organization be designed or structured in order for more effective innovations and faster innovation processes to occur? While a mechanistic structuring would generate too few ideas and proposals, an organic structuring would promote a slow and conflict-ridden implementation.

Clearly, the traditional solution to this innovation design dilemma had been to let the initiation stage take place in one organizational unit—say, a planning or R&D group, and then implement the innovation in another organizational unit—say, a manufacturing or marketing department. Such a solution, however, involved permanent *differentiation in space*. For smaller organizations, this was a realistic alternative only when linkages to other organizations were involved. In this case, the task of managing the innovation process became a task of managing an *inter*-organizational process.

Gradually, a second alternative, *differentiation in time* emerged as an organization design device. Shepard (1967) had discussed how a military raiding unit during World War II used structural forms alternating over time:

> The planning before a raid was done jointly by the entire unit— the private having as much opportunity to contribute to the planning as the colonel. During the raid, the group operated under a strict military command system. Following each raid, the unit returned to the open system used in planning for purposes of evaluation and maximizing learning from each raid. (pp. 474–5)

Similarly, a group or organization in civil life might shift its structure while moving through the various stages of the innovation process. With two major stages in the process, this calls for a dual structure (Zaltman and Duncan 1977). The existence of iteration and feedback loops between the stages imply the possibility of switching back and forth between an organic and mechanistic form—between an initiating and implementing mode of problem solving.

Duncan (1973) found, on the basis of empirical data, that the same decision-making group utilized different structures for making non-routine and routine decisions. An organic (non-bureaucratic) structure was used for non-routine decision situations, in which the needs for new information was high. A mechanistic (bureaucratic) structure was used and found more effective for routine decision situations, in which the needs for new information was low. Such a change between non-routine and routine decision situations, or between ill- and well-structured problem situations, is equivalent to a change between initiation and implementation in the innovation process.

IS THE DILEMMA LESS REAL THAN IT USED TO BE?

We have emphasized the relevance and importance of the innovation design dilemma, i.e. that the initiation of innovation is facilitated through the use of organic structures, while the implementation of innovation is facilitated with the use of mechanistic structures. Recent studies, however, suggest that there may be exceptions to these relationships (cf. Daft and Becker 1978; Daft 1982; Zmud 1982). The exception seems to occur in the case of administrative innovations, i.e. innovations relating to the social structure of the organization. More specifically, it is claimed that initiation of administrative innovations is facilitated with mechanistic rather than organic structures. Our interpretation of the works producing these results coincides with that of Huber (1984): the exception is an apparent exception, not a real one. The appearance probably stems from what we will refer to as the problem of measurement location.

A problem with some innovation studies is that "organic" vs. "mechanistic" are viewed primarily as characteristics of the organization as a whole. An overall measure, however, will frequently hide differences between submits. As noted by several authors, the degree of bureaucratization vs. organicity may differ considerably across the various parts of a single organization. Mintzberg (1982), for instance, has emphasized that:

> Although we can (and will) characterize certain organizations as bureaucratic or organic overall, none is uniformly so across its entire range of activities. (p. 37)

In a rich and insightful study of school organizations, Daft and Becker (1978) have shown that different units or groups within an organization tend to take the responsibility for initiating different kinds of innovations. It is, we believe, these initiating units or groups that will tend to be organically designed, irrespective of any overall measure. Indeed, such an interpretation of results is in accordance with the principle of parsimony.

Having noted that different kinds of innovations tend to be initiated by different units and groups in the organization, Daft and Becker (1978) further show that educational and administrative innovations follow "entirely different processes". Educational innovations are said to "percolate up" in the organizations, while administrative innovations "trickle down" (cf. Evan 1966). Although these differences are indeed of interest, the processes share—according to our interpretation—a basic similarity in relation to the innovation design dilemma.

Our interpretation of Daft and Becker's study, with respect to innovation process similarities, runs as follows: In schools, technical or *educational* innovations are proposed or initiated by teachers— primarily teachers ranking high on educational professionalism (an organic measure). Due to lack of authority among teachers, these ideas will have to "trickle up" ("percolate up") from an organic unit to the level of administrators, where an adoption decision is eventually made. On the other hand, *administrative* innovations are proposed or initiated by administrators at top levels, since the managers here are the experts and have highest administrative professionalism (an organic measure) with regard to administrative arrangements. Administrative innovations will also be approved (adopted) near the top of the hierarchy. Next, adopted administrative ideas "trickle down" for implementation, in a context of mechanistic arrangements. Thus, if measures of organicity are not taken at the level or in the unit of initiating administrators, one may possibly make the conclusion that "administrative innovations are facilitated with mechanistic structures". This, indeed, is likely to be correct, but only as far as implementation is concerned.

It appears, then, that both a "percolating up" and a "trickling down" innovation process are linked to the constraints of the innovation design dilemma: Initiations occur primarily in organic units or groups—wherever these are located (internally or externally), while implementations occur primarily in mechanistic surroundings. In our view, therefore, the innovation design dilemma is just as real as it used to be. This is a conclusion with respect to the issue of *relevance* of the innovation design dilemma. We have been concerned here with whether innovation *initiation*—under specific conditions—primarily occurs in *mechanistic* contexts, and our conclusion is that this does not seem to be the case. The other side of the coin in the case of the relevance issue, is whether innovation *implementation*—under specific conditions—primarily occurs in *organic* contexts. Again, our conclusion is in the negative, but we will return to this issue when discussing the "organic organization".

THE INNOVATION PROCESS IN MORE DETAIL

In the previous presentation of the innovation design dilemma, our conception of the innovation process was a simple one. The process was said to consist of two major subprocesses or stages: initiation and implementation. This simple conception was sufficient to highlight the basic idea involved in the innovation design dilemma. However, empirical research on innovation design, as well as the practical design experience of managers, suggests that the specification of a more detailed innovation process is required.

The total innovation process, according to Zaltman et al. (1973), consists of five substages: (1) knowledge–awareness, (2) formation of attitudes, (3) decision, (4) initial implementation, and (5) continued–sustained implementation. One advantage of this process is its flexibility: While being essentially a rational model of the innovation process, like the more specific activity-stage models (Saren 1984), it does give room for response to unexpected findings as well as to political decision making.

Feedback is an important aspect of the Zaltman et al. process—guiding and controlling the actual sequencing and performance of the process. Feedback cycles may take care of unexpected findings and key uncertainties emerging in the process. For instance, a complex innovation may require several returns to previous phases in the initiation subprocess; next, an unexpected problem occurring during implementation may require a return to the problem—solving phases of initiation. This iterative or "circular" element in the process will remind the researcher and manager that the initiation as well as the implementation stage, may consist of a *multitude* of decisions (cf. Cooper and More 1979).

The Zaltman et al. process is also flexible in the sense that it can be easily linked to other models. For instance it may be linked to the normative model of participation in decision making described by Vroom and Yetton (1973). This could be a useful link in many instances, since the Vroom and Yetton model might suggest how various degrees of participation in decision making and problem solving should be applied in different innovative situations. It can further be noted that the Zaltman et al. process makes no limiting assumptions about political variables like those of authority relationships and coalition building. This implies a potential for linking the process to models of political decision making, like that proposed by Elkin (1983).

While we consider flexibility to be a main advantage of the Zaltman et al. process, we are not going to explore the many implications of that flexibility in the present article. Our intention here is primarily to make some notes on alternative solutions to the innovation design dilemma. For this purpose, we have found the Zaltman et al. process a useful framework for presentation.

So far in this article, we have made the implicit assumption that the tasks of innovation initiation and innovation implementation should be differentiated in space or time between substages 3 and 4 in the Zaltman et al. process. Apparently, the tasks related to knowledge-awareness (1), formation of attitudes (2) and decision (3) should be performed in an organic unit, while the tasks related to initial implementation (4) and continued-sustained implementation (5) should be performed in a mechanistic unit.

There are, however, two problems with such an assumption. First, the "decision" substage may in some instances have more mechanistic than organic characteristics (cf. Pierce and Delbecq 1977). While decisions in the early rounds of initiation are likely to be of an organic kind, with varied knowledge inputs and a high degree of participation—especially in the case of complex innovations—the decisions in the latter rounds are sometimes of a more mechanistic kind, say, when "technical" decision rules can be applied. Indeed, as Vroom and Yetton (1973) have pointed out, the kind of decision process to apply for problem solving may fruitfully be determined in view of situational characteristics.

The second problem with implicitly assuming a differentiation between substages 3 and 4, is an emerging impression of a "best" differentiation or transfer point between what Galbraith (1982) has termed an "innovative" organization unit and an "operating" organization unit. However, as empirical research and practical experience have shown, the "best" transfer point is determined by situational factors, as we shall see in a later section on some specific Dilemma solutions. First, however, we are going to take a look at the more basic solutions.

BASIC TYPES OF DILEMMA SOLUTIONS

Two types of solutions to the innovation design dilemma were briefly commented upon in a previous section: (1) differentiation in *space*, i.e. initiation and implementation are performed by different people in separate organization units; (2) differentiation in *time*, i.e. initiation and implementation are performed by the same people during separate intervals in time. A third "solution" to the Dilemma is to ignore it: (3) the organic organization—a quasi-solution. These three basic types of solution to the innovation design dilemma will be discussed next. (Combining the two first ideal type solutions, i.e. differentiating in space *and* time, makes possible a multitude of *hybrid* solutions characterized by multiple activities both in series and in parallel; cf. Cooper 1983).

Differentiation in Space

In principle, differentiation in space as an innovation dilemma solution occurs in two modes: (i) *internal* differentiation in space, i.e. both initiation and implementation take place in the same organization, and (ii) *external* differentiation in space, i.e. either initiation or implementation tasks are performed in the focal organization, while implementation or initiation tasks, respectively, are performed in other organizations.

Internal differentiation in space can be designed both horizontally and vertically. With *horizontal* differentiation, the traditional innovation dilemma solution is to let the initiation stage take place in one organizational unit like a planning or R&D group, and then implement the innovation in another organizational unit like a manufacturing or marketing department. But an innovation formulated in one place and implemented in another, involves a set of assumptions which are valid mainly under conditions of stability (cf. Mintzberg 1983). When research and development or planning units are segregated, their innovative proposals may be ignored if they deviate too much from the status quo.

Both emotional and cognitive issues are involved here. Incremental innovations like productivity improvements, fine-tuning of the organization, and small changes as part of a more major change may be relatively easy to implement. However, when innovation radicalness is increasing, a major problem with the differentiation-in-space solutions is that "the users are not the choosers". The result is frequently resistance to change, nurtured by a "not invented here" syndrome. This means that innovations initiated by "them" will not be implemented by "us". On the cognitive level, new insights are frequently difficult to transfer in writing. Increasing radicalness of innovations further involves increasing uncertainty, and thereby an increasing need for processing new information en route (Galbraith 1977). Also, problems during implementation may require a return to the problem-solving phases of innovation. The more non-routine (radical) an innovation is, then, the stronger the need for creating horizontal linking mechanisms between initiating units and implementation units.

Galbraith (1977) has identified the most relevant of the linking mechanisms ("lateral processes"), and has listed these in a sequence determined by increasing ability to handle new information and cost to the organization: (1) direct contact between managers, (2) creation of liaison role, (3) creation of task forces, (4) use of teams, (5) creation of integrating role, (6) change to managerial linking role, and (7) establishment of the matrix form. Following this sequence from (1) to (7) generally involves an increasing potential for creating overlap between "choosers" and "users" (initiators and implementors). The sequence, therefore, implies that we are moving into a *hybrid* area of innovation design solutions, where differentiation-in-space merges with differentiation-in-time solutions.

As noted at the beginning of this section, internal differentiation-in-space can be designed both horizontally and vertically. The most common *vertical* differentiation is probably that between the top management group and the rest of the organization. With reference to our previous emphasis upon the importance of measuring "organic" vs. "mechanistic" at the *subunit* level of organizations, we would expect high overall centralization of the organization to go along with high organicity in the top management *group*, when the need for innovation is high. Low organicity in the top management group would, under these circumstances, produce ineffective initiation tasks.

In some instances the roles and tasks of an initiating top group become aggregated in a single individual, say, a top manager of a small firm or local government agency. In such cases, characteristics of the initiating individual appear to be compatible or consistent with characteristics of organicity at the group level. For instance, the individual top manager is likely to take better care of the function of initiating innovations when he or she has higher professionalism (Daft and Becker 1978) and stronger cosmopolitan orientation (cf. Kimberly 1981). Both of these characteristics are strongly related to scanning the external environment for new developments, thus removing what Rhenman (1973) found to be the major obstacle to innovation in top management: insensitivity to the environment.

As already noted, a major problem with the differentiation-in-space solutions is that "the user is not the chooser". This creates transfer problems with respect to understanding as well as motivation. For horizontal differentiation-in-space, this problem is partly solved by applying horizontal integration or linking mechanisms. For vertical differentiation-in-space, the problem may partly be solved by means of vertical integration or *participation*—by applying a "collaborative strategy" across levels (cf. Guest and Knight 1979; Lyngdal 1987).

External Differentiation in Space
An innovation dilemma solution of increasing importance is what Miles and Snow (1986) have called the "dynamic network". This solution involves two or more different organizations. Business functions such as product design and development, manufacturing, marketing, and distribution, typically conducted within a single organization, are performed by independent organizations within a network. Signs of this organizational form are linked to concepts such as joint ventures, subcontracting, licensing, and leasing.

Each organization (component) in the network can be seen as complementing rather than competing with the others. Miles and Snow (1986) state:

> Complementarity permits the creation of elaborate networks designed to handle complex situations, . . . which cannot be accomplished by a single organization. It also permits rapid adjustment to changing, competitive conditions . . . (p. 65)

For the individual organization (firm), the primary benefit of participation in the network is the opportunity to pursue its particular competence. Within an innovation perspective, an organization may choose to perform initiation tasks rather than implementation tasks—or vice versa. We shall see in a later section that such choices may lead to, for instance, a "new-style joint venture" (Roberts 1980) or an "operating adhocracy" (Mintzberg 1979). In both of these cases, one organization performs initiation and initial implementation tasks, while another organization performs continued-sustained implementation tasks.

Differentiation in Time

The innovation dilemma solution that involves "differentiation in time", has traditionally been the solution of smaller organizations. Interestingly, however, it has also—in modified versions—been increasingly applied in recent years by larger organizations. With this solution, initiation and implementation are performed by the same people during separate intervals in time, i.e. the "users" are also the "choosers". It should be noted that "separate intervals in time" does not only imply a simple, linear sequence of the two major stages. Since the innovation process is iterative, the differentiation-in-time solution has also been referred to as "periodicity", meaning a capacity to use alternating organizational forms (cf. Rowe and Boise 1974).

A prime example of an organization applying the differentiation-in-time solution is the small firm. Mintzberg (1983) applies the notion of "Simple Structure" in this case, and characterizes it as "an organic structure" (p. 255). However, while the overall organicity of a small firm may be relatively high, Holbek (1977) has pointed out—with reference to empirical findings—that this organicity does vary over the life-cycle of the innovation. Typically, there is a relatively high periodicity during the initiation stage, with people switching back and forth between informal, unstructured discussions—sometimes using simple creativity techniques, on the one hand, and more regimented and focused task performance when an alternative is (temporarily) worked on and tested, on the other. Having finally made an adoption decision, rules, procedures and relationships are gradually specified, making the activity context increasingly mechanistic. During implementation, periodicity occurs less and less frequently, and only when some unexpected problem occurs.

Earlier, we quoted Shepard (1967), who provided an example of a differentiation-in-time solution: a military raiding unit, whose members employed an organic system for planning, but reverted to a mechanistic command system during the raid. Row and Boise (1974) give an example from public agencies with line responsibilities:

> . . . many recreational programs at the neighbourhood level are likely to be planned in relatively open-ended climates by those who will later operate effectively in rational climates in order to carry out the plans. (p. 292)

In a later section on "hybrid" solutions to the innovation design dilemma, we shall see that differentiation in time is frequently used in tandem with differentiation in space. This is particularly the case in organizations where a "parallel organization" is applied to supplement a mechanistic base organization (bureaucracy).

Rowe and Boise (1974) have hypothesized that the greater the number of individuals in an organization who can operate in both an organic and a mechanistic structure/climate, the greater the potential for innovation in that organization. Implied in this statement is the assumption that the ability to function effectively under the conditions of high periodicity varies between organizations as well as between individuals (organization members). In order to increase its innovation potential, therefore, an organization might usefully apply a combination of recruitment and organization development policies aimed at increasing the number of members being comfortable with the level of periodicity required. A parallel organization might be a useful learning device for this purpose (cf. Zand 1974; Stein and Kanter 1980).

The Organic Organization—a Quasi-solution

We have earlier discussed a challenge to the general relevance of the innovation design dilemma: Should initiation tasks—under specific conditions—primarily take place in mechanistic units? Our answer to this question was in the negative. The "organic organization" concept challenges the other side of the relevance coin: Should implementation tasks—under specific conditions—primarily take place in organic contexts? An affirmative answer to this question would mean that the organic organization could be viewed as a conditional solution to the innovation design dilemma.

In a frequently quoted study of health and welfare organizations, Aitken and Hage (1971) found empirical support for the notion that the organic organization (as identified by overall, organization-wide measures) has characteristics that facilitate innovation: In particular, a diversity of occupations, high involvement in professional association, high intensity of scheduled communications, and high intensity of unscheduled communications with those of higher status, both within and between departments, are highly related to the degree of innovation. These results are not necessarily inconsistent with the basic notion of the innovation dilemma. This is implicit in the following statement from the two authors:

> . . . so many proposals for innovation are made in such (organic) organizations that, even though only a small proportion are actually adopted, the absolute number of adoptions may still be greater in such organizations. (p. 71)

In other words: high organicity is so favorable to initiation that it outweighs the disadvantages of high organicity to adoption and implementation. It appears, however, that such an outweighing of disadvantages is situation specific. Indeed, Aitken and Hage themselves have made statements which are relevant in this respect. The two authors note that " . . . if the rate of innovation is too rapid, more rigid, inflexible mechanisms of social control might have to be used to implement innovation" (p. 76). This statement may suggest, first, that innovations can become too easily adopted in organic surroundings (cf. Miller and Friesen 1982), and, second, that implementation in the context of an organic structure may become too time-consuming and costly. In any event, more mechanistic implementation devices are called for, when time and cost is of importance.

It is instructive at this point to take a brief look at the two organic structures described by Mintzberg (1979). Within his "structure in fives" framework, Mintzberg has identified two organic and three more mechanistic structures or "natural configurations". The latter are highly efficient in stable environments, have standardization of work, skills or outputs, and are best at performing implementation tasks (particularly "continued–sustained implementation"). On the other hand, the two organic structures—called "Simple Structure" and "Adhocracy"—are adaptive and effective in dynamic environments, and are—we might assume—best at performing initiation tasks. What is of concern to us here is whether these two latter structures can function effectively at both major stages of the innovation process.

Mintzberg's "Simple Structure" is best exemplified by the small entrepreneurial firm, and as we have already seen, this type of organization frequently uses "differentiation-in-time" solutions. Although the overall level of organicity may be high, some mechanistic devices are used when possible for efficiency or productivity purposes.

Mintzberg's "Adhocracy" has two basic types: the *operating* adhocracy innovates and solves one-of-a-kind problems directly on behalf of its clients. "The organization cannot easily separate planning and design of the operating work—in other words, the project—from its actual execution" (Mintzberg 1981, p. 112). The organization, then, emerges as "an organic mass". Mintzberg's second type of adhocracy, the *administrative* adhocracy, undertakes projects on its own behalf. This organization has a two-part structure internally: a highly organic administrative component, and a highly mechanistic operating component. The latter is separated or truncated, "so that its need for standardization will not interfere with the project work" (Mintzberg 1981, p. 112).

It appears, then, that among the innovative organizations of Mintzberg, only the operative adhocracy emerges as truly "organic". It can do so for a very good reason, however: the continued-sustained implementation substage of the innovation process is left to the client.

The heading of this subsection suggested that we are viewing the organic organization as a "quasi-solution". We have noted that very few organizations are evenly organic overall, and that those who are have either let other organizations take care of some part of the innovation process (external truncation), or have low needs for speed and efficiency. We are not saying that organizations should not increase their overall organicity; on the contrary, such increase is generally considered crucial in dynamic and unstable environments. What we do argue, however, is that an increased organicity cannot do away with the need for a "mechanistic shift" towards the late phase of the total innovation process. Clearly, the importance of this need, and thus the importance of the innovation design dilemma, does vary. But even when the need for speed and efficiency is low, a mechanistic shift—say, a differentiation in time—will certainly *reduce* conflicts, time-consumption and costs.

Having considered now the *basic* types of Dilemma solutions, we will next take a brief look at some of the more *specific* solutions. We will do this with due reference to some of the "innovation designs" for larger organizations described in the literature. Since larger organizations are forced to solve the innovation dilemma by means of differentiation-in-space solutions, transfer of the innovation must take place at "transfer points" between different organization units. Our focus in the following overview will be on the location of such transfer points.

SOME SPECIFIC DILEMMA SOLUTIONS

In the present context, a transfer point marks the differentiation in space between organization units performing different tasks over the life-cycle of an innovation. As used here, a transfer point is synonymous with the notion of an interface:

> Interfaces are those boundaries where two systems meet, such that the output of one system is the input to the other. These boundaries can be internal or external to the system itself. (Athery 1982, p. 101)

Identification of transfer points is not always an easy matter.

Feedback loops in the innovation process will frequently blur the interfaces. Also, when "the users are not the choosers", it may be of importance to design the hand-off between initiation and implementation so that it is almost invisible:

> before the baton changes hands, the runners should have been running in parallel for a long time. (Leonard-Barton and Kraus 1985)

On the other hand, there is a need for "innovators" to form a self-contained group of their own with considerable autonomy (Child 1984). Particularly in the case of major or radical innovations, there is a need for an organizational structure that effectively insulates innovative developments from the prevailing corporate culture (Littler and Sweeting 1985).

While the location of transfer points may vary considerably between organizations and innovation processes, the majority of locations are to be found between substages 3 and 4, and between 4 and 5 in the Zaltman et al. process, i.e. between the substages of "decision" (to adopt) and "initial implementation", or between "initial implementation" and "continued-sustained implementation". In Figure 1 some common examples found in the literature are presented; "initiation" is a simplified expression for substages 1–3 in the Zaltman et al. process. We believe the examples suggest, however rudimentary, that the innovation design dilemma can be used as a rather generic conceptual underpinning for a number of innovation practices encountered in today's formal organizations.

The first two examples in the left part of Figure 1 are from Souder (1983). "Modern-Interfacial" is an organization design said to be "ideally suited to handling the initiation stage of the innovation process" (p. 36). That stage is taken care of in organic surroundings by creating interdisciplinary project groups composed of people drawn from across the functional areas. The "Modern-Interfacial" structure may be well suited for handling a high need for incremental innovations. However, it is capable of handling somewhat more radical innovations in an episodic fashion only. The reason for this is the inability of the structure to handle the implementation stage of the radical innovation process. There is no structural provision for commercializing the innovations (products) that come from completed projects. Operating divisions or departments are reluctant to use new products from the projects unless they are "fully developed", well-established in their market, highly profitable and relatively riskless" (Souder 1983, p. 36).

Some specific solutions to the innovation design dilemma: alternative locations of transfer point(s).

Major stages of the innovation process (Zaltman et al. 1973).	Some specific design solutions				
	"Modern-Interfacial" design (Souder. 1983)	"Modern Integrative" design (Souder. 1983; Frontini and Richardson. 1984)	"Innovative form" design (A & B. 1971). Internal venturing design	"New-style joint venture" design (Roberts. 1980; Hlavacek et al. 1977).	Planning agency design. Operating adhocracy (Mintzberg. 1979).
Initiation	Project group	Project group	Innovation group team	Smaller developing firm	R & D/ Planning agency.
Initial implementation	Operating departments	New enterprise group			
Continued-sustained implementation		Business group (division)	Current business group	Larger established firm	Client organization.

Note:
A number of situational factors will have to be considered in order to choose an optimal Dilemma solution. Among these factors are: variability and complexity of the **environment** (Souder. 1983; Mintzberg. 1979). performance needs of the **organization** (Huber. 1984; Ansoff and Brandenburg. 1971). and attributes of the **innovation** (Daft and Becker. 1978; Zaltman et al.. 1973)

Key:

- High organicity
- Medium organicity
- Low organicity
- ◇ External transfer point
- ⬖ Internal transfer point
- A & B Ansoff and Brandenburg

The weak part of the "Modern-Interfacial" structure is initial implementation—a substage which may be virtually lacking in many instances. This means that operating departments with mechanistic structures must take care of the initial implementation tasks, implying reduced steady-state efficiency. For more radical innovations, assuming internal development, two common improvements of the Modern-Interfacial design exist: "Modern Integrative" and the "Innovative Form" (internal venturing). In the former, one more transfer point has been added, and in the latter, the transfer point has been moved so that it occurs between "initial implementation" and "continued-sustained implementation".

"Modern-Integrative" design has three different groups taking care of different innovation stages: A *project group* serves as a source of new ideas and projects for a *new enterprise group*, which temporarily takes care of ventures until they have achieved a break-even level of profits. Currently profitable products and established product lines are housed in *business groups* (divisions), where emphasis is upon steady-state efficiency. In order to provide the necessary linkages across interfaces, the division manager, the new enterprise manager, and the projects manager meet periodically as a new products committee. Emerging conflicts are resolved by top management.

What distinguishes the "Modern-Integrative" design is an explicit emphasis upon initial implementation. Several authors have stressed the importance of this substage, which should be at least moderately organic (cf. Figure 1). Frontini and Richardson (1984) have highlighted a "design and demonstration" function. Such a function, they believe,

> gives a mature industrial enterprise an opportunity to pilot the more risky steps of launching an innovation without hurting other (more mechanistic) aspects of the business. (p. 39)

Other types of organizations are also in need of demonstration and "piloting". Delbecq and Mills (1985) conclude that in the case of health services organizations as well as high technology firms, high-innovation organizations began with a small, controlled pilot study. In the case of more radical innovations,

> pilot studies help maintain the attitude that the initial implementation is clearly experimental and the results will require further modifications before direct, large-scale implementation can occur (p. 33)

As noted, the "Modern-Integrative" design has two transfer points. However, an additional transfer point means another barrier of understanding and motivation in the innovation process (cf. Myers 1986). It may in many instances be advantageous, therefore, to maintain one transfer point only, like in the "Modern-Interfacial" structure, and rather move the transfer point to the interface between initial implementation and continued-sustained implementation. This gives a design which Ansoff and Brandenburg (1971) have called the "Innovative form", consisting of two major groups: (i) an organically structured *innovation group*, in which the development of new product-market positions is placed, and (ii) a mechanistically structured *current business* group, in which currently profitable, established product-markets are gathered.

In this design, the innovative group is responsible for the project until its commercial feasibility has been established; this is done e.g. through pilot production and market tests. Entering the current business group, the project may become a part of an *existing* division or form the nucleus of a *new* division. Ansoff and Brandenburg (1971) note that the transfer between the two groups may include all of the personnel and facilities, or just the innovation (product and/or technology):

> The former mode has considerable merit, because it exposes managers to operations in both the innovative and "steady state" environment and provides a valuable exchange of information and experience. In some firms, such transfer is only temporary and the innovation oriented people return to the innovative group after a tour of duty (pp. 725–6)

It is worth while to emphasize here that this transfer of personnel between the innovation group and the current business group represents a differentiation-in-time solution *added to* the differentiation-in-space solution created by the interface between the innovation group and the current business group. (We will return to this point in the next section).

The innovation group in the "Innovative form" design is frequently referred to in the literature as a "venture group". The venture concept is in most cases based on the assumption that radical innovations, say, a radically new product, stand a better chance of being successfully developed in larger companies, if a condition devoid of "bureaucracy" can be created (a small business development condition). While the lists of characteristics said to distinguish successful venture groups are long and varied (see, for instance, Hill and Hlavacek (1972) and MacMillan (1986)), almost every author emphasizes the importance of establishing such groups as separate structures.

Venture groups are regarded as having great potential, but their record is mixed. While the reasons for this are many, one major problem with a radically new product emerging from a separated venture group is to link that product to existing systems of the current business for continued-sustained implementation. Roberts (1980) and Hlavacek et al. (1977) have argued that in many instances the larger firm may more easily link itself to another, smaller firm whose new product or technology fits reasonably well into its existing production and marketing organization. In this "New-style joint venture", the smaller company (with relatively high organicity) has already performed initiation and—in low-risk situations—initial implementation. The larger company, next, takes care of continued-sustained implementation in more mechanistic surroundings.

From the point of view of the smaller company in this joint venture, "separation in time" and a mechanistic shift from initiation to initial implementation may have been successfully performed. Smaller firms do have problems with continued-sustained implementation in many industries, however, so that a coupling to a larger firm with compatible implementation characteristics may appear attractive.

The New-style joint venture provides an interesting example of an innovation design where the transfer point is "external" in the sense that it is located at the interface between two different organizations. Another example is the "Operating adhocracy" (Mintzberg 1981). This type of organization

> . . . carries on innovative projects directly on behalf of its clients, usually under contract, as in a creative advertising agency, a think tank consulting firm, or a manufacturer of engineering prototypes (p. 112)

Here, the content of the contract will determine, implicitly, the location of the transfer point. In public organizations, an external transfer point is frequently the rule. Among local governments, for example, a planning agency formulates plans, which are subsequently implemented by one or more other agencies or client organizations. No initial implementation is performed by the planning agency (see Figure 1).

What we have briefly presented in this section is some of the "innovation designs" for larger organizations described in the literature. The presentation is not exhaustive. It does demonstrate, however, that the *variation* of designs can be described and understood in terms of coping with the innovation design dilemma.

Our location of transfer points in Figure 1 serves analytical purposes. More exact transfer point location will have to be determined for the individual organization. Kilmann (1981) has presented an approach where dependency concepts from Thompson (1967) are applied for practical differentiation purposes. Further, the issue of decoupling (Stinchcombe 1985) should be seriously considered, both in general and for more specific applications (see also Mintzberg (1979) on "truncation").

The question of internal vs. external transfer points, as well as the question of transfer point location in the innovation process, await further research. Answers to such questions are clearly situation specific. Recent contributions that look promising for the study of larger companies have been made by Burgelman (1984)—studying internal entrepreneurship, and by Roberts and Berry (1985)—studying alternative diversification strategies.

HYBRID SOLUTIONS—BALANCED DIFFERENTIATIONS IN SPACE AND TIME

"Differentiation in space" is a necessary solution to the innovation design dilemma in larger organizations. However, with the exception of incremental innovations, it is hardly a sufficient solution. On both theoretical and empirical grounds, it can be hypothesized that differentiation-in-time solutions should be *added to* differentiation-in-space solutions—resulting in various *hybrid* solutions that might improve the effectiveness of innovative endeavors. More specifically, it can be hypothesized that increasing innovation radicalness implies increasing importance of hybrid solutions, as compared to "ideal type" solutions of differentiation in *either* space *or* time.

With reference to Figure 1 and our previous presentations of alternative locations of transfer points, the importance of adding differentiation-in-time solutions is possibly most striking in the case of the "Innovative form", and we will use this as an example. Here, the innovative group or venture group was supposed to take care of two important substages, initiation and initial implementation. For an effective functioning of such groups, a differentiation in time should take place—however gradually—between the two sub-stages. If such a differentiation, with concomitant mechanistic shift, does not occur, members of the groups are more likely to accept unsolicited alternatives, to have slower processing during initial implementation, and to produce output with more errors or lower quality (cf. Zand 1974).

Discussing the hand-over between the innovation group and current business group—the latter taking care of continued-sustained implementation, Ansoff and Brandenburg (1971) note that the transfer may include both personnel and innovation (product or technology). They further emphasize that personnel transfer provides a valuable exchange of information and experience, and increases the likelihood of a sustained implementation. But transfer of personnel across spatial interfaces in the innovation process requires a differentiation in time for the people involved. Here, for efficient routinization of the innovation to occur, people involved in the transfer must switch to task performance under conditions of low organicity, i.e. to task performance within the context of a mechanistic structure.

A very useful notion, when applying "differentiation in time" as an organization design device in large and medium-sized organizations, is the one of *parallel organization* (Stein and Kanter 1980) or collateral organization (Zand 1974). Flat and flexible, the parallel organization is supposed to form an organic mechanism for managing change and building environmental responsiveness into bureaucratic organizations on a continuing basis. The parallel organization is composed of a set of project (pilot) groups, a steering committee, and possibly an advisory group. Employees are strongly involved, since a major purpose of the parallel organization is to "energize the grass roots" (Kanter 1983; Cooke 1979).

The parallel organization should be

> . . . loose enough to allow for flexibility and some trial-and-error, yet connected enough to the line organization that the lessons learned could easily be seen as relevant to the larger setting and ultimately incorporated into ongoing operations. (Kanter 1983, p. 195)

Since the outputs of the parallel organization are inputs to the base organization (the "bureaucracy"), the ultimate value of the former depends on successfully linking it to the latter. The parallel organization has a high potential for successful linkage, in that the people who work in its project groups are also members of the base organization. This suggests that differentiation in time will be a recommended solution in most instances: having performed the initiation tasks in the parallel organization, implementation can be completed within the confines of the mechanistic, base organization. This movement between two "home" organizations will necessarily imply a mechanistic shift for people engaged in the total process. In the case when initiating members of a project group belong to a minority among implementors, the solution might be considered a *partial* differentiation-in-time solution. Such a designation would serve to underline that an important people transfer has taken place, thus serving both integration and extended learning purposes.

Hybrid solutions are innovation dilemma solutions combining the advantages of differentiation in space and differentiation in time. "Space" solutions provide specialization advantages, and "time solutions bring integration or linkage advantages. The optimal balance between the two as far as emphasis is concerned will depend upon the situation at hand. We have already hypothesized that increasing innovation radicalness should lead to stronger emphasis upon "time" solutions: on the other hand, specific characteristics of the organization's industry would be likely to encourage stronger emphasis upon "space" solutions. A systematic inquiry of these issues, however, is outside the scope of this article.

FINAL NOTE

In focus throughout this article has been the innovation design dilemma, its relevance and some of its solutions for today's formal organizations. The innovation dilemma is not just another problem to be solved in order for organizations to renew or change themselves. The Dilemma is basic, we believe, to a thorough understanding of the dynamics of innovation, and to an understanding of organization structuring and design.

One implication of the Dilemma for research methodology is clear: measurement of organicity or bureaucratization must be made for the relevant organization *units* as well as for the organization as a whole. Overall measures only, will not uncover variations of the type presented in Figure 1. Another issue of relevance for research is a systematic mapping of content in the concepts termed "organic" and "mechanistic".

Kimberly (1981) is clearly right in stating that the innovation design issue involves *more* than questions of organic vs. mechanistic structures. This being said, however, those same questions do emerge as important for "innovation designs", and thus to the practicing manager. On a more theoretical level, the innovation design dilemma appears useful as a major conceptual underpinning for a design approach to innovation.

REFERENCES

Aiken, M. & Hage, J. 1971: The organic organization and innovation. *Sociology,* 5 (1), January, 63–82.

Ansoff, H. J. & Brandenburg, R. G. 1971: A language for organization design. *Management Science, 17* (12), August (B-705–B-731).

Athery, T. H. 1982: *Systematic systems approach.* Englewood Cliffs, N.J.: Prentice-Hall, Inc.

Burgelman, R. A. 1984: Designs for corporate entrepreneurship in established firms. *California Management Review* (3), Spring 154–66.

Burns, T. & Stalker, G. M. 1961: *The management of innovation.* London: Tavistock Publications.

Child, J. 1984: *Organization: A guide to problems and practice.* London: Harper & Row, Publishers.

Cooke, R. A. 1979: Managing change in organizations. In G. Zaltman (ed.), *Management principles for non-profit agencies and organizations.* New York: AMACON (American Management Association).

Cooper, R. G. 1983: The new product process: an empirically-based classification scheme. *R&D Management, 13* (1), 1–13.

Cooper, R. G. & More, R. A. 1979: Modular risk management: an applied example. *R&D Management, 9* (2), 93–9.

Daft, R. L. 1982: Bureaucratic versus nonbureaucratic structure and the process of innovation and change. In S. G. Bacharach (ed.), *Research in the sociology of organization,* Vol. 1. Greenwich, Conn.; Jai Press, Inc. 129–66.

Daft, R. L. & Becker, S. W. 1978: *Innovation in organizations: innovation adoption in school organizations.* New York: Elsevier.

Delbecq, A. L. & Mills, P. K. 1985: Managerial practices that enhance innovation. *Organizational Dynamics,* 24–34.

Duncan, R. B. 1973: Multiple decision-making structures in adapting to environment uncertainty: the impact on organizational effectiveness. *Human Relations, 26,* 273–91.

Duncan, R. B. 1976: The ambidextrous organization: designing dual structures for innovations. In R. H. Kilmann, L. R. Pondy & D. P. Slevin (eds.), *The management of organization design, Vol. I.* Amsterdam: North-Holland, 167–88.

Elkin, S. L. 1983: Towards a contextual theory of innovation. *Policy Sciences, 15,* 367–87.

Evan, W. M. 1966: Organizational lag. *Human Organization, 25,* Spring, 51–3.

Frontini, G. F. & P. R. Richardson 1984: Design and demonstration: the key to industrial innovation. *Sloan Management Review,* Summer, 39–49.

Galbraith, J. 1977: *Organization Design.* Reading, Mass.: Addison-Wesley.

Galbraith, J. 1982: Designing the innovating organization. *Organizational Dynamics, 10* (3), Winter, 5–25.

Grønhaug, K. & Fredriksen, F. 1981: Resources, environmental contact and organizational innovation. *Omega, 9* (2), 155–62.

Guest, D. & Knight, K. (eds.) 1979: *Putting participation into practice.* Westmead, Engl.: Gower Press.

Hill, R. M. & Hlavacek, J. D. 1972: The venture team: a new concept in marketing organization. *Journal of Marketing, 36,* 44–50.

Hlavacek, J. D., Dovey, B. H. & Biondo, J. J. 1977: Tie small business techology to marketing power. *Harvard Business Review*, (1), January–February, 106–16.

Holbek, J. 1977: Innovasjon i småforetak (Innovation in small firms); paper delivered at the yearly conference of professional economists, Bergen, Norway.

Huber, G. P. 1984: The nature and design of postindustrial organizations. *Management Science, 30* (8), August 928–51.

Kanter, R. M. 1983: *The change masters.* New York: Simon & Schuster.

Kilmann, R. H. 1981: Organization design for knowledge utilization, *Knowledge: Creation, Diffusion, Utilization, 3,* (2), December, 211–31.

Kimberly, J. R. 1981: Managerial innovation. In P. C. Nystrom & W. H. Starbuck (eds.), *Handbook of organizational design, Vol. 1: Adapting organizations to their environments.* New York: Oxford University Press, 84–104.

Leonard-Barton, D. & Kraus, W. A. 1985: Implementing new technology. *Harvard Business Review, 63* (6), November–December, 102–10.

Littler, D. A. & Sweeting, R. C. 1985: Radical innovation in the mature company. *European Journal of Marketing, 19* (4), 33–44.

Lyngdal, L. E. 1987: Deltakelse i byråkratiske organisasjoner. (Participation in bureaucratic organizations.) *Norsk statsvitenskapelig tidsskrift, 3.* årg. (1), 49–64.

MacMillan, I. C. 1986: Progress in research on corporate venturing. In D. L. Sexton & R. W. Smilor, *The art and science of entrepreneurship.* Cambridge, Mass.: Ballinger Publishing Co., 241–64.

Mohr, L. B. 1982: *Explaining organizational behavior.* San Francisco: Jossey-Bass Publishers.

Miles, R. E. & Snow, C. C. 1986: Organizations: new concepts for new forms. *California Management Review, XXVIII* (3), Spring 62–73.

Miller, D. & Friesen, P. H. 1982: Innovation in conservative and entrepreneurial firms: two models of strategic momentum. *Strategic Management Journal, 3* (1), January–March, 1–25.

Mintzberg, H. 1979: *The structuring of organizations: a synthesis of the research.* Englewood Cliffs, N.J.: Prentice-Hall, Inc.

Mintzberg, H. 1981: Organization design: fashion or fit? *Harvard Business Review,* January–February, 103–16.

Mintzberg, H. 1983: *Structures in fives: designing effective organizations.* Englewood Cliffs, N.J.: Prentice-Hall, Inc.

Myers, D. D. 1986: How many champions will an innovation cycle support? In D. L. Sexton & R. W. Smilor, *The Art and Science of Entrepreneurship.* Cambridge, Mass.: Ballinger Publishing Co., 211–22.

Pierce, J. L. & Delbecq, A. L. 1977: Organization structure, individual attitudes and inovation. *Academy of Management Review, 2,* January, 27–37.

Rhenman, E. 1973: *Organization theory for long-range planning.* London: John Wiley & Sons.

Roberts, E. B. 1980: New ventures for corporate growth. *Harvard Business Review, 58* (4), July–August, 134–42.

Roberts, E. B. & Berry, C. A. 1985: Entering new business: selecting strategies for success. *Sloan Management Review,* Spring, 3–17.

Rogers, E. M. 1983: *Diffusion of innovation.* New York: The Free Press (3rd ed.).

Rowe, L. A. & Boise, W. B. 1974: Organizational innovation: current research and evolving concepts. *Public Administration Review,* May–June, 284–93.

Sapolsky, H. 1967: Organizational structure and innovation. *Journal of Business, 40* (4), 497–510.

Saren, M. A. 1984: A classification and review of models of the intra-firm innovation process. *R&D management, 14* (1), 11–24.

Shephard, H. A. 1967: Innovation-resisting and innovation-producing organizations. *Journal of Business, 40* (4), 470–7.

Souder, W. E. 1983: Organizing for modern technology and innovation: a review and synthesis. *Technovation, 2,* 27–44.

Stein, B. A. & Kanter, R. M. 1980: Building the parallel organization: toward mechanisms for permanent quality of work life. *Journal of Applied Behavioral Science*, *16* (3), 371–88.

Stinchcombe, A. L. 1985: Project administration in the North Sea. In A. L. Stinchcombe & C. A. Heimar, *Organization theory and project management.* Oslo: Norwegian University Press.

Thompson, J. D. 1967: *Organizations in action.* New York: McGraw-Hill.

Vroom, V. H. & Yetton, P. W. 1973: *Leadership and decision making.* Pittsburgh: University of Pittsburgh Press.

Wilson, J. Q. 1966: Innovation in organization: notes toward a theory. In J. D. Thompson (ed.), *Approaches to organizational design.* Pittsburgh: University of Pittsburgh Press 193–218.

Zaltman, G. & Duncan, R. 1977: *Strategies for planned change.* New York: John Wiley & Sons.

Zaltman, G., Duncan, R. & Holbek, J. 1973: *Innovations and organizations.* New York: John Wiley & Sons.

Zand, D. E. 1974: Collateral organization: a new change strategy. *Journal of Applied Behavioral Science*, *10* (1), 63–86.

Zmud, R. W. 1982: Diffusion of modern software practices: influence of centralization and formalization. *Management Science*, *28* (12), December, 1421–31.

10

A Process Model of Internal Corporate Venturing in the Diversified Major Firm[1]

ROBERT A. BURGELMAN

This paper examines the management of new ventures in a firm of the "diversified major" or "related business" type. Such firms are large agglomerates of widely diverse yet related businesses grouped into divisions whose general managers report to corporate management. In recent years, a substantial literature has emerged on the relationships between strategy, structure, degree of diversification, and economic performance in the divisionalized firm (Chandler 1962; Williamson 1970; Wrigley 1970; Rumelt 1974; Galbraith and Nathanson 1979; Caves 1980). The actual processes of corporate entrepreneurship and strategic change, however, remain less well understood. This is probably because these processes in such firms are complex and are difficult and costly to research. While large, diversified firms are clearly not representative of business organizations in general (Aldrich 1979), they represent such a large proportion of the total industrial activity in the developed economies that efforts to construct a theory of corporate entrepreneurship would seem valuable (Arrow 1982).

The research reported here investigates the process through which a diversified major firm transforms R&D activities at the frontier of corporate technology into new businesses through internal corporate venturing (ICV). These new businesses enable the firm to diversify into new areas that involve competencies not readily available in the operating system of the mainstream businesses of the corporation (Salter and Weinhold 1979). Previous systematic research of ICV has not clearly distinguished between new product and new business development and has investigated the ICV development process only up to the "first commercialization" phase (von Hippel 1977). The present study specifically examines the relationship between project development and business development showing how new organizational units developed around new businesses become integrated into the operating system of the corporation either as new freestanding divisions or as new departments in existing divisions. The rationale for studying projects utilizing new technologies is that the strategic management problems involved in corporate entrepreneurship are likely to be most accentuated and most identifiable in projects in which innovative efforts are radical (Zaltman, Duncan, and Holbek 1973).

Ansoff and Brandenburg (1971) discussed the strategic management problems of diversification through internal development in the divisionalized firm, and proposed that corporations create separate units within the corporate structure to facilitate new venture development. During the seventies, many large corporations adopted the new venture division (NVD) design (Hanan 1976; Hutchinson 1976). Fast (1979), however, showed that new venture divisions often occupy a precarious position within the corporate structure because of erratic changes in corporate strategy or in the political position of the NVD in the corporate context. Argyris and Schön (1978) provided anecdotal evidence of the various problems that impede the effectiveness of the NVD in divisionalized firms. The present study further elucidates the management problems inherent in internal corporate venturing.

Frohman (1978), Quinn (1979), and Maidique (1980) suggested categories of specialized roles to conceptualize the innovation process in organizations. The present study uses a different approach, documenting the key activities of persons on different hierarchical levels within the organization. The flow of these interlocking activities is represented in a process model of internal corporate venturing. Such a model is useful to elucidate the "generative mechanisms" (Pondy 1976) of corporate entrepreneurship. It indicates how the entrepreneurial activities of individuals combine to produce entrepreneurship at the level of the corporation, as well as how forces at the level of the corporation influence the entrepreneurial activities of these individuals.

METHODOLOGY AND RESEARCH DESIGN

A qualitative method was chosen as the best way to arrive at an encompassing view of ICV. Concerns of external validity were traded off against opportunities to gain insight into as yet incompetely documented phenomena. The caveats pertaining to field methods described by Kimberly (1979) are in order.

ICB project development has a ten-to-twelve-year time horizon (Biggadike 1979), and a truly longitudinal study was thus beyond the available resources. Instead, a longitudinal-processual approach (Pettigrew 1979) was adopted. The ICV process was studied exhaustively in one setting. Data were collected on six ongoing ICV projects that were in various stages of development. The historical development of each case was traced and the progress of each case during a fifteen-month research period was observed and recorded. These materials formed the basis for a comparative analysis of the six projects. This approach should not be confused with the so-called "comparative method" of early sociology, which used, often selectively, cross-sectional data to support a priori theories—most aptly called metaphors—of stages of development (Nisbet 1969). No such theory guided the present research, nor is one proposed as a result of it.

In fact, because of the exploratory nature of the study and the objective of generating a descriptive model of as yet incompletely documented phenomena, Glaser and Strauss's (1967) strategy for the discovery of "grounded theory" was adopted. This strategy requires the researcher ". . . at first, literally to ignore the literature of theory and fact on the area under study, in order to assure that the emergence of categories will not be contaminated by concepts more suited to different areas" (Glaser and Strauss 1967, p. 37). It also requires joint collection, coding, and analysis of the data. Data must be collected until patterns have clearly emerged and additional data no longer add to the refinement of the concepts.

The lack of previous research at the ICV project level of analysis made it fairly easy to follow these guidelines. By the same token, great uncertainty existed as to what conceptual framework would emerge from the data. Throughout the research period, idea booklets were used to write down new insights and interpretations of data already collected. These ongoing, iterative conceptualization efforts resulted in the creation of a new set of terms for the key activities in ICV and provided the bits and pieces out of which the conceptual framework finally emerged.

Research Setting

The research was carried out in one large, U.S.-based, high-technology firm of the diversified major type which I shall refer to as GAMMA. GAMMA had traditionally produced and sold various commodities in large volume, but it had also tried to diversify through the internal development of new products, processes, and systems so as to get closer to the final user or consumer and to catch a greater portion of the total value added in the chain from raw materials to end products. During the sixties, diversification efforts were carried out within existing corporate divisions, but in the early seventies, the company established a separate new venture division (NVD). Figure 1 illustrates the structure of GAMMA at the time of the study.

Data were obtained on the functioning of the NVD. The charters of its various departments, the job descriptions of the major positions in the division, the reporting relationships and mechanisms of coordination, and the reward system were studied. Data were also obtained on the relationships of the NVD with the rest of the corporation. In particular, the collaboration between the corporate R&D department and divisional R&D groups was studied. Finally, data were also obtained on the role of the NVD in the implementation of the corporate strategy of unrelated diversification to help explain why it had been created, how its activities fit in the corporation's Strategic Business Unit system, and how it articulated with corporate management. These data describe the historical evolution of the structural context of ICV development at GAMMA before and during the research period. The bulk of the data was collected in studying the six major ICV projects in progress at GAMMA at the time of the research.

- *Fermentation Products* was in the earliest stage of development. The new business opportunity was still being defined and no project had been formally started. Five people from this project were interviewed, some several times, between November 1976 and August 1977.
- *Fibre Components* was a project for which a team of R&D and business people were investigating business opportunities and their technical implications. Five people in this group were interviewed between January 1977 and May 1977.
- *Improved Plastics* had reached a point where a decision was imminent as to whether the project would receive venture status and be transferred from the corporate R&D department to the venture development department of the NVD. Seven people from this project were interviewed, some several times, between February 1977 and April 1977.
- *Farming Systems* had achieved venture status, but development had been limited to the one product around which it had been initially developed. Efforts were being made to articulate a broader strategy for further development of the venture. This

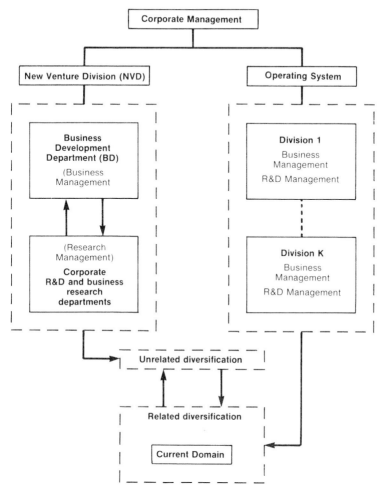

Fig. 1. The structure of GAMMA corporation.

was achieved during the research period and an additional project was started. Seven people were interviewed, some several times, between November 1976 and August 1977.

- *Environmental Systems* had also achieved venture status, but was struggling to deal with the technical flaws of the product around which its initial development had taken place. It also was trying to develop a broader strategy for further development. It failed to do so, however, and the venture was halted during the research period. Six people from the project were interviewed between March 1977 and June 1977.
- *Medical Equipment* was rapidly becoming a mature new business.

It had grown quickly around one major new product, but had then developed a broader strategy that allowed it to agglomerate medically related projects from other parts of the corporation and to make a number of external acquisitions. After the research period, this venture became a new freestanding division of the corporation. Eleven people were interviewed, some several times, between June 1976 and September 1977.

Data Collection

In addition to the participants in the six ICV projects, I interviewed NVD administrators, people from several operating divisions, and one person from corporate management. All in all, sixty-one people were interviewed. Table 1 indicates the distribution of persons interviewed over job categories.

The interviews were unstructured and took from one and a half to four and a half hours. Tape recordings were not made, but the interviewer took notes in shorthand. The interviewer usually began with an open-ended invitation to tell about work-related activities, then directed discussion toward three major aspects of the ICV development process: (1) the evolution over time of a project, (2) the involvement of different functional groups in the development process, and (3) the involvement of different hierarchical levels in the development process. Respondents were asked to link particular statements they made to statements of other respondents on the same issues or problems and to give examples, where appropriate.

A major benefit from this approach was that it was possible to interview more people than originally planned. Respondents mentioned names of relevant actors and were willing to help set up interviews with them. It was thus possible to interview the relevant actors in each of the ICV cases studied and to record the convergence and divergence in their views on various key problems and critical situations throughout the development process. In some cases, it was necessary to go back to a previous respondent to clarify issues or problems, and this was always possible. After completing an interview, the interviewer made a typewritten copy of the conversation. All in all, about 435 legal-size pages of typewritten field notes resulted from these interviews.

The research also involved the study of documents. As could be expected, the ICV project participants relied little on written procedures in their day-to-day working relationships with other participants. One key set of documents, however, was the set of

Table 1. Distribution of persons interviewed, by job title.

Top management of the New Venture Division (NVD)	
Director of NVD	2
Director of corporate R&D Department	1
Director of Business Research Department	1
Director of Business Development Department	2
Participants from corporate R&D Department	
R&D managers	4
Group leaders	10
Bench scientists	6
Participants from Business Research Department	
Business managers	2
Business researchers	4
Participants from Business Development Department	
Venture managers	5
Business managers	1
Technology managers	3
Group leaders in venture R&D group	3
Marketing managers	4
Marketing researchers	2
Operations managers	4
Project managers	1
Administration of NVD	
Personnel managers	1
Operations managers	1
Participants from other operating divisions	
R&D managers	1
Group leaders	2
Corporate management	
Executive staff	1
Total	61

written corporate long-range plans concerning the NVD and each of the ICV projects. After repeated requests, I received permission to read the plans on site and to make notes. These official descriptions of the evolution of each project between 1973 and 1977 were compared with the interview data.

Finally, occasional behavioral observations were made, for example when other people would call or stop by during an interview or in informal discussions during lunch at the research site. These observations, though not systematic, led to the formulation of new questions for further interviews.

A PROCESS MODEL OF ICV

A Stage Model

As the research progressed, four stages of ICV development were identified—a conceptual, a pre-venture, an entrepreneurial, and an organizational stage. Table 2 indicates the stages reached in each project, the number of projects observed for each stage, and the number of real time observations of each stage.

Table 2. Stages of development reached by six ICV projects.

Project	Conceptual	Pre-venture	Stages Entrepreneurial	Organizational
Medical equipment	*	*	*	*
Environmental systems	*	*	*	
Farming systems	*	*	*	
Improved plastics	*	*		
Fibre components	*	*		
Fermentation products	*			
Projects observed	6	5	3	1
Real time observations	1	2	2	1

Note: An asterisk indicates that the project reached this stage prior to the conclusion of the study.

This research design thus resulted in seven case histories. At the project level, the comparative analysis of the six ICV cases allowed the construction of a grounded stage model that described the sequence of stages and their key activities. At the level of the corporation, the research constituted a case study of how one diversified major firm went about ICV and how the corporate context influenced the activities in each stage of development of an ICV project.

A stage model describes the chronological development of a project. It provides a description of the development activities and problems in a series of stages, which is convenient for narrative purposes. Such a model, however, is somewhat deceptive because it does not capture the fact that strategic activities take place at different levels in the organization simultaneously as well as sequentially and, sometimes, in a different order than would be expected.

ICV Process

The process-model approach proposed by Bower (1970) for strategic capital investment projects permits one to connect the project and corporate level of analysis and to depict simultaneous as well as sequential strategic activities. Subsequent research has established the usefulness and generalizability of the process-model approach for conceptualizing strategic decision making in and around projects other than capital investment in large, complex firms (Hofer 1976; Bower and Doz 1979).

The inductively derived process model for ICV at GAMMA presented below shows how managers from different generic levels in the organization got involved in the development of ICV projects. The first step was to map the stages of ICV development onto the *definition and impetus* processes of the model. The definition process encompassed the activities involved in articulating the technical-economic aspects of an ICV project. Through the impetus process, it gained and maintained support in the organization. Definition and impetus were identified as the *core* processes of ICV.

The second step was to map the corporate-level findings onto the *strategic context* and *structural context* determination processes, which make up the corporate context in which ICV development takes shape. Structural context refers to the various organizational and administrative mechanisms put in place by corporate management to implement the current corporate strategy. It operated as a selection mechanism on the strategic behavior of operational and middle-level managers. Strategic context determination refers to the process through which the current corporate strategy was extended to accommodate the new business activities resulting from ICV that fell outside the scope of the current corporate strategy. Strategic and structural context determination were identified as the *overlaying* processes of ICV.

The third step was the documentation of the managerial activities that constitute these different processes.

Figure 2 maps the activities involved in ICV onto the process model. It shows how the strategic process in and around ICV is constituted by a set of key activities (the shaded area) and by a set of more peripheral activities (the nonshaded area). These activities are situated at the corporate, NVD, and operational levels of management.

Figure 3, which can be superimposed on Figure 2, shows how these different activities interlock with each other, forming a pattern of connections. The relative importance of activities is indicated by the different types of line segments. The data also suggested a sequential flow of activities in this pattern, as indicated by the numbers in Figure 3.

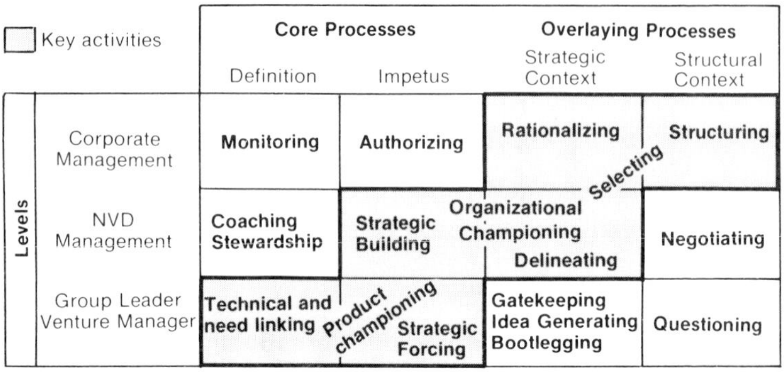
Key activities
Core Processes
Overlaying Processes
Definition
Impetus
Strategic Context
Structural Context
Levels
Corporate Management
NVD Management
Group Leader Venture Manager
Monitoring
Authorizing
Rationalizing
Structuring
Selecting
Coaching Stewardship
Strategic Building
Organizational Championing
Delineating
Negotiating
Technical and need linking
Product championing
Strategic Forcing
Gatekeeping
Idea Generating
Bootlegging
Questioning

Fig. 2. Key and peripheral activities in a process model of ICV.

Figure 3 shows that ICV is primarily a bottom-up process and depicts the key role performed by middle management. Looking at Figure 3, entrepreneurial activities at the operational and middle levels (1, 2, 3) can be seen to interact with the selective mechanisms of the structural context (5). These selective mechanisms can be circumvented by activating, through organizational championing (6), the strategic context, which allows successful ICV projects to become retroactively rationalized by corporate management in fields of new business delineated by the middle level (7, 8). These parts of the pattern, represented by the full line segments in Figure 3, constitute the major forces generated and encountered by ICV projects.

The finely dotted lines in Figure 3 (4, 9) represents the connection between the more peripheral activities in the ICV process and their linkages with the key activities. Corporate management was found to monitor the resource allocation to ICV projects. Middle-level managers managed these resources and facilitated collaboration between R&D and business people in the definition of new business opportunities; however, these activities seemed to support, rather than drive the definition process. In the same fashion, authorizing further development was clearly the prerogative of corporate management, but this was a result, not a determinant of the impetus process. In the strategic context determination process, gatekeeping, idea generating, and bootlegging activities by operational level participants were all found to be important in developing a basis for further definition processes but seemed to be more a result of the process than a determinant of it. In the process of structural context determination, questioning of the structural context by

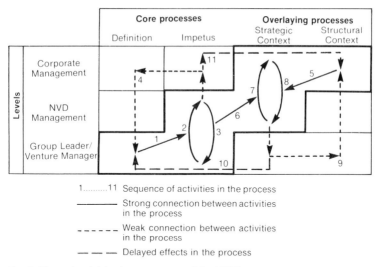

1.........11 Sequence of activities in the process

———————— Strong connection between activities
in the process

— — — — — Weak connection between activities
in the process

— — — Delayed effects in the process

Fig. 3. Flow of activities in a process model of ICV.

operational level participants and efforts by middle managers to negotiate changes in it seemed to be reactive rather than primary.

The broken line segments in Figure 3 (10, 11) indicate two important delayed effects in the ICV process. First, the successful activation of the process of strategic context determination encouraged further entrepreneurial activities at the operational level, thus creating a feedforward loop to the definition process (10). Second, corporate management attempted to influence the ICV process primarily through its manipulations of the structural context. These manipulations appeared to be in reaction to the results of the previously authorized ICV projects. This created a feedback loop (11) between the core and overlaying processes.

Figures 2 and 3 and the preceding overview of the process model can now serve as a road map for detailed examination of the interlocking key activities that constitute the major driving forces in the four processes—definition, impetus, strategic context determination, and structural context determination—that together constitute ICV.

DEFINING NEW BUSINESS OPPORTUNITIES

The case data of the present study suggest that the definition process of an ICV project encompasses the conceptualization and pre-venture stages of the development process. As the definition process takes shape, an idea for a new business opportunity evolves into a concrete new product, process, or system around which a pre-venture team of R&D and business people is formed. As a result of the successful technical and market development efforts of this pre-venture team, a project grows into an embryonic business organization. These stages take place in the context of the corporate R&D department. Critical for the definition of new business opportunities are *linking processes* and *product-championing* activities.

Linking Processes

In all of the cases studied, the initiation of the definition process involved a double linking process. Technical linking activities led to the assembling of external and/or internal pieces of technological knowledge to create solutions for new, or known but unsolved, technical problems. Need linking activities involved the matching of new technical solutions to new, or poorly served, market needs.

In five out of six cases, the definition of the new business opportunity had its origin in technical linking activities in the context of ongoing research activities in the corporate R&D department. In the Fibre Components case, the idea came from a business-oriented manager, but once the idea was to be made concrete, technical linking activities began to dominate the definition process there, too. This suggests "technology first" (Schön 1967) as the dominant mode of conceiving of a new venture. However, the case data also suggest that the continued viability of a project depended to a very great extent on the integration of technical and marketing considerations in the definition process.[2]

An important characteristic of ICV project definition was its autonomy from current corporate strategy. ICV project initiators perceived their initiatives to fall outside the current strategy, but felt that there was a good chance for them to be included in future strategic development if they proved to be successful. For instance, in the Improved Plastics project, SURF was a process through which cheap plastics—a major business of GAMMA—could be given certain properties of expensive plastics. However, since knowledgeable and influential people at GAMMA were convinced that SURF could not work because it was too violent a process, it was very difficult to obtain formal support for work in this area.[3] The leader of the efforts in SURF persisted, however, and was capable of developing an application of the process with plastic aerosol bottles. Later on, it turned out that they had focused their efforts on the wrong size bottles for commercial application, but in the meantime a basis for corporate support had been demonstrated.

The key position in the definition process turned out to be that of group leader, a first-line supervisory position, in the corporate R&D department. This person had sufficient direct involvement in the research activities to perform technical linking activities, sufficient contact with the business side to be aware of market needs and start the need linking activities, and sufficient experience of the corporate tradition to know what might be included in corporate strategy.[4] Fermentation Products, Improved Plastics, Farming Systems, and Medical Equipment all clearly illustrated the importance of the group leader in the definition process. Fibre Components and Environmental Systems involved higher levels of management in a very superficial way in the initiating phase, but it was the group leader who was able to perform the concrete linking activities, and the higher level involvement soon became very remote even in these two cases.

Product Championing: Linking Definition and Impetus

Because group leaders were most deeply involved in the definition process, they tended to take on the product-championing activities (Schön 1967) that formed the connection between the definition and impetus processes. Product championing was required to turn a new idea into a concrete new project in which technical and marketing development could begin to take shape. These activities required the ability to mobilize the resources necessary to demonstrate that what conventional corporate wisdom had classified as impossible was, in fact, possible. To overcome difficulties in resource procurement resulting from this conventional wisdom, product champions acted as scavengers, reaching for hidden or forgotten resources to demonstrate feasibility. SURF, for instance, demonstrated the validity of its need for pumps by using modified pumps from the corporate reserve list.[5]

Product championing also set the stage for the impetus process by creating market interest in the new product, process, or system while, from the corporate point of view, it was still in the definition process. To do so, the product champion sometimes cut corners in corporate procedures, as in a case where unauthorized selling efforts were started from the R&D site before the project had become an official venture.[6] ICV projects of the nature investigated in this study thus had to be fought for by their originators. Hiding their efforts until they could show positive results clearly had survival value for product champions. Once such positive results were available, however, pressure began to build to give a project venture status and to transfer it to the business development department, where the impetus process took further shape.

The importance of product championing was especially clear in the cases where it was lacking. In the Fibre Components case, a product champion had not yet emerged, and this hampered the momentum of the project. The more careful balance between the technical and business considerations fostered in this case seemed to make the emergence of a champion more difficult. In Improved Plastics, the original product champion returned to more basic research, and the subsequent reorganization of the pre-venture team with greater balance between R&D and business people made the emergence of a new product champion more difficult. In the Farming Systems, Environmental Systems, and Medical Equipment cases, however, a product champion was able to develop a single product or system around which an embryonic business organization could be formed.

IMPETUS

The impetus process of an ICV project encompasses the entre-preneurial and organizational stages of development. Major impetus was received when a project was transferred with venture status to the business development department. At this time it acquired its own organization, general manager, and operating budget, thus becoming an embryonic new business organization in the department. In the course of the impetus process, the embry-onic business grew into a viable one-product business and then, possibly, into a more complex new business with several products. The impetus process reached its conclusion in the decision to integrate this new unit into the operating system of the corporation as a freestanding new division or as a major new department of an existing division. The data indicate that there were no clear general criteria that guided the decisions to transfer projects to the business development department. Although formal screening models existed and the participants in all cases were very able in quanti-tiative analysis, there was little reliance on formal analytical tech-niques in the ICV process. This is understandable, since each project was unique and could not easily be judged by prior experi-ence. Not surprisingly, the transfer decision thus tended to be greatly influenced by the success of the product-championing activi-ties. The latter allowed a project to reach a threshold level of commercial activity which, in turn, created pressure for it to be given venture status. Farming Systems, Environmental Systems, and Medical Equipment all manifested this pattern. The data on these cases also indicate that after a project was transferred, its further development was highly dependent on the combination of *strategic forcing* and *strategic building* activities and their corollary forms of *strategic neglect*. These activities together give shape to the impetus process.

Strategic Forcing

In the first phase of the impetus process, product-championing activities were transformed into strategic forcing by the entrepreneurial venture manager. This transformation happened naturally, because, in the cases studied, the product champion had become the venture manager. Even though normative theory might question this practice, there were very strong pressures to let the technically oriented product champion become the venture manager. These pressures were in part motivational, because product champions were attracted by the opportunity to become general managers, but they also resulted because there was nobody else around who could take over and maintain momentum. Strategic forcing required that the venture manager concentrate his efforts on the commercialization of the new product, process, or system. In particular, it required a narrow and short-term focus on market penetration.

The Medical Equipment case illustrates successful strategic forcing. Under the impulse of a product champion/venture manager, this ICV project doubled its sales volume each year for five consecutive years. This created the beachhead for further development into a new, mature business.[7] Such successful strategic forcing created a success-breeds-success pattern that allowed the new venture to maintain support from top management and facilitated collaboration from people in other parts of the corporation who liked to be part of the action of a winner. In addition, the success of strategic forcing allowed the emerging venture organization to acquire substantial assets that could not easily be disposed of, thus committing the corporation.[8]

The Environmental Systems case, on the other hand, illustrates unsuccessful strategic forcing. In this case, premature commercialization caused strategic forcing to degenerate into mere selling, and technical people were forced to spend their time correcting the technical flaws of systems already sold. The resulting failure-breeds-failure pattern led first to a reduction of the control of the product champion/venture manager, then to management-by-committee, then to the termination of the venture.

The corollary of successful strategic forcing, however, was strategic neglect of the development of the administrative framework of the new venture. Strategic neglect refers to the more or less deliberate tendency of venture managers to attend only to performance criteria on which the venture's survival is critically dependent, that is, those related to fast growth. To carry out the strategic forcing efforts, the entrepreneurial venture manager attracted or was assigned generalist helpers who usually took care of more than one of the emerging functional areas of the venture organization. This was inexpensive and worked sufficiently well until the volume of activity grew so large that operating efficiency became an important issue. Also, as the new product, process, or system reached a stage of maturity in its life cycle, the need for additional new product development was increasingly felt. To deal with the operating problems and to maintain product development, some of the generalists were replaced with functional specialists who put pressure on the entrepreneurial venture manager to pay more attention to administrative development. In the cases studied, this led to severe friction between the venture manager who continued to be pressured by forces in the corporate context to maintain a high growth rate, and the functional specialists.[9]

In the successful Medical Equipment venture, the venture manager neglected the administrative development of the venture and experienced increasingly strong conflicts with the professional functional managers brought in to replace the generalists. This became a problem especially in manufacturing. The venture manager also neglected to maintain close relationships with the corporate R&D group and focused everything on development efforts related to the original product. The venture R&D group, seeking its own identity, sealed itself off from corporate R&D.[10] One of the problematic results of this was that the flow of new product development never got under control. Eventually, the organizational problems and the difficulties in new product development required the replacement of the venture manager.

This study of ICV thus reveals an important dilemma in the process of radical corporate innovation. Successful strategic forcing is required if a project is to gain and maintain impetus in the corporate context. Yet, the very success of strategic forcing seems to imply strategic neglect of the administrative development of the venture. This, in turn, leads to the ironic result that the new product development may become a major problem, and to the tragic result that the entrepreneur may become a casualty in the process of gaining a beach head for the venture.

Strategic Building

Successful strategic forcing was a necessary, but not sufficient, condition for the continuation of the impetus process. Strategic forcing had to be supplemented by strategic building activities if the project was to overcome the limitations of a one-product venture and maintain the growth rate required for continued support from corporate management. Strategic building took place at the level of the business development (BD) department manager (the venture manager's manager). Thus, consistent with Kusiatin's (1976) and von Hippel's (1977) findings, the present study identifies the venture manager's manager as a key position in the ICV process.

Strategic building involved the articulation of a master strategy for the broader field of new business development opened up by the product champion/venture manager and the implementation of this strategy through the agglomeration of additional new businesses with the original venture. This involved negotiating the transfer of related projects from other parts of the corporation and/or acquisition of small companies with complementary technologies from the outside.

The Medical Equipment case illustrates successful strategic building. From year to year, the written long-range plans showed an increase in depth of understanding of what the real opportunity was. Strategic plans grew more specific, and there was a progression in identifying problems and solving them. Based on this articulation of the principles underlying success, the BD manager negotiated the transfer of one major medically related project from one of the divisions and was able to identify suitable acquisition candidates and convince top management to provide the resources to get them.

Strategic building was iterative in nature. The evolving master strategy reflected the learning-by-doing that resulted from the assessment of the success of the strategic forcing efforts of the venture manager. The BD manager learned to understand the reasons for the success of these efforts and used this insight to further articulate the strategy. This, in turn, increased his credibility and provided a basis on which to claim further support of the venture.[11]

The Environmental Systems case illustrates how failure to understand the nature of the opportunity prevented further progress. Over a five-year-period, the long-range plans remained vague about what the opportunity was. There was no progress in terms of identifying and then solving problems. An acquisition was actually made, but it turned out to be as much technically flawed as the original system around which the venture was formed.

The Farming Systems case illustrates how the impetus received from fairly successful strategic forcing can slow down, and even halt, when strategic building is lacking. Only after a new BD manager took over and an analysis was made of the underlying principles of the business opportunity did the impetus process pick up again. The new BD manager discarded the original product, which had been the vehicle for strategic forcing, and articulated a new master strategy that led first to the redirection of the R&D efforts and then to the acquisition of two small companies with complementary technology.

Strategic building, like strategic forcing, was accompanied by strategic neglect in the Medical Equipment case. Because forces in the corporate context emphasized fast growth, the BD manager got absorbed in the search and evaluation of companies that could be acquired, in negotiations with divisions to transfer related projects, and in courting top management. The coaching of the venture manager was, again more or less deliberately, neglected, which seemed to suit the venture manager. As a result, the emerging administrative problems in the venture organization deteriorated from petty and trivial to severe and disruptive, and some high-quality people left the venture.

The personal orientations of the venture managers further reinforced this tendency in the cases in my study. The venture manager of Medical Equipment complained about a lack of guidance from the BD manager, but he also pointed out that the situation gave him leeway for his mistakes. Furthermore, he pointed out that because the venture was growing very fast, there was little time for coaching. He also admitted that his style was probably considered a bit "adversarial" by the BD manager, and that this did not facilitate the coaching process.

The venture manager of Environmental Systems also complained about a lack of guidance.[12] This manager, however, admitted that he had been eager to get the venture manager's job in spite of his lack of experience. Others in the venture organization pointed to this manager's stubbornness and lack of responsiveness to others' inputs.

The present study thus suggests a second important dilemma in the strategic management process. The BD manager can spend more time trying to guide the impetuous venture manager, but this may both interfere with the strategic forcing efforts of the venture manager and limit the time available to the BD manager for strategic building activities. Or, he can leave the venture manager alone and let him run his course until the problems in the growing venture organization require his replacement, but by that time the venture itself should have reached a viable position in terms of commercial activity. The data suggest that the forces exerted by the corporate context—the emphasis on fast growth—seem to favor the second of these possibilities.

Successful strategic forcing and strategic building created a new business organization with several products and a sales volume of about 35 million dollars in the case of Medical Equipment, but important managerial problems remained to be solved. First, the effects of the strategic neglect of the administrative framework of the venture became particularly pronounced. This administrative instability was exacerbated by the fact that there was not yet a strong common orientation, and there was still a lot of opportunistic behavior on the part of some key participants in the venture organization, who seemed to work more to improve their resumes to get a better position elsewhere than for the overall success of the venture. Also, the delayed effects of the strategic neglect of new product development in the original area of business manifested themselves. Furthermore, strategic building efforts had led to the creation of a complex new business organization, where growth could no longer be maintained solely by the hard work of the venture manager. New strategies for the different business thrusts had to be generated by the organization, but this required that people work in a strategic planning framework in which the concerns of the different new business thrusts could be traded off and reconciled, and the participants were still learning to do this.

In addition to these internal managerial problems, this new venture also had to cope with the problem of securing its position in the corporation. The venture's size made it visible in the external and internal environments, and corporate management became increasingly aware of the differences in modus operandi between the new business and the rest of the corporation and of the effects of these differences on the corporate image. NVD management thus was faced with the problem of convincing corporate management that the new venture was compatible with the rest of the corporation and was moving toward institutionalization.

STRATEGIC CONTEXT

For institutionalization to take place, an area of new venturing must become integrated into the corporation's concept of strategy. Adaptation of corporate strategy at GAMMA involved complex interactions between managers of the NVD and corporate management in the process of *strategic context* determination.

Strategic context determination refers to the political process through which middle-level managers attempt to convince top management that the current concept of strategy needs to be changed so as to accommodate successful new ventures. Strategic context determination constitutes an internal selection mechanism that operates on the stream of autonomous strategic behavior in the firm. The key to understanding the activation of this process is that corporate management knows when the current strategy is no longer entirely adequate but does not know how it should be changed until, through the selection of autonomous strategic initiatives from below, it is apparent which new businesses can become part of the business portfolio.[13]

Critical activities in this process involve *delineating* new fields of business development and *retroactive rationalizing* of successful new venture activities. The link between the process of strategic context determination and the impetus process of a particular new venture is constituted by *organizational championing* activities.

Organizational Championing: Linking Impetus and Strategic Context Determination

The case data indicate that during the impetus process, organizational championing activities became the crucial link between the emerging new business organization and the corporate context. Organizational championing involved the establishment of contact with top management to keep them informed and enthusiastic about a particular area of development. This, in turn, involved the ability to articulate a convincing master strategy for the new field, so as to be able to communicate where the development was leading and to explain why support was needed for major moves. These activities were also performed at the level of the business development manager.

Organizational championing was, to a large extent, a political activity. The BD manager committed his judgment and put his reputation on the line. Astute organizational champions learned what the dispositions of top management were and made sure that the projects they championed were consistent with the current corporate strategy. More brilliant organizational champions were able to influence the dispositions of top management and make corporate management see the strategic importance of a particular new business field for corporate development.

Organizational championing required more than mere political savvy, however. It required the rare capacity to evaluate the merit of the proposals and activities of different product champions in strategic rather than in technical terms. Thus, in the Medical Equipment venture, a sound master strategy for the new venture and corresponding strategic building moves allowed the organizational champion to convince top management that the medical field was an attractive and viable one for the corporation. In the Environmental Systems venture, on the other hand, the failure to come up with a master strategy prevented the organizational champion from obtaining the resources needed to straighten out the technological problems of the new venture and prevented him from engaging in strategic building. His organizational championing was limited to gaining more time, but eventually top management concluded that the opportunity just wasn't there. Finally, in the Farming Systems venture, new impetus was developed as a result of the involvement of the same person who was the organizational champion in Medical Equipment.

Delineating

Through organizational championing based on strategic building middle-level managers were capable of delineating in concrete terms the content of new fields of business development for the corporation. It is a critical finding of this study that these new fields became defined out of the agglomeration of specific commercial activities related to single new products, processes, or systems, developed at the level of venture projects rather than the other way around. Delineating activities were thus iterative and aggregative in nature. This was clearly reflected in the written long-range plans of the NVD in 1975, which stated: "Instead of dealing with an ever-growing number of separate arenas, the NVD should henceforth focus its attention on a critical few major fields, within each of which arenas may be expanded, grouped together, or added."

Retroactive Rationalizing

To be sure, corporate management, too, got involved in the process of strategic context determination. Top management gave indications of interest in venture activity in certain general fields and expressed concern about the fit of ongoing ICV activities with corporate resources and strategy. In the final analysis, however, corporate management's role was limited to rejecting or rationalizing, retroactively, the ICV initiatives of lower-level participants in fields delineated by middle-level management.

These findings corroborate and extend the findings of previous research. They confirm the critical role of middle-level managers in shaping the strategy of internal development in the diversified major firm (Kusiatin 1976). More generally, these findings also extend Kimberly's (1979) observation of the paradox that the success of a new nonconformist unit creates pressures in the larger organizational context toward conformity, thereby affecting the very basis of success. Entrepreneurial and institutional existence seem to be inherently discrete states and middle-level management needs to bridge the discontinuity.

STRUCTURAL CONTEXT

Given the limited substantive involvement of corporate management in the process of strategic context determination, how do they try to exert control over the ICV process? The present study suggests that they did so by *structuring* an internal selection environment.

Structuring

As in the situation studied by Bower (1970), corporate management relied on the determination of the structural context in its attempts to influence the strategic process concerning ICV. The structural context includes the diverse organizational and administrative elements whose manipulation is likely to affect the perception of the strategic actors concerning what needs to be done to gain corporate support for particular initiatives. The creation of the NVD as a separate organizational unit, the definition of positions and responsibilities in the departments of the NVD, the establishment of criteria for measuring and evaluating venture and venture-manager performance, and the assignment of either entrepreneurially or administratively inclined managers to key positions in the NVD all seemed intended to affect the course of ICV activity.

The corporate level seemed dominant in the determination of structural context. Corporate management's manipulations of the structural context seemed to be guided primarily by strategic concerns at their level, reflecting emphasis on either expansion of mainstream businesses or diversification, depending on perceptions at different times of the prospects of current mainstream businesses.

These changes in structural context did not reflect a well-conceived strategy for diversification, however, and seemed to be aimed at consolidating ICV efforts at different levels of activity rather than at guiding and directing these efforts. The NVD was created in the early seventies because people in the divisions had been engaging in what some managers called a "wild spree" of diversification efforts. Corporate management wanted to consolidate these efforts, although at a relatively high level of activity. Key managers involved in those earlier decisions pointed out that the direction of these consolidated efforts was based on preceding lower level initiatives that had created resource commitments, rather than on a clear corporate strategy of diversification.

The lack of a clear strategy for directing diversification was also evident in 1977, when significant changes in the functioning of the NVD took place. The newly appointed NVD manager pointed out that corporate management had not expressed clear guiding principles for further diversification beyond the emphasis on consolidation and the need to reduce the number of fields in which ICV activity was taking place.

Selecting

Structural context determination thus remained a rather crude tool for influencing ICV efforts. It resulted in an internal selection environment in which the autonomous strategic initiatives emerging from below competed for survival. In all the ICV cases, strong signals of fast growth and large sizes as criteria for survival were read into the structural context by the participants. This affected the process, if not so much the specific content of their behavior. The importance of product championing, strategic forcing, strategic building, and the corresponding forms of strategic neglect would seem to indicate this. The inherent crudeness of the structural context as a tool for influencing the ICV process provided, of course, the rationale as well as the opportunity for the activation of the strategic context determination process discussed earlier.

CONCLUSIONS AND IMPLICATIONS

The preceding discussion of a process model of ICV does not, to be sure, treat the entire range of phenomena associated with new ventures (Roberts 1980) and corporate entrepreneurship (Peterson

1981). Reasons of focus as well as space constraints prevent discussion of issues such as management of the interfaces between business and R&D people and structural and managerial innovation associated with the separate new venture division.

The purpose here has been to construct a grounded model and to use this model as a framework for insights into the generative mechanisms of one form of corporate entrepreneurship in one type of large business organization. Verification is necessary to identify the generalizable relationships embedded in the process model generated in this paper and to identify the contingency factors that might explain variance across organizations in these relationships. The major insights gained from this exploratory study of the ICV process are recapitulated below and some major implications are briefly discussed.

First, the findings suggest strongly that the motor of corporate entrepreneurship resides in the autonomous strategic initiatives of individuals at the operational levels in the organization. High-technology ventures are initiated because entrepreneurially inclined technologists, usually at the group-leader level, engage in strategic initiatives that fall outside the current concept of corporate strategy. They risk their reputations and, in some cases, their careers, because they are attracted by the perceived opportunity to become the general manager of an important new business in the corporation. This stream of autonomous strategic initiatives may be one of the most important resources for maintaining the corporate capability for renewal through internal development. It constitutes one major source of variation out of which the corporation can select new products and markets for incorporation into a new strategy.

Second, because of their very nature, autonomous initiatives are likely to encounter serious difficulties in the diversified major firm. Their proponents often have to cope with problems of resource procurement, because they attempt to achieve objectives that have been categorized by the corporation as impossible. Because such initiatives require unusual, even unorthodox, approaches, they create managerial dilemmas that are temporarily resolved through the more or less deliberate neglect of administrative issues during the entrepreneurial stage. The success of the entrepreneurial stage thus depends on behaviors that, paradoxically, have a high probability of eliminating the key actors from participation in the organizational stage. There seems to be an inherent discontinuity in the transition from entrepreneurial to institutionalized existence, as well as a possible asymmetry in the distribution of costs and benefits for the actors that may underlie the myth of the entre-preneur as tragic hero in the large corporation.

Third, the study of ICV elucidates the key role of middle-level managers in the strategy-making process in the diversified major firm. The venture manager's manager performs the crucial role of linking successful autonomous strategic behavior at the operational level with the corporate concept of strategy. Both the continuation of the impetus process of a particular ICV project and the change of the corporate strategy through the activation of the process of strategic context determination depend on the conceptual and political capabilities of managers at this level. The importance of this role seems to confirm the above-mentioned discontinuity between entrepreneurial activity and the mainstream of corporate activity.

Fourth, corporate management's role in the ICV process seems to be limited to the retroactive rationalization of autonomous strategic initiatives that have been selected by both the external environment at the market level and the internal corporate environ-ment. Top management's direct influence in the ICV process is through the manipulation of structural context. These manipu-lations, however, seem to be predicated less on a clearly formulated corporate strategy for unrelated diversification than on concerns of consolidation. Ironically, from this perspective, the estab-lishment of a separate new venture division may be more a mani-festation of corporate management's uneasiness with autonomous strategic behavior in the operating system than the adaptation of the structure to implement a clearly formulated strategy. The present study thus suggests that the observed oscillations in ICV activity at GAMMA may have been due to the lack of articulation between these manipulations of the structural context and a cor-porate strategy for unrelated diversification. It also provides further corroboration for the similar findings of Fast (1979) on the unstable position of NVDs in many corporations and for Peterson and Berger's (1971) suggestion that top management may view cor-porate entrepreneurship more as insurance for coping with per-ceived environmental turbulence than as an end in itself.

Implications for Organization Theory and Strategic Management

The research findings presented in this paper can be related to the current discussions in organization theory of the validity of rational versus natural selection models to explain organizational growth and development (Pfeffer and Salancik 1978; Aldrich 1979; Weick

1979). Relatively successful, large diversified major firms like GAMMA would seem to be representative of the class of organizations that have sufficient control over their required resources to escape, to a great extent, the tight control of external selection and to engage in strategic choice (Child 1972; Aldrich 1979). The detailed, multilayered picture of the strategic management process presented in this paper suggests, however, that these strategic choice processes, when exercised in radical innovation, take on the form of experimentation and selection rather than strategic planning. This is fundamentally different from the view that administrative systems "program" their own radical change (Jelinek 1979).

Further research is needed to establish the conditions under which different systems for innovation in organizations can be adequate. The limited evidence of the present study, however, suggests that the tight coupling implied in the institutionalized approach may be inadequate for organizations with multiple, mostly mature technologies in their operating system. In an organization like GAMMA, there seems to be relatively little opportunity for generating radical innovation from within the operating system through the imposition of a strategic planning approach.

Large, complex business organizations have separate variation and selection mechanisms. Previously unplanned, radically new projects at the product/market level are generated from the relatively unique combination of productive resources of such firms. Not all of these projects are adopted, not so much because the market may turn out to be unreceptive but because they must overcome the selection mechanisms in the internal administrative environment of the firm, which reflect, normally, the current strategy of the corporation, i.e. the retained wisdom of previously selected strategic behavior. Thus, the experimentation and selection model draws attention to the possibility that firms may adopt externally unviable projects or may fail to adopt externally viable ones and provides a clue to why firms occasionally produce strange innovations.[14] This analysis posits a conceptual continuity between internal and external selection processes, analogous to Williamson's (1975) analysis of external and internal capital markets, to explain the existence of the conglomerate form of the divisionalized firm. Because corporate entrepreneurship, as exemplified by the ICV activities in this paper, seems to differ from traditional individual entrepreneurship, as well as from traditional organizational economic acvivity, it may be necessary to devise different arrangements between the corporate resource providers and their entrepreneurial agents. Further research, both theoretical and empirical, would seem useful here.

The insights generated by the present study also have some implications for further research on the management of the strategy-making process in general. Comparative research studies of a longitudinal-processual nature, carried out at multiple levels of analysis, are necessary to document and conceptualize the multi-layered, more or less loosely coupled network of interlocking, simultaneous, and sequential key activities that constitute the strategy-making process. Following Bower (1970), the present study has found it useful to focus the research on a particular strategic project rather than on the strategy-making process in general. This is consistent with Quinn's (1980, 52) observation that top managers "deal with the logic of each subsystem of strategy formulation largely on its own merits and usually with a different subset of people." A concrete focus, it would seem, is more likely to produce data on the vicious circles, dilemmas, paradoxes, and creative tensions that are embedded in the strategy-making process.

Comparative analysis of process models of various strategic projects could produce grounded concepts and categories that would initially be somewhat rudimentary and evocative. Hopefully, these would stimulate the imagination of other scholars and provide the base for more formal and precise concepts of managerial activity in the strategy-making process. Eventually, this could lead to a general theory of the management of strategic behavior in complex organizations and to the conceptual integration of content and process, formulation and implementation.

The present study may then be viewed as an attempt to augment the substratum of rudimentary and evocative concepts and categories. One result of this attempt is the identification of the new concepts of autonomous strategic behavior and strategic context determination and categories of key strategic activities. Further research along these lines may be able to provide a clearer understanding of the interactions between strategy, structure, and managerial activities and skills.

NOTES

1 This paper is based on the author's doctoral dissertation, which received a Certificate of Distinction for Outstanding Research in the field of Strategic Management, Academy of Management and General Electric Company, 1980. L. Jay Bourgeois, Arthur P. Brief, David B. Jemison, Leonard R. Sayles, Stephen A. Stumpf, and Steven C. Wheelwright have made useful comments on earlier drafts of this paper. The constructive comments of three anonymous

ASO reviewers have contributed significantly to this final version. Support from the Strategic Management Program of Stanford University's Graduate School of Business is gratefully acknowledged. My thanks also to Barbara Sherwood for excellent administrative assistance.

2 This is how the originator of the medical equipment venture recounted a story that illustrated the importance of integrating technical and marketing considerations in the definition process: "In 1968, we had a think tank session in Connecticut. A scientist from our government-sponsored lab, I found out, was working on a new way to handle and transfer blood samples, an entirely new concept . . . but the scientist had very fixed ideas about how the product should look as a commercial product . . . . An outside group also had discovered the existence of the scientist's idea and followed closely his recommendations. It was a small company, with a sales volume of some eight million dollars. I decided not to make 'Chinese copies' of their approach. I insisted on doing market research, and actually spent two months full-time doing this. We ended up with a radical departure from the scientist's approach; we used only the nucleus of his physical concepts. We had found out some advisable product characteristics from our market research, which led, for instance, to a broader-sized 'reader'. We also combined the analyzer with a computer." A further discussion of how this integration is achieved and of the issues related to the collaboration between R&D and business people in the definition process is provided in Burgelman (1980).

3 In the words of the group leader: "As with most new ideas, people would give little time to it. People "knew" that SURF was "unpractical," so the divisions did not really get involved, except in an informal way.

4 Argyris and Schön (1978:214) noted a similar phenomenon in their Mercury case, in which key participants were those who could recognize "a Mercury problem." In the present study, however, initiators were more concerned with avoiding the work on projects that would be perceived by top management as *not* a GAMMA problem. Projects were avoided in those areas in which there had been failures in the past, in those where there might be risk to the corporation's image, or in areas having special legal liabilities.

5 Said the group leader: "But these pumps are costly, and people at the management level are afraid to commit themselves to such outlays. At that time, however, an engineer came on the project. He knew of the corporate surplus lists and got some old pumps. We rebuilt them and showed that we could pump 35 percent to 49 percent solutions. Having showed that, we could now get the pumps we needed."

6 As the product champion in this case explained: "When we proposed to sell the ANA product by our own selling force, there was a lot of resistance, out of ignorance. Management did numerous studies, had outside consultants on which they spent tens of thousands of dollars; they looked at XYZ Company for a possible partnership. Management was just very unsure about its marketing capability. I proposed to have a test marketing phase with 20 to 25 installations in the field. We built out own service group; we pulled ourselves up by the 'bootstrap.' I guess we had more guts than sense."

7 In the words of the venture manager: "We were convinced that we could develop simultaneously domestically and internationally. We were fearless, and, management being ignorant, we just started to do it. What we did was, in fact, a parallel international new development. That made our sales 55 percent larger and allowed a larger profit fraction. If we had not done this, we might have lost the business."

8 In the words of one of the key participants in a venture: "The mechanism is to double each year your size. The next step is then to acquire assets that are not easily disposed of. Then management cannot get rid of you that easily, and you can relax if you have a bad year."

9 Arrow (1974) uses "salutory neglect" to denote the situation in which problems

for which there are no satisfactory solutions are not placed on the agenda of the organization. Strategic neglect, independently observed in the present study, has a similar meaning. Arrow points out that neglect is never productive. In the long run, and from the perspective of the larger system, this may be true, and of course the larger system will, in time, correct for neglect. From the perspective of the entrepreneurial actor, however, strategic neglect of administrative issues was the necessary cost of forcing growth.

10 In the words of one person who was transferred from corporate R&D to the venture: "We were, at the time, basically separated from the group in the venture. The group there wanted to identify itself. They did it to such an extent that they put a wall between themselves and us . . . In a way, it was ironic. We were funded by the venture, and the technology that we developed was not accepted by them!"

11 Explaining his approach, one BD manager said: "First, I look for demonstrated performance on an arbitrarily chosen—sometimes not even the right one—tactic. For instance, developing a new analyzer may not be the right move, but it can be done and one can gain credibility by doing it. So, what I am really looking for is the ability to predict and plan adequately. I want to verify your claim that you know how to predict and plan, so you need a 'demonstration project' even if it is only an experiment. The second thing that I look for is the strategy of the business. That is the most important milestone. The strategy should be attractive and workable. It should answer the questions where you want to be in the future and how you are going to get there . . . And that, in turn, allows you to go to the corporation and stick your neck out."

12 Right after his replacement, this manager observed: "I should have gotten help from my management—counseling and education. Most venture managers tend to come from the technology side because these ventures require a lot of high technology input. But in the technology area there is relatively little need for broad general management skill development. I was lacking that kind of judgment."

13 The identification of the process of strategic context determination leads to a major extension of the process model. It suggests that the corporate context is more complex than was revealed by Bower's (1970) study of strategic capital investment projects. These projects were situated in the operating system of the corporation. Even though they were clearly strategic because of the large amounts of resources involved, they did not require a change in the business portfolio of the corporation. These projects fell within the scope of and were induced by the current concept of strategy of the corporation.

14 In the course of the present study, anecdotal evidence for the emergence of very unusual projects was amply available. In one case, a scientist pulled out a file with a whole series of such abortive projects, e.g. the mining of gold from sea water.

REFERENCES

Aldrich, Howard E. 1979: *Organizations and environments.* Englewood Cliffs, N.J.: Prentice-Hall.

Ansoff, H. I. & Brandenburg, R. G. 1971: A language for organization design, Part II. *Management Science, 17*: B717–B731.

Argyris, Chris & Schön, Donald A. 1978: *Organizational learning.* Reading, MA: Addison-Wesley.

Arrow, Kenneth J. 1972: *The limits of organization.* New York: Norton.

Arrow, Kenneth J. 1982: *Innovation in large and small firms.* New York: Price Institute for Entrepreneurial Studies.

Biggadike, Ralph 1979: The risky business of diversification. *Harvard Business Review, 56,* 103–11.

Bower, Joseph L. 1970: *Managing the resource allocation process*. Boston: Graduate School of Business Administration, Harvard University.

Bower, Joseph L. & Doz, Yves 1978: Strategy formulation: A social and political view. In Dan E. Schendel & Charles W. Hofer (Eds.), *Strategic management*, 152–166. Boston: Little, Brown.

Burgelman, Robert A. 1980: Managing innovating systems: A study of the process of internal corporate venturing. Unpublished doctoral dissertation, Columbia University.

Caves, Richard E. 1980: Industrial organization, corporate strategy, and structure. *Journal of Economic Literature, 18*, 64–92.

Chandler, Alfred D., Jr. 1962: *Strategy and structure*. Cambridge, MA: MIT Press.

Child, John 1972: Organization structure, environment, and performance: The role of strategic choice. *Sociology, 6*, 1–22.

Fast, Norman D. 1979: The future of industrial new venture departments. *Industrial Marketing Management, 8*, 264–79.

Frohman, Alan L. 1978: The performance of innovation: Managerial roles. *California Management Review, 20*, 5–12.

Galbraith, Jay R. & Nathanson, Daniel A. 1979: The role of organizational structure and process in strategy implementation. In Dan E. Schendel & Charles W. Hofer (Eds.), *Strategic Management*, 261–286. Boston: Little, Brown.

Glaser, Barney J. & Strauss, Anselm L. 1967: *The discovery of grounded theory*. Chicago: Aldine.

Hanan, Mack 1976: *New venture management*. New York: McGraw-Hill.

Hofer, Charles W. 1976: Research on strategic planning: A survey of past studies and suggestions for future efforts. *Journal of Business and Economics, 28*, 261–86.

Hutchinson, John 1976: Evolving organizational forms. *Columbia Journal of World Business, 11*, 48–58.

Jelinek, Mariann 1979: *Institutionalizing innovation*. New York: Praeger.

Kimberly, John R. 1979: Issues in the creation of organizations: Initiation, innovation, and institutionalization. *Academy of Management Journal, 22*, 435–57.

Kusiatin, Ilan 1976: The process and capacity for diversification through internal development. Unpublished doctoral dissertation, Harvard University.

Maidique, Modesto A. 1980: Entrepreneurs, champions, and technological innovations. *Sloan Management Review, 21*, 59–76.

Nisbet, Robert A. 1969: *Social change and history*. New York: Oxford University Press.

Peterson, Richard A. 1981: Entrepreneurship and organization. In Paul Nystrom & William Starbuck (Eds.), *Handbook of Organizational Design*, 65–83. New York: Oxford University Press.

Peterson, Richard A. & Berger, David G. 1971: Entrepreneurship in organizations: Evidence from the popular music industry. *Administrative Science Quarterly, 16*, 97–106.

Pettigrew, Andrew M. 1979: On studying organizational cultures. *Administrative Science Quarterly, 24*, 570–81.

Pfeffer, Jeffrey & Salancik Gerald R. 1978: *The external control of organizations*. New York: Harper & Row.

Pondy, Louis R. 1976: Beyond open system models of organizations. Paper presented at the Annual Meeting of the Academy of Management, Kansas City, August 12.

Quinn, James B. 1979: Technological innovation, entrepreneurship, and strategy. *Sloan Management Review, 20*, 19–30.

Quinn, James B. 1980: *Strategies for change*. Homewood, IL: Irwin.

Roberts, Edward B. 1980: New ventures for corporate growth. *Harvard Business Review, 57*, 134–42.

Rumelt, Richard P. 1974: *Strategy, structure, and economic performance*. Boston: Graduate School of Business, Harvard University.

Salter, Malcolm S. & Weinhold, Wolf A. 1979: *Diversification through acquisition.* New York: Free Press.

Schön, Donald A. 1967: *Technology and change.* New York: Delacorte.

von Hippel, Eric 1977: Successful and failing internal corporate ventures: An empirical analysis. *Industrial Marketing Management, 6,* 163–74.

Weick, Karl E. 1979: *The social psychology of organising.* Reading, MA: Addison-Wesley.

Williamson, Oliver E. 1970: *Corporate control and business behavior.* Englewood Cliffs, N.J.: Prentice-Hall.

Williamson, Oliver E. 1975: *Markets and hierarchies.* New York: Free Press.

Wrigley, Leonard 1970: Divisional autonomy and diversification. Unpublished doctoral dissertation, Harvard University.

Zaltman, Gerald, Duncan, Robert L., & Holbek, Jonny 1973: *Innovations and organizations.* New York: Wiley.

11

Organizational Effects on Creativity: The Use of Roles in a Technical Audit of Organizations

MORRIS I. STEIN

Industrial organizations are established and capital is invested in them for profits. To achieve this goal existing goods and services must be provided that are competitive in price and quality with those already in the marketplace. Or, new and unique goods and services must be provided which satisfy societal needs and consumer demands.

In the first instance we deal primarily with a production orientation; in the second with an orientation that involves innovation and creativity. Our concern in this paper is with the second orientation and specifically with the use of the concept of roles to study organizational effects on creativity.

Historically, innovations in industry originated in various sources. They originated with inventors who not only originated a product but also used their own capital or capital they raised from others to produce, distribute and sell their inventions. Innovations were also introduced by venture capitalists who sponsored inventors or who bought up inventions and then organized companies to produce them. Institutes and universities also contained within them sub-units charged with the specific responsibility for the generation of new ideas or for the transformation of existing ones for the development of new products and processes. It was a still later development in which production companies began to contain within them Research and Development (R&D) units charged with the generation and development of novelty. A still more recent development is the investment by private companies in university professorships or academic research in exchange for early access to information that might give them the edge over their competitors in getting new products and processes to the marketplace.

Throughout these developments management's primary experience was with production. And, in such an orientation it regarded its personnel as "primarily *passive instruments*, capable of performing work and accepting direction, but not initiating action or exerting influence in any significant way" (March and Simon 1958). They also viewed their employees as individuals who "have to be motivated or induced to participate in the system or organization of behavior; that there is incomplete parallelism between their personal goals and organization goals; and that actual or potential goal conflicts make power phenomena, attitudes and morale centrally important in the explanation of organizational behavior" (March and Simon 1958).

Creative persons, however, are about as far as one can get from the "passive instruments" that production management was accustomed to. Creative persons are actively involved in transforming existing materials into new combinations; they find problems where no one else even thinks they would exist and then proceed to solve them; they look at existing order, develop disorder out of it and then create a newer and better (Barron 1958). They are autonomous, independent, inclined to take risks, playful but also achievement oriented, etc. They are rather complex psychologically but their very complexity may be critical in producing elegant problem solutions.

Management has had its problems in working with creative persons and one could hardly blame some managements for seeking corporate growth through acquisitions and mergers rather than through the novel contributions of their own scientists and engineers. But, management alone is largely to blame for the problems it experienced in this area. It was hardly as innovative in its management style as it might have been. It spent little if any time trying to understand its creative people and to boot managed them in terms of principles that from their point of view stood them in good stead with their production people.

Creative persons wanted the freedom to select their own problems and the opportunity to explore possible solutions as they saw fit. They broke out of their niches and frequently required the breaking down of corporate barriers. And when it came to rewards, yes money was important but it was not all-important since other rewards like sabbaticals, increased opportunity to select their own problems and increased responsibility for their solution, were equally or even more desirable.

What did they get? Instead of freedom or problem selection they experienced problem assignment. And those assigned were selected because competitors were working on them and not because they were original. Instead of flexible corporate structure they experienced regimentation and constant admonition to "go through channels". And, instead of rewards that would be in line with their scientific and engineering interests and development, they were treated like everyone else with periodic small monetary rewards.

Management's interaction with its creative personnel has frequently been reminiscent of how air routes were determined when airplanes were first used for commercial flights. Instead of determining routes on the basis of the airplane's characteristics—maneuverability, speed, etc.—they were determined by what already existed—railroad lines.

Creative individuals, if they are to be managed effectively, have to be understood. They have to be understood in terms of their psychological makeup and in terms of their relationships to the organizations in which they work. Treatment of their psychological makeup is treated elsewhere in this book. Our purpose here is to raise our consciousness about the organizations in which creative individuals work by providing a set of role variables which when combined with other variables and parameters can be of help in the understanding of the current organization and in the selection of individuals who can probably function best and be most creative in that organization. All of this has been combined into a *Technical Audit*. [The Stein *Technical Audit* is available by writing to the Mews Press, Drawer D, Amagansett, N.Y. 11930.]

Before presenting the parameters and variables a few words need to be said about our conceptualization of organizations that are most likely to be conducive to the development of creative ideas, processes and products (1974, 1975).

THE ORGANIZATION AS A SYSTEM

The R&D organization, if it is to be a viable organization, is an open system involved in the transformation of existing information for the development of new information. It combines and integrates both maintenance and growth functions. Maintenance functions consist largely of support systems designed to keep a company's ongoing functions operating smoothly and efficiently. But in an open system they cannot be absolutely fixed. They need to be so developed and structured that they are sufficiently flexible, adaptable and receptive to changes in other parts of the system—especially those parts designed to fulfill growth functions.

Organizational growth functions are manifest in how the organization generates new information. It is also manifest in the transport of information within substructures within the organization and between these or the total organization and other structures outside the organization. Growth functions vary in time and content. Some are short-term while others are long-term; some are concerned with basic processes and others are focused applied products.

In a smoothly running organization, maintenance functions have to be closely coordinated with growth functions. Personnel departments, with their sources for employees to fill ongoing needs, may have to develop completely new sources on rather short notice when there is a breakthrough in some scientific or engineering development. The Comptroller's office typically bound up with tried and tested techniques for the allocation and disbursement of funds may suddenly find itself confronted with new and unexpected demands for large sums of money to carry out some exciting and promising project. Secretaries in the typing pool accustomed to set working hours may suddenly find themselves coming much earlier to work and staying longer at work than has been the case heretofore.

Just as maintenance functions have to accommodate to growth functions in a smoothly running R&D organization, so it is also true that growth functions have to accommodate to maintenance functions. The bench chemist cannot always get his/her work done as rapidly as they wish in the pilot plant which has its own schedule. Marketing people do not always have the opportunity to "get with it" and test new products on demand.

In a smoothly runnning organization accommodations work out over time and both maintenance groups and growth groups "contricipate" (Stein 1984) in the total process—they *contribute* to the total process directly or they *participate* in the process by facilitating its development or appreciating its end results. Accommodations are not always smooth and easy integration is not always possible. Perfection is impossible and conflict, frustration, and discord are not infrequent. A tension-less living system is an impossibility if the organization is to maintain its steady state.

Organizations that are living systems are not expected to get maximal scores on a morale survey or a technical audit. That is an erroneous expectation that stems from a testing tradition in which a high score on an intelligence test is better than a low score. But, while living systems are not expected to get maximal scores they are expected to get *optimal* scores. Just what these "optimal scores" are frequently has to be determined empirically for each organization.

Admittedly this is an all too brief description of our conceptualization, but our purpose was only to set the stage for what is to follow and to call attention to the fact that the kind of setting we are concerned with and the kinds of people we are interested in are called upon to perform both maintenance functions and growth functions.

THE CONCEPT OF ROLES

Roles, as the sociologists use this term, refer to the expectations that one has for another. As used in the context of this paper, roles refer to the expectations management has for its scientists and engineers—what kinds of behaviors do the former expect of the latter.

Drawing on our experience with R&D organizations, we focused on five roles and developed a relatively small number of questionnaire items (13) with which to gather our information.

Roles in Industrial Research Organizations

The five roles which we focused on are: the Scientist Role, the Professional Role, the Administrator Role, the Employee Role, and the Social Role. Brief definitions for each role follow and their institutional imperatives or constraints are presented in Table 1.

The Scientist Role

In the scientist role the individual is expected to "discover, systematize, and communicate knowledge about some order or phenomena" (Hughes 1952). As a scientist the individual undertakes activities not because they will be of direct and immediate benefit to anyone who may be considered his client, but because these activities will result in more knowledge. "Scientists, in the purest case, do not have clients" (Hughes 1952).

The Professional Role

The professional, like the scientist, is trained in a specific tradition and "only members of the profession are treated as qualified to interpret the tradition authoritatively and, if it admits of this, to develop and improve it" (Parsons 1954, p. 372). This statement holds true for both the professional and the scientist. What distinguishes the two is that while the latter is concerned primarily

Table 1. Roles of the researcher in the industrial organization.

Scientist role

i. *"Universalism"* Claims for truth "are to be subjected to *pre-established impersonal criteria*: consistent with observation and with previously confirmed knowledge. The acceptance or rejection of claims entering the lists of science is not to depend on the personal or social attributes of their protagonist: his race, nationality, religion, class and personal qualities are as such irrelevant. Objectivity precludes particularism.

ii. *"Communism"* Science's substantive findings "are a product of social collaboration and are assigned to the community. They constitute a common heritage in which the equity of the individual producer is severely limited. The scientist's claim to 'his' intellectual 'property' is limited to that of recognition and esteem, which, if the institution functions with a modicum of efficiency, is roughly commensurate with the significance of the increments brought to the common fund of knowledge (Merton 1949, p. 312)". Furthermore "secrecy is the antithesis of this norm: full and open communication its enactment" (Merton 1949, p. 312).

iii. *"Disinterestedness"* Science demands objectivity and has no place for the personal and subjective motivations of the individual. To make certain of this "scientific research is under the exacting scrutiny of fellow-experts. Otherwise put . . . the activities of scientists are subject to rigorous policing, to a degree perhaps unparalleled in any other field of activity. The demand for disinterestedness has a firm basis in the public and testable character of science and this circumstance, it may be supposed, has contributed to the integrity of men of science (Merton 1949, p. 314)". "Although the scientist is not accountable to a client he is accountable to his colleagues who will impose sanctions upon him if he has personal motives to influence the course of his work and results".

iv. *"Organized Skepticism"* This imperative "involves the suspension of judgment until 'the facts are at hand' and the detached scrutiny of beliefs in terms of empirical and logical criteria . . . (Merton 1949, p. 315)".

Professional role

i. *Particularism* Although principles of universalism hold with respect to the content of the work whether it gets incorporated into the mainstream of the organization's activity may also depend on a variety of particularistic factors that involve power relationships between individuals.

ii. *Limited Communism* Results are to be shared only with certain selected individuals in the organization. Their number may vary but they have been "cleared" to receive non-public information. Results are shared with outsiders only when all legal rights are secure. Rights to work in the same field as one's employer after one has left the company may also be signed away for a period of time.

iii. *Focussed Truth* Attention is focussed on and energy is devoted to problems and products of interest to the organization. Freedom to investigate problems is limited. Guidelines are unclear for there are instances where "bootleg" work has paid off and so some proportion of time is allowed for free exploration of one's own choosing in some instances.

iv. *Selflessness* Intense motivation may be an important ingredient of quality work but when work pressures on other matters are intense one may have to interrupt one's work and be ready to accept an assignment elsewhere on some other work. Moreover the decisions may not be one's own but one's client-management.

v. *Communication in Client's Language* Clients must be communicated with in terms they understand. Technical language must be translated into lay language; lengthy and weighty experiments and procedures must be condensed in "one page or less" or else they are not read; and the bottom line must always contain a dollar figure.

vi. *Vested Interest* One must always be loyal to and maintain the interests of one's company, division, department, section and/or work group. One needs to be constantly alert to how one's group or organization can better service the client. Competitiveness with other organizations outside in the marketplace is an important factor. Within the organization competition and rivalry with other subgroups for personnel, space, and money are crucial for one's survival.

Administrative role

i. *Scientist/Professional* By virtue of the status position as an administrator, it is also assumed that the individual in this role has the knowledge and experience to give counsel and advice to other scientists and professionals in the organization. The information cannot be dated but must be current. It has to be broad in scope and also with sufficient depth that subordinates in the area respect the knowledge.

ii. *Leader/Manager of People* Proper selections of personnel at all levels and for a variety of activities must be made. Implicit or explicit predictions are to be made that they will perform properly at present as well as in the future. One must be capable of providing proper reviews and evaluations; opportunities for training into new positions; allocate space and serve as mentor and model.

iii. *Financial Responsibility* In this role one carries responsibility for the allocation of funds for ongoing work and for salaries, equipment and the like. One prepares budgets and seeks more funds as necessary.

iv. *Venture Capitalist* One seeks funds and argues for the support of new research endeavors that are of an original and unique nature. One serves as an intermediary and gatekeeper deciding between conflicting and competitive programs. One has a risk taking orientation as well as the capacity to deal with ambiguity.

v. *Entrepreneur* One evaluates existing products to determine how they might be improved to increase their marketability. One is constantly alert to profitability.

Social role

i. *Representative of the Organization* The scientist/engineer is a representative of the organization in the sense that what he does and how he behaves will reflect on the organization. Consequently proper dress and comportment must be observed at all times. In some organizations form and style in dress may follow some specific norm.

ii. *Social Sensitivity* The scientist/engineer needs to be sensitive to the various status relationship in the organization and outside of it so that each relationship gets its due. The scientist/engineer's place must be observed. Relationships here refer to both formal status relationships as well as informal power relationships.

iii. *Social Participation* The scientist/engineer must be willing and able to participate in various social activities both within the organization and outside of it. These may vary from the social activities in the organization to social, athletic, charitable, educational, etc. activities outside of the organization. In these activities knowledge of the social amenities is a requirement and must be practised.

iv. *Social Priorities* Social awareness must be such that the scientist/engineer is capable of knowing when social factors must be given priority over non-social and intellectual factors.

Employee role

i. *Consistent Productivity* Work must be produced consistently. Difficult problems may show little if any progress, consequently the scientist/engineer usually undertakes more than one problem at a time. If one is a laggard the other may show more progress.

ii. *Financial Awareness* The time to sit and think costs money and it is charged against the appropriate unit like equipment and personnel. One must always ask oneself whether and to what extent one's work will eventually "ring the cash register".

iii. *Efficiency* Time is an important ingredient in all undertakings, efficiency is therefore mandatory. From an economic point of view the best idea is one that requires a minimum of re-tooling and a minimum of re-allocation of personnel and equipment. Major personnel and material requirements must go through many stages of approval and risk being turned down.

iv. *Accepting Status Position and Adjusting to Authority* The limitations of one's status have to be accepted. Moreover the status of others has to be accepted and their requests fulfilled. One must learn the proper channels for getting things done.

v. *Regularity and Flexibility* Rules and regulations are followed. Jobs are attended regularly and during prescribed working hours. Special permission is required for work after working hours. Work records have to be kept carefully and consistently. Reports have to be submitted with regularity and flexibility has to be shown if changes are necessary.

with increasing knowledge and in communicating with his colleagues, the professional earns his livelihood by giving what Hughes has called "esoteric service" to a client. "The essence of the matter appears to be that the client is not in a position to judge for himself the quality of the service he receives. He comes to the professional because he has met a problem which he cannot himself handle . . . He has some idea of the result he wants; little of the means or even the possibility of attaining it" (Hughes 1952, p. 442). The client of the industrial researcher is "the company". By accepting a position with a company, the researcher both implicitly and explicitly accepts the task of working on problems related to the products that the company produces. But the company is not only the researcher's client; it is also his patron, providing him with financial security, equipment, personnel, etc., to carry out his work. It is the client-patron role of the company vis-à-vis the scientist and engineer that puts certain constraints on the fulfilment of the scientists role for the industrial researcher.

The Administrator Role

Sooner or later the scientist/engineer has to take on administrative responsibility. The number and types of persons and activities involved in administration may vary. In general, the activities may be separated into two categories—Administration *of* Research and Administration *for* Research. The former involves activities of a scientific and professional nature and the latter involves paper-pushing, meetings, "putting out fires", etc.

The Employee Role

This role does not add new information to the system. It maintains the system, limits the noise in the system, and facilitates the flow of already existing information in the system. The imperatives and constraints of this role are not unique to the scientist and engineer in the organization. They are found among others in the organization also.

The Social Role

This role refers to the rules of behavior and conduct within the organization and outside of it. Like the Employee Role, the expectations here are also designed to keep noise in the system at a minimum and to make for smooth and pleasant working relationships. The constraints of this role vary as a function of administrative status. Its prescriptions, in general, are neither codified nor verbalized. Social "rules" are "discovered" through experience or the advice of friends and mentors. When the rules are questioned, they may even be denied as non-existent and certainly as "irrational".

Use of Role Analysis
For Job Demands and Job Fit

These five roles for the scientist/engineer in the industrial research organization may be used in several ways for assessment purposes. They may be used to characterize the characteristics and demands of a scientist's and engineer's job. As we shall see shortly, by using the roles to characterize the job demands at various administrative levels, one can get a picture of how job requirements change as a function of rising administratively. One can also use the roles to learn of their relative salience for being creative or for being successful in an organization. Then by comparing a man's perception of his job requirements with his perception of what makes for creativity and success in the organization, one can get a fairly good idea of whether the person sees his position as allowing for creativity; or whether he sees himself in a job with a dead end. By the same token, by comparing the man's perception of what makes for success and what makes for creativity with the perceptions of the same matters by the managers in the organization, one can get a picture of inter-adminstrative level agreements or disagreements. Where disagreements appear it may be a matter of lack of information or it may be a matter of some rather serious problems in the organization.

For Hiring

Role analysis may also be used in the hiring process. A manager requesting a new-hire does not limit himself to the verbiage of the traditional job description, he describes the job in terms of role demands. Consequently it is then a simple matter of comparing the rank order of the role items as they describe the job with the man's perception of his own skills and abilities to determine whether he is or is not a good fit for the job.

For Periodic Review

The roles may also be used in the periodic review process. The review process contains many questions but in our procedure the reviewer also uses the roles to characterize the reviewee's skills and abilities. The reviewee does the same for himself with the same items. Then, at review time, it is a simple matter to put both sets of rankings side by side to determine degree of agreement. Items on which there are disagreements are items that make for good discussion during review meetings.

Illustrative Data

The five roles specified above are represented by 13 items in the *Audit*. The items are presented under various conditions and in terms of different questions as indicated above. The person answering the Audit rank orders the items from 1 to 13 from most important to least important.

To illustrate the data obtained with this kind of analysis and the kinds of factors that may affect creativity of persons in industrial organizations I will not present the items as they appear in the *Audit* to preserve their future usefulness. Rather I shall limit myself to the core meaning of the item. Material will be presented on how the roles vary as a function of managerial or administrative level in the R&D organization and then on how role requirements also vary as a function of managerial level for success or creativity in the organization. In reading these data, the reader needs to bear in mind that the results are not likely to be applicable to all situations. Organizations vary. Therefore when one talks about the relationships of organizational factors to creativity, the precise nature of the relationship has to be determined for each situation. Hopefully, over time as more of us share our data, we may be able to say something about the generalizability of the results. At the moment the data are limited to the organizations studied.

The data presented were collected from 157 scientists and engineers in four major R&D organizations in the United States. They all had their Ph.D. degrees. On the average (median figures) they were employed an average of 5.6 years in the companies studied. But the range of years of company employment was large, from 1 to 23, which was to be expected from the range of managerial levels studied (described below). The average number of years of full-time employment before coming to their companies was 1.3 years. On the average (median) the men studied had published 3.5 papers and had 1.1 patents.

They came from five different supervisory levels. Level 1 is that of the "bench" researcher (N = 49). At this level the man has no subordinates, is usually a member of a group/team/section, and works on supervisor-assigned projects. Level 2 is that of the Group Head (N = 45). This researcher may have one new Ph.D. and/or one or more non-Ph.D.'s working for him. Level 3 is that of Section Head (N = 39) and this person has responsibility for two or more groups. A Department Head (N = 11) manages several sections

and a Division Head (N = 13) has two or more departments under
him.

Variations in Roles as a Function of Managerial/Administrative Status

The relative importance of the different role items as a function of
managerial/administrative status are presented in Figure 1.

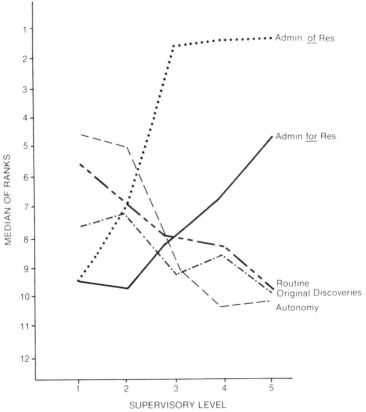

Fig. 1. Variations in role items for skills and abilities required by the job as a
function of supervisory level.

Looking at variations in five skills and abilities as a function of
supervisory level we find, as one would expect, that the two
administrative items show marked increases. The Administration
of Research becomes the most important requirement of the job. It
starts off low and shows a steady and marked increase in importance
from the lowest level to the Section Head level, and it stays at that

point up to the highest managerial position (Division Head) in Figure 1.

Administration *of* Research allows managers at the higher administrative levels to keep involved in their scientific and engineering interests. However, these activities come at some sacrifice in the direct involvement in the Scientific Role and in making original discoveries on their own. Thus, as administrators rise they obviously give up independence in their work (which is related to both Scientific and Professional roles). They have to delegate their work to others. But they also do not have to be much involved with routine (an Employee Role item).

The distribution of time devoted to different roles at the higher administrative levels also reflects the frustration of top level managers. True, the fact that they have risen administratively reflects their success in the organization. But our interviews with them indicate that they are also frustrated by the fact that they do not get directly involved with the work. Some of it refers to their own enjoyment of the work. But some of it also reflects their feeling that they could probably do the work better than their subordinates. In essence, managers when they allow themselves to think of the excitement they had in doing their own work at the bench are sometimes inclined to envy the positions of their subordinates, but one wonders if they would in fact trade positions with them.

It is at the first two positions in the administrative hierarchy (at the level of the Bench Researcher and the Group Head) that, relatively speaking, one finds greater opportunity for independence and for fulfilling Scientific and Professional roles. Administrative demands are low and because the Employee Role is of relative greater importance there is a greater requirement for routine work.

The distribution of skills and abilities at the lower administrative positions is consistent with our interview data with persons at these positions. One of them captured the feeling of the group when he said, "It takes four years to get the academic Ph.D. But it takes another four to get the industrial Ph.D." The person with the academic Ph.D. is fitted for the requirements of the jobs at the lower administrative levels, but in time he will have to learn how to rise administratively. This is a rather complicated matter and well worthy of the diploma—the industrial Ph.D.

Roles and the Rewards for Creativity

One of the lessons on the way to achieving the industrial Ph.D. that the researcher has to learn is what kinds of activities are to be rewarded if one's creativity is to be fostered. To gather data on this matter we asked the managers at administrative levels 4 and 5 to rank order the skills and abilities that made up the roles in terms of how they should be rewarded if creativity is to be promoted. The persons at the first three administrative levels were asked to make the same rankings. In essence the question we were asking was: To what extent do managers and the workers agree or disagree as to what kinds of skills and abilities should be rewarded if creativity is to be promoted in the organization?

The data revealed that there was fairly good agreement between the two groups. The median rank order correlation between the two groups was 0.79. However, a study of the items which provided the largest differences between the groups indicates important areas of disagreement.

Both managers and the researchers agree that rewarding Social Role items is unimportant if creativity is to be promoted. They are in agreement, however, that originating new products and processes (a Professional Role item) is of first rank in importance. But, among the next three items one finds disagreement. Administrators put more emphasis on the importance of the Administration *of* Research and the ability to communicate with others. The researchers call attention to the importance of theory and science and independence of thought and behavior.

The organization's values, as reflected in the orientation of the managers, call attention to the crucial importance of Professional and Administrative role behaviors for creativity. The researchers share these views but also call attention to the importance of more academically and scientifically oriented values. When they get their industrial Ph.D.'s they will probably change their minds.

Roles and the Rewards for Success

To be successful in an organization and to be creative in an organization may or may not involve the same skills and abilities. Hypothetically, an organization that values creativity should reward it in a manner that a creative individual is also successful. On the other hand, where skills and abilities necessary for creativity and those necessary for success are dissimilar, then one might well be in a situation that is characterized by stress and conflict.

Since stress and conflict may diminish the energy an individual has available for creative pursuits, this matter was investigated with the role items. Both managers and workers were asked to rank order the skills and abilities involved in the roles for their relative importance if one was to be successful in the organization. Then one could compare the rankings for the two groups to judge the level of agreement or disagreement between managers and the researchers. This would provide data for *inter-group* stress or conflict. Then we shall turn to the relationships between success and creativity for the researchers themselves and for the managers themselves. This would reflect possible *intra*-personal stress and conflict.

Analysis of the data indicates that there is fairly good agreement between the managers and the researchers as to what makes for success in the organization. The median intercorrelation is 0.61. It should be noted, however, that it is lower than the 0.79 obtained for creativity.

Looking at the variations in individual items in both groups we find that administrators and researchers agree that the more academic skills and abilities and autonomy are less important for success than are the more Professional Role items which involve the development of useful products and processes. Among the more important items for success are to be found the two administrative role items plus the ability to communicate with one's salesmen and customers and the ability to get along socially with one's colleagues.

The difference between the researchers and the managers is reflected in a more cynical attitude on the part of the researchers. For success, researchers place more emphasis on knowing the right people in the organization (a Social Role item) and the Administration *for* Research than do the managers. They also regard independence and theoretical work as less important for success than do the managers.

In essence then, to be successful in the organizations studied an individual has to learn how to integrate Professional and Administrative roles—if one assumes that the managers are seeing the situation correctly. For the researchers, if *they* see the situation correctly, they have to continue with the adult socialization process and learn how to move from their academic to their newer industrial situations and, somehow, in the process, combine their scientific knowledge with the demands of the company. As they do this they have to learn how to take on more and more administrative responsibility.

Roles and the Relationships Between Success and Creativity—Intra-personal Stress (?)

The data obtained from managers and researchers in answer to the success questions may also be used in another way. One may correlate the managers' rankings of the role items for success with their rankings of the same items for promoting creativity. One can do the same thing for the researchers themselves. In this instance a high correlation would indicate that the manager or the researcher sees little, if any, difference between what makes for success and what makes for creativity. A low correlation would mean that there is a difference. If there is a large difference it may reflect potential for intra-personal stress—stress that would stem from what is essentially a value conflict and stress which could put constraints on creative work.

Analysis of the data indicated that the median rank order for the administrators was 0.80. In other words, they saw that which was involved in creativity and that which was involved in success as rather similar. The data obtained from the researchers, however, revealed another story. The median rank order correlation for their two rankings was only 0.34. In other words, scientists and engineers studied felt that which would make for success in the organization was quite different from that which made for creativity.

Turning to the managers' data it will be recalled that their orientation to promoting creativity was described as stemming from an orientation that regarded research activity within an organized framework and therefore they placed less emphasis on the autonomy of the researcher and the Scientific Role items. When managers were asked to rank the items for success they again regarded the items within an organized framework and on this basis the two sets of item rankings did not differ from each other. Nevertheless, a study of the managers' rankings indicates that they themselves regard the Scientific Role items as more important for creativity than for success.

Analysis of the researchers' data tell a somewhat different story. As indicated previously, their average rank order for success and creativity was 0.34. A study of their rankings indicates that they regard the Scientific Role item that would make for growth in scientific knowledge and, another item regarding autonomy in one's work, as more important for creativity than for success. But then they believe that a Social Role item, involving getting to know the right people in the organization, is more important for success than it is for creativity. For creativity this item was among the least important. Another item that the scientists regarded as least important for creativity but more important for success was the Administration *for* Research. For the researchers, an individual with skill in "pushing papers around" has at least one characteristic in his favor insofar as being successful in the organization.

These data on success and creativity for managers and researchers indicate that somehow the managers have integrated the two values and find congruence between the skills and abilities necessary for success and those necessary for creativity. The men lower in the administrative hierarchy, who are also the younger men, are still in the throes of the conflict and in all probability are still struggling with the stress involved—all of which may detract from the energies and time they have available for creative work. One wonders if specially designed programs for persons at these administrative levels might not alleviate much of their stress and release more energy for creative pursuits.

CONCLUSION

Creativity is a process which occurs in a social context. To be able to understand it and predict it one must not only comprehend the intra-individual psychological factors that are related to creativity, but also the social or inter-individual factors. Insofar as the latter are concerned one needs to consider the effects on creativity of geographical, political, economic, religious, and other social factors which operate at the macrolevel. But at the microlevel one needs to focus on the effects of organizational characteristics on individuals' creative processes. This paper focused on such matters.

The paper highlighted the importance of studying organizations with specific concern for how they effect the creativity of their scientists and engineers. Organizations can be studied from different vantage points and for different purposes. The aim of this paper was to present a specifically designed procedure, a *Technical Audit* which was based in theory and which was constructed with scientists and engineers in mind.

The *Technical Audit* considers a number of organizational characteristics that may affect the creative process but for our purposes here only that part of it that concerned itself with the organization's expectations of the scientist and engineer in industry, or his roles, was presented. Our study of industrial research (R&D) organizations indicated that scientists and engineers in industry were expected to fulfill five roles: Scientist, Professional, Administrator, Employee, and Social. These roles may be used to specify skills and abilities in different job situations, for hiring purposes, for review purposes and for general assessment purposes.

The data presented in this paper focused on the variations in roles as a function of administrative status and pointed out how the assumption of administrative responsibility was a mixed blessing for managers. Managers may enjoy their higher status in the organization but they seem to miss their direct involvement in research and engineering. Their satisfactions are constrained by the fact that they may have to delegate responsibility for solving problems to others while they, at the same time, believe that they themselves might be the best persons for this job.

Men lower in the administrative hierarchy have more opportunity for the pursuit of scientific and professional activities but if they find they also have to be successful in the organization they have to shift emphasis from Scientific to Professional roles and they will have to learn how to be managers.

Life in the organization for scientists and engineers is a continuation of the adult socialization and educational process. "It may take four years to obtain the academic Ph.D.", one man said, "but it may take an additional four years to obtain the industrial Ph.D." In this process the shift to administration and professionalism has to occur. Also, while one learns how to bring academic values into the industrial organization, one may have to adapt to the stresses and strains of conflicts that arise because of the lack of congruence between the pictures of that which makes for success and that which makes for creativity in the organization.

In-house programs can be organized around the data presented here. For older administrators there might still be some opportunities available to them to engage in research activities. For younger scientists and engineers R&D organizations might well profit from designing training or discussion groups for them in which they could learn how to select research problems which would pay off both for their scientific interests and professional role for the company. In this regard experience in various parts of the company, which is usually not available to them, could go far. Management can profit from knowledge of the stresses and strains within their company for then they can develop a climate in which creativity can be encouraged and facilitated. Toward these ends, the role analysis presented here and the total *Technical Audit* can be a valuable instrument. It is also of considerable value in the hiring, placement, and review of scientists and engineers who are expected to be both creative and successful in their organizations.

REFERENCES

Barron, F. 1958: The needs for order and for disorder as motives in creative activity. In C. W. Taylor (Ed.), *The second (1957) conference on the identification of creative scientific talent*, pp. 119–28. Salt Lake City: University of Utah Press.

Hughes, E. C. 1952: Psychology: science and/or profession. *American Psychologist*, 7, 441–3.

March, J. G. & Simon, H. A. 1958: *Organizations*. New York, Wiley.

Merton, R. K. 1949: *Social theory and social structure*. Glencoe, Ill.: Free Press.

Parsons, T. 1954: A sociologist looks at the legal profession. In *Essays on sociological theory* (rev. ed.). Glencoe, Ill.: Free Press.

Stein, M. I. 1974: *Stimulating creativity: Individual procedures (Vol. 1)*. New York: Academic Press.

Stein, M. I. 1975: *Stimulating creativity: Group procedures (Vol. 2)*. New York: Academic Press.

Stein, M. I. 1984a: Homo transformare. In S. Gryskiewicsz (Ed.), *Creativity week VI: Blueprint for innovation*. Greensboro, North Carolina: Center for Creative Leadership.

Stein, M. I. 1984b: *Making the point: Anecdotes, poems and statements about the creative process*. 149 York Street, Buffalo, New York: Bearly Limited.

Stein, M. I. 1985a: *Audiotape: Homo transformare*, Greensboro, North Carolina: Center for Creative Leadership.

Stein, M. I. 1985b: *Videotape: On creativity*. Greensboro, North Carolina: Center for Creative Leadership.

12

Entrepreneurship and Strategic Management: Synergy or Antagony?

KJELL GRØNHAUG AND TORGER REVE

> If there is a hell for planners, over the portal will be carved the term "Cash Cow". (*Business Week*, Sept. 17, 1984, p. 53)

This paper raises the question whether innovative and entre-preneurial activities on the one hand and strategic management on the other represent synergy or whether the two sets of activities are in antagony when combined in an organizational setting. The background for this question is illuminated, basic concepts are defined and contrasted, the impact of organizational properties on the various activities is discussed, and suggestions how to overcome the entrepreneurship—strategy dilemma are offered.

Many firms and industries are continually being confronted with increasing environmental turbulence and competitive pressure threatening economic prosperity and survival (cf. Ansoff 1984). Major European industries are losing market shares in their domestic and foreign markets.

The former American challenge (cf. Servan-Schreiber 1968) no longer represents the major competitive threat. In fact important parts of American industry are facing problems similar to those facing their European counterparts (cf. Lawrence and Dyer 1983). New competitors are in the process of taking the lead role. The Japanese challenge (Kahn and Pepper 1979) recognized decades ago is now number one position (Vogel 1979). Other new entrants with major competitive advantages are also entering the scene. For years European shipyards—facing declining markets—have been outperformed by competitors from the newly industrialized countries (NICs)—particularly South Korea. The Hong Kong car producers, now test-marketing in Canada, are being carefully watched by headquarters in Detroit.

Keener competition, declining markets, drops in market shares and increasing environmental turbulence challenge present management practice and paradigms. The new signals have made decision makers at the firm, industry and governmental levels fully aware that "something has to be done" (cf. Thurrow 1984).

Strategy and Entrepreneurship

Two broad categories of activities—strategic management and entrepreneurship/innovation—seem to be the proposed answers for bringing back the competitive edge. By letting Ansoff's (1965) seminal book represent the start of the strategy area, it is not unfair to say that strategic planning/management has been a major subject in business schools as well as in business for almost two decades, even though specific aspects of strategic planning are now under fire (cf. *Business Week* 1984). [This does not imply that strategic thinking did not exist before that time. In fact, strategic thinking is probably as old as man himself, and courses in strategy were also taught at various business schools before World War Two.] The quest towards renewing European and American industries (cf. Lawrence and Dyer 1983; Selvik 1984; OECD 1978) may be interpreted as demand for and belief in strategic thinking at both industrial and corporate levels. In a similar vein, there seems to exist a deep-rooted belief that innovation and entrepreneurship are important for solving the crisis (Drucker 1985b).

Innovation and entrepreneurship have for long been considered a major force of economic growth (cf. Schumpeter 1939). Belief in entrepreneurship and innovation as a major source of growth and success is found at firm, industrial as well as governmental levels; this is underscored by the emphasis on product development and various programs for restoring innovation and entrepreneurship—and thus bringing back firms'/industry's competitive edge (cf. *Newsweek* 1979; Kamien and Schwartz 1982; OECD 1979).

Due to the basic role of firms (organizations) in an economy, we will—by conceiving firms as a subset of organizations—direct our attention to the activities under scrutiny in an organizational context.

BASIC CONCEPTS

Before discussing the major concepts constituting the basis for the discussion to follow, it should be stressed that our discussion relates mainly to relatively *large* organizations. By this we are *not* saying that entrepreneurship and strategy are unimportant to small firms; we in fact hold the belief that entrepreneurship/innovation and strategy are of equal, if not higher, importance in such a context. Our attention towards larger organizations is dictated rather by the complexity and specific characteristics of such organizations as well as the economic impact of the single large organization compared to a small one. Thus, revitalizing large business organizations remains the core issue in the paper.

Strategy

The meaning of *strategy* was originally shaped in military settings. Strategy meant the development and use of military forces and material to achieve specific objectives. Business people have found it convenient to use military terms to describe their competitive situations (cf. Kotler and Singh 1981).

Companies engage in price "wars", "border clashes", and "skirmishes". As noted by Day (1984), however, the military analogy is most insightful "when the objective of a strategy is interpreted as achieving a better state of peace rather than annihilating the competition" (p. 15). Or, stated in another way, firms seek to enter and stay in positions which give them specific relative *advantages*. The firms want to be in high growth markets, keeping the competitors away, and earning high profits (return on investments).

There is no single definition of strategy agreed upon. According to Porter (1980) competitive strategy involves "*positioning* to maximize the value of the capabilities that distinguish it (the firm) from its competitors" (p. 47). Ansoff (1984) maintains that strategy is a set of decision-making *rules* for guidance or organizational behavior (p. 31), while Day (1984) claims strategy to be "the *direction* the organization will pursue within its chosen *environment*" (p. 1).

The following set of underlying assumptions may be deduced from these definitions:

- organizational resources are scarce;
- organizations are conceived as open systems;
- efficient resource allocations are important, but difficult to attain;
- market positioning is important;
- rules and guidelines are needed to direct the organizational resource allocation.

Strategy is often thought of as a *steering* device. An organization without a strategy is compared with a sailing ship without a rudder—condemned to wander aimlessly in response to winds, currents, and outside events (cf. Ross and Kami 1973).

Strategy is often conceived as a *process*, and various descriptions and prescriptions of the strategic planning process are offered throughout the literature (e.g. Ansoff 1984; Day 1984; Pearce and Robinson 1985). Agreement seems to exist that assessment of the present situation is a useful starting point. This definitely makes sense. In order to choose direction and find a specific niche, knowledge about the present position is required. The analogy of using map and compass is stumbling close. There are, however, some differences. When a compass is being used, the destination is assumed known. Moreover, there are specific procedures to follow to point out the direction from A to B. A strategy, however, has to be based on some conception. Based on the assessment of the present situation (including analyses of market needs, competitors' and one's own strengths and weaknesses and so on), the strategy has to be created. In other words, it is the conception—the creative act based on the various pieces of information—which represents the basis for a strategy.

Strategy creation is the starting point only. The strategy has to be specified, and plans for what to do, when and where and by whom have to be worked out. The necessary resources have to be allocated, and the various actions subsumed under the strategy have to be activated. The strategy has to be implemented, evaluated and controlled. Thus strategy implies *commitment to action*, where the actions should be coordinated in the direction dictated by the focus of the intended strategy.

Innovation and Entrepreneurship

The concept of *innovation* subsumes several meanings and is used in partly different ways across disciplines. In some disciplines, the *creation* of something *new* is stressed (e.g. in psychology and economics). Other disciplines emphasize the *adoption* of something which is *new* to the adopter (e.g. in marketing and organization science). In addition, focus has been placed on the *diffusion* process, i.e. how the adoption of a new product, process or idea is spread in a social system over time (cf. Rogers 1983).

Novelty, however, is implied in creation as well as in adoption of something "new". Moreover, to create as well as to adopt something new requires *deviation* from past practice. When Kodak created their instamatic camera, it represented a new conception in the process of photography. For the adopters this also represented a new concept, a new solution which was associated with perceived advantages. It should also be noted that many innovations have to be modified by the adopters. In order to adopt a specific process, a new computer system, etc., the innovation has to be modified to the specific needs of the adopters, or, as emphasized by Rogers and Agarwala-Rogers (1976), " . . . many innovations go through extensive revisions, essentially announcing to 'reinvention', in the process of their adoption and implementation" (p. 139). Thus, adopting a new product or a new business idea will require creative activities and new conceptions. And "creative imitation"—to use Drucker's term (1985b)—often has the largest commercial potential.

Entrepreneurship for one thing requires something new, a new idea, a new product, or a new business idea. Besides the idea, which may be created by the entrepreneur him-/herself or borrowed (bought, adopted), entrepreneurship also implies that the idea is *realized*. Entrepreneurship requires that the new product or business idea has proved viable. Thus, entrepreneurship requires focus, resources and actions beyond the creation of the basic idea. However, entrepreneurship is limited to some *initial* phases of the process of implementing and realizing the new idea.

Strategy and Innovation/Entrepreneurship Contrasted

Figure 1 may serve as our point of departure in contrasting strategy and innovation/entrepreneurship.

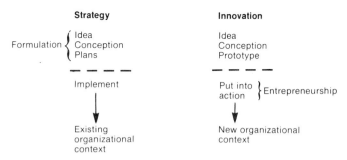

Fig. 1. Strategy and innovation contrasted.

Strategy and innovation both start with some underlying idea and conception. Plans are the outcome of the strategy formulation phase, while a prototype may be seen as the end of the innovation process. Strategic plans have to be implemented, and the new idea has to be put into action. Thus, there are stumbling similarities between strategy formulation and innovation, as well as between strategy implementation and entrepreneurship. It should be noted, however, that strategy—in most cases—is implemented in the existing organizational context, while entrepreneurship typically implies the creation of a new organizational context.

Several writers on strategy emphasize the importance of identifying threats to avoid and opportunities to pursue (cf. Day 1984, p. 2), thus strategy should—when needed—imply entrepreneurship as well as basically new ideas and conceptions. The fact that few—if any—organizations live forever, demonstrates the difficulties in doing so. The many business failures in fact imply that most organizations only possess modest ability to adapt to changing environmental requirements (Starbuck and Hedberg 1977), a perspective which is fully reflected in the natural selection paradigm (cf. Freeman 1982). Empirical studies reported by Mintzberg and his associates also show that few firms change their strategies (Mintzberg and Waters 1982). Organizations operate in a dynamic world. As the surrounding environments change as new threats and opportunities appear—or should be discovered—the organizational strategy should be modified, and—in some cases—be replaced by a completely new one. Observations show, however, that firms find it difficult to change and to maintain competitive advantages. A high fraction of the excellent companies praised by Peters and Waterman (1982) had lost their excellency just two years later (cf. *Business Week*, Nov. 5, 1984).

NEED FOR STRATEGY CHANGES

Strategy implies some conception of own relative advantages, competitive forces, a clear understanding of market needs, the part of the market to serve, how to do it, and commitment to do so.

When looking at companies like McDonalds, Levi Strauss, Coca Cola and Kellogg, the basic conceptions behind their strategies are all old, and apparently still viable. In a true Kotlerian sense they have adopted the marketing concept, emphasizing customers' needs and wants (cf. Kotler 1984, p. 22). This, however, is only part of the story. Throughout the years, these organizations have introduced several new products and undergone several changes, but mainly within the frame of the same strategy concept. And it has worked. But will such a determined focus always work?

The companies mentioned above all offer *non-durables*. Their markets all exhibit repeat buying. In fact, very much of the basic idea is based on satisfied customers who will buy their products over and over again. Directing our attention to *durables* may, however, change this perspective. When the product durability is long, say 10–20 years, and the market (i.e. number of potential buyers) limited, the market may be emptied. And as the potential adopters are turned into adopter, the remaining number of potential adopters approaches zero. [This is easily seen by inspecting the first derivative of the equation for the diffusion curve.] One may argue that this represents a hypothetical case. True, but when looking at markets such as shipbuilding, production equipment for agriculture (tractors, harvesters, etc.) and train production, the market situation is almost as described above, i.e. limited number of potential buyers, long product durability, and decreasing total demand. [This may be interpreted as a specific stage of the life cycle of the niche. See Brittain and Freeman (1980) for relevant strategies in such cases.]

Repeat buying of durables often implies that the last product bought differs from the former. A Ford 1985 possesses some novelties compared to a Ford 1984. An important issue is the degree of needed change. In order to stay in the computer business, great changes in product offerings are often needed, requiring innovation, entrepreneurship, and the ability to unlearn previous conception (cf. Starbuck 1983). Examples offered by Levitt (1960) on "myopic marketing" demonstrate, however, that this is hard to attain.

New competitors may enter the market. Moreover, the new competitors may outperform the organizations presently dominating the market. The observation made by Lawrence and Dyer (1983) that many American firms and industries fail in their maturity, may partly be due to the entry of competitors possessing more efficient and more cost-saving production processes. New offerings from the entering competitors may also change buyers' perceptions of acceptable solutions not being handled adequately by the previous dominant firms. *Substitutes*, too, may change the market's perception of acceptable solutions, as detergent represented a new way to solve various tasks of cleaning. In sum, changes are often needed. The changes needed may vary in degree of *deviance* from the activities and solutions suggested by the present strategy. Moreover, the needed changes may arise discontinuously, conveyed by signals difficult to detect (cf. Ansoff 1984).

STRATEGY AND ORGANIZATIONAL STRUCTURING

It is a common observation that organizations are often less adaptable than usually assumed (cf. Lawrence and Dyer 1983; Freeman 1982; Starbuck 1983), and that the organizational strategy often remains stable (cf. Mintzberg and Waters 1982). But why is this so?

Structuring

Organizations consist of people embedded in hierarchical structures. Their tasks are more or less specialized. In most cases they hold limited perspectives on the total set of activities taking place within the organization (cf. Wilkins and Ouchi 1983).

In order to achieve goals, the organizational activities have to be *directed* and *coordinated*. In applying the transaction cost approach to organizational coordination, Ouchi and his associates have identified three different types of governance structure, i.e the market, the bureaucracy and the clan (cf. Ouchi 1980; Wilkins and Ouchi 1983). Strategy implementation implies for one thing internal coordination. The various activities subsumed under the strategy have to be specified. What to do and how to do it, has to be learnt. The claimed "commitment to action" requires allocation of resources needed, devotion of attention and time—also scarce resources—from members at every level of the organization (March and Olsen 1984). Direction of the organizational activities and commitment to action requires that the individual member knows what to do, how to do it—and that he/she behaves as expected. The shapening-up and "renewal" of SAS which started a few years back, may partly be seen as focusing and directing the organizational activities, including the motivation of individual members of the SAS organization to do what they were supposed to do! Such activities imply *socialization* of the individuals within the organization (Pascal 1985), which implies learning of values and skills (cf. Ziegler and Child 1968) and, as maintained by Wilkins and Ouchi (1983), this also implies costs. Such direction, commitments and socialization activities all attribute to the *structuring* of the organization (cf. Mintzberg 1979), exerting impact of the actions performed, how they are performed, and under which conditions.

The structuring of the organization and the subsequent routinized behavior will have a stabilizing effect on the organization and its strategy. This for one thing directs attention. The known is more easily seen than the unknown. Moreover, the routinized and well known more easily gets access to the limited *attention* capacity of the organization member, and is more easily allocated time and other resources. Gresham's law, formulated several decades ago, stresses the importance of this argument (cf. March and Simon 1958), and contributes to explaining why organizations remain stable sticking to the same strategy.

Organizational characteristics such as high degree of job specialization, formalization, and emphasis on professional values attribute to a particularistic ego-centered perspective on the activities performed, hampering a more wholistic what-is-good-to-the-company perspective as demonstrated by several well-run companies (cf. Ouchi 1980, 1981). In sum, the structuring forces described above contribute to maintaining established routines and will thus impact future strategies and actions. [Over the years the strategy-structure sequence has been subject to considerable amount of discussion and research (cf. Galbraith and Nathanson 1978 for overview). There is little doubt, however, that the present structure exerts impact on future activities (cf. Hall and Sais 1980).]

"Top-down" Planning

Strategic planning and strategic management are often considered a major—if not *the* major—management task (cf. Ansoff 1984; Day 1984; Schendel and Hofer 1979); this is reflected in widely adopted "top-down" planning procedures. Some authors definitely claim the necessity of combining the "top-down" with a "bottom-up" planning process. The former remains, however, the dominant planning process.

Entrepreneurs have been subject to considerable amounts of research (cf. Collins and Moore 1970). Among the common findings are personality characteristics demonstrating high degree of independence, self-orientation, and the disliking of formality and authoritive dictating (cf. Steiner 1965 for overview). Thus the hierarchical organizational structure seems to contradict entrepreneurial personality, supported by several observations. Several of the most recent innovations originate from small organizations (*The Economist* 1983). Despite the enormous R&D spendings of IBM (amounting to more than $2.5 billions in 1983, i.e. the second largest R&D-spender in the US., cf. *Business Week* 1984), the IBM organization still continues to acquire entrepreneurial ideas developed elsewhere.

Organizational Resistance

Suggesting something new is to break with conformity. The entrepreneur has to break the established pattern, at the risk of considerable personal cost. He or she has to impose a new perspective on other organizational members, which is not easy (cf. Kidder 1981), and the resisting organization may easily strangle the entrepreneurial spirit, or may even force the innovative persons to leave the organization, which in the last few years has been given considerable attention (see Drucker (1985a) and the *Economist* (1983) for detailed discussions).

Planning Activities

Corporate strategy which by its very nature should enhance adaption to environmental change, also becomes a hindrance to internal organizational change. The reasons are the very activities associated with strategic planning—strategy formulation and strategy implementation. The Swedish Facit case—once claimed to have the best strategic planning system—is an eclatant example of the blindfolding which may occur in large corporate planning units (cf. Starbuck and Hedberg 1977). The Norwegian Reksten shipping case is an example of the tyranny of strategy implementation as new super tankers were built even after the oil crisis which ruined the market for oil transportation. Middle managers became highly efficient implementers of the wrong ideas. Planning may thus strangle innovation, but the lack of planning hampers industrial activity and adaption. This is the *administrative paradox* faced in business organizations. Moreover, corporate entrepreneurship is by definition in conflict with corporate strategy. Thus corporate entrepreneurship can be defined as strategic initiatives which *break with* the established corporate strategy. Anyone who shows corporate entrepreneurship runs the risk of being expelled from the inner circles—and even from the organization itself. Thus the *incentives* in most organizations run counter to stimulating innovation and entrepreneurship.

CAN THE STRATEGIC AND ENTREPRENEURIAL TASKS BE COMBINED?

According to Thompson (1967) organizations simultaneously seek *certainty* and *flexibility*. Certainty is needed in order to get the work done, and flexibility is required to adapt to a changing environment. Both are needed for the firm to be profitable and to survive. Management—according to Thompson—is the act of *combining* the search for certainty with the search for flexibility. This is the dialectic component of management.

How can organizations combine the search for firm strategies with the search for innovations and entrepreneurship? Or is this one of those insolvable dilemmas which is nice for dialectic analysis, but weak in practical recommendations?

The Quest for Internal Renewal

A first question is whether the organization should base its renewal function on various types of acquisitions, or rely on internal entrepreneurship. In acquisitions, the organization buys new ideas, new products or acquires "spin-offs", i.e. small entrepreneurial firms (cf. *The Economist* 1983). There are several reasons why we would hesitate to propose a renewal strategy based on acquisition of spin-offs only. First, if all entrepreneurial/innovative activities are put aside, the creative and entrepreneurial people will tend to leave the organization. Thus such a strategy will result in brain-drain.

Second, to *see* new opportunities also requires innovative skills. Thus, being without the creative elements makes the organization less sensitive to perceive opportunities and threats.

Third, adoption of new ideas also requires creative skills (cf. reinnovation). Thus the brain-drain to follow the acquisition-only strategy will make the organization less capable of adjusting to new solutions necessary to perform well.

Some Suggestions

There are, however, ways of reducing some of the negative effects of the administrative paradox posed above. The organization should be prepared to favor change, or at least to get a better understanding of the necessity of change and renewal. This may be done in several ways.

- *Job rotation* is one such device. This has been applied to a greater extent in Japanese than in American and European organizations (cf. Ouchi 1981; Reve 1983). Job rotation can be an effective device for creating a more wholistic perspective of the organizational tasks, and thus pave the way for acceptance of change and new ideas put forth by individual members.
- In the same way a *broader*, general background of the employees may attribute to a more wholistic understanding of the organizational needs.
- Intensive use of various informational activities in order to create some degree of shared understanding of the organizational tasks

and requirements, may also favor pro-change attitudes. So called *"internal marketing"* activities may serve as examples of the emphasis on shared understanding or organizational tasks among the members in order to coordinate activities, but also as a means of adopting new perspective, and thus adaption to new requirements (cf. Carlzon and Hubendick 1983).

- Creating an organizational climate which stimulates *experimentation* and plays down the failures may in a similar vein contribute to more corporate entrepreneurship.

- Increased *participation* in planning and decision making may contribute to consensus and increased sensitivity to new ideas. Thus, by implementing the true "bottom-up" perspective, new ideas will more easily be elicited and accepted. Quality circles represent a relevant example.

- In addition, the tight, hierarchical structure should be changed in more *informal* directions to impact organizational communication and the climate for change. [Kagono et al. (1983) demonstrated such characteristics to be valid for Japanese firms in a large cross-cultural research effort.] This is the familiar agreement first advocated by Burns and Stalker (1961).

- In order to guarantee that innovation and entrepreneurship get attention, various types of *agenda-building* devices may also be applied (cf. Cobb and Ross 1976). Innovation and entrepreneurship can be made a permanent issue in the strategic plan by establishing specific procedures allowing these topics to enter the agenda.

- Similarly, entrepreneurial values can be stressed in *organization culture* building activities (Wilkins and Ouchi 1983).

- *Project groups* may be established to solve specific entrepreneurial tasks. Such groups should be recruited from all the departments involved in order to create a broader perspective on the entrepreneurial tasks, which may ease the acceptance of new ideas.

- Recently, considerable attention has been given to *internal corporate venturing* (Burgelman 1983), which may be conceived as a device for strengthening the innovativeness and the renewal ability of the organization. The focus in internal corporate venturing (IVC) is broader than product development or business development. In fact, the basic idea is to bring new ideas to the stage where they have proven viable, which may include product development, strategy building, and implementation—including the organizational structuring needed to get the new idea flying.

Something—even breaking organizational rules—is necessary to attain entrepreneurial results.

- Internal corporate venturing needs not only product champions as known from R&D. Such ventures also need *organizational champions*, who are able to make the ventures succeed in the organizations. From a managing point of view, the major job is to build the *strategic argument* to make the new venture consistent with existing corporate strategy, and thus pave the way for the new idea.

The list is not exhaustive, but the common thread is to find ways to stimulate *"intrapreneurship"*—ways to regain entrepreneurial activities within large, mature organizations. Spin-offs, split-ups, or fusions represent one form of visible entrepreneurial avenue (*Business Week*, July 1985). Intrapreneurship represents another, and probably more important, entrepreneurial revitalization.

CONCLUDING COMMENTS

Strategic management and entrepreneurship/innovation are of crucial importance to business organizations. Both sets of activities are needed to prosper and survive.

The structuring of the organization and the planning activities as such tend to strangle innovation and weed out entrepreneurship. In fact, many organizations actively stand in the way of innovation. Such activities represent disturbances to the rules, procedures and commitments towards getting the work done.

Business organizations are, however, confronted with increasingly turbulent environments. Environmental changes occur faster, and the consequences of such changes are more dramatic than ever before. Thus the need for organizational flexibility is accelerating. The increasing need for organizational flexibility imposes new requirements on organizational design and management. To bring back and promote the entrepreneurial spirit will be an important, if not the major management task in the years to come. The emerging trend towards "splitting up" (*Business Week* 1985) may be seen as one way of handling the need for flexibility and entrepreneurship, so may be the escape of members possessing new ideas and entrepreneurial spirit (*The Economist* 1983). In order to keep the business organization viable the question of how to balance the need for innovation and flexibility with the requirements for certainty and planning has to be emphasized—and research should be devoted to this problem to bring forth workable solutions.

REFERENCES

Ansoff, H. I. 1965: *Corporate strategy*. Harmondsworth: Penguin Books.
Ansoff, H. I. 1984: *Implementing strategic management*. Englewood Cliffs, N.J.: Prentice-Hall International.
Brittain, I. W. & Freeman, J. H. 1980: Organizational proliferation and density-selection: Organizational evolution in the semiconductor industry. In J. R. Kimberly & R. H. Miles (Eds.), *The organizational life cycle*, pp. 291–338. San Francisco: Jossey-Bass.
Burgelman, R. A. 1983: A process model of internal corporate venturing in the diversified major firm. *Administrative Science Quarterly, 28* (June), 223–44.
Burns, T. & Stalker, G. M. 1961: *The management of innovation*. London: Tavistock.
Business Week 1984: A deepening commitment to R&D. *Business Week*, July 9, 64–78.
Business Week 1984: The new breed of strategic planners. *Business Week*, Sept. 17, 52–7.
Business Week 1984: Who's excellent now?. *Business Week*, Nov. 5, 46–55.
Business Week 1985: Splitting up. July 1, 50–5.
Carlson, J. & Hubendick, E. 1983: Intern marknadsföring som ett ledningsinstrument vid stora förändringar. In J. Arndt & A. Freeman (Eds.), *Intern marknadsföring*, Malmö: Liber Förlag.
Cobb, R. & Ross, M. H. 1976: Agenda building as a comparative political process. *American Science Review, LXX*(1), 126–38.
Collins, O. & Moore, D. G. 1970: *The Organization Maker*. New York: Appelton-Century-Crofts.
Day, G. S. 1984: *Strategic market planning*. St. Paul: West Publishing Co.
Drucker, P. F. 1985a: The discipline of innovation. *Harvard Business Review, 85* (May–June), 67–72.
Drucker, P. F. 1985b: *Innovation and entrepreneurship*. New York: Harper & Row.
Economist 1983: The new entrepreneurs. *Economist, 289*, Rec. 24, 59–71.
Freeman, J. H. 1982: Organizational life cycles and natural selection processes. B. M. Staw & L. L. Cummings (Eds.), *Research in Organizational Behavior, 4*, Greenwich, CT.: JAI Press, 1–32.
Galbraith, J. R. & Nathanson, D. A. 1978: *Strategy implementation: The role of structure and process*. New York: West Publishing Company.
Hall, D. T. & Sais, M. A. 1981: Strategy follows structure. *Strategic Management Journal, 1*, (2), 149–64.
Kagano, T. 1983: An evolutionary view of organizational adoption. Kobe University, Working paper.
Kahn, H. & Pepper, T. 1979: *The Japanese challenge*. New York: Crowell.
Kamien, M. I. & Schwartz, N. L. 1982: *Market structure and innovation*. Cambridge, Mass.: Cambridge University Press.
Kidder, T. 1981: *The soul of a new machine*. Boston: Little, Brown & Company.
Kotler, P. 1984: *Marketing management* (5th ed.). Englewood Cliffs, N.J.: Prentice Hall.
Kotler, P. & Singh, R. 1981: Marketing warfare in the 1980s. *Journal of Business Strategy, 1* (Winter), 20–41.
Lawrence, P. R. & Dyer, D. C. 1983: *Renewing American industry*. New York: The Free Press.
Levitt, T. 1960: Marketing Myopia. *Harvard Business Review, 38*, (Sept.–Oct.), 26–44, 173–81.
March, J. G. & Simon, H. A. 1957: *Organizations*. New York: John Wiley & Sons.
March, J. G. & Olsen, J. P. 1984: Garbage can models of decision making in organizations. In J. G. March & R. Weissinger-Baylen (Eds.), *Ambiguity and Command: Organizational Perspectives on Military Decisions Making*.

Mintzberg, H. 1979: *The structuring of organizations.* Englewood Cliffs, N.J.: Prentice Hall.

Mintzberg, H. & Waters, J. A. 1982: Tracking strategy in an entrepreneurial firm. *Academy of Management Journal, 25*, (3), 465–99.

Newsweek 1979: Innovation: Has America lost its edge? *Newsweek*, June 4, 58–65.

OECD 1978: *Policies for the stimulation of industrial policies.* Paris: OECD.

Ouchi, W. G. 1980: Markets, bureaucracies, and clans. *Administrative Science Quarterly, 25*, 129–41.

Ouchi, W. G. 1981: *Theory Z. How the American business can meet the Japanese challenge.* Reading, MA.: Addison-Wesley.

Pascal, R. 1985: The paradox of "Corporate Culture": Reconciling ourselves to socialization. *Californian Management Review, XXVII*, (2) (Winter), 26–41.

Pearce, J. A. II & Robinson, R. B., Jr. 1985: *Strategic management:* Strategy formulation and implementation. Homewood, Ill.: R. D. Irwin.

Peters, T. J. & Waterman, R. H. 1982: *In search of excellence: Lessons from America's best-run companies.* New York: Harper & Row.

Porter, M. E. 1980: *Competive strategy.* New York: The Free Press.

Reve, T. 1983: Organisasjon og strategi. *Bedriftsøkonomen*, No. 5, 228–232.

Rogers, E. M. 1983: *Diffusion of innovations* (3rd ed.). New York: The Free Press.

Rogers, E. M. & Agarwala-Rogers, R. 1976: *Communication in organizations.* New York: The Free Press.

Ross, J. E. & Kami, M. J. 1973: *Corporation in crisis: Why the mighty fail.* Englewood Cliffs, N.J.: Prentice-Hall.

Schendel, D. H. & Hofer, C. W. (Eds.) 1979: *Strategic management.* Boston: Little Brown & Company.

Schumpeter, J. 1939: *Business cycles, 1*, New York: McGraw-Hill.

Selvik, A. (red.) 1984: *Omstilling: Erfaringer og utfordringer.* Bergen: Industriøkonomisk Institutt.

Servan-Schreiber, J. J. 1968: *The American challenge.* New York: Atheneum.

Starbuck, W. H. 1983: Organisations as action generators. *American Sociological Review, 48* (February), 91–102.

Starbuck, W. H. & Hedberg, B. L. 1977: Saving an organization from a stagnating environment. In H. B. Thorelli (Ed.), *Strategy + Structure = Performance.* Bloomington: Indiana University Press.

Steiner, G. A. 1971: *The creative organization.* Chicago: University of Chicago Press.

Thompson, J. D. 1967: *Organizations in actions.* New York: McGraw-Hill.

Thurrow, L. 1985: Revitalizing American industry; Managing in a competitive world economy. *California Management Review, XXVII* (1) (Fall), 9–44.

Vogel, E. F. 1979: *Japan as number one: Lessons for America.* New York: Harper & Row.

Wilkins, A. L. & Ouchi, W. G. 1983: Efficient cultures: Exploring the relationship between culture and organizational performance. *Administrative Science Quarterly, 28*, 468–81.

Ziegler, E. & Child, I. L. 1968: Sozialization. In G. Lindzey & E. Aronson (Eds.), *Handbook of social psychology*, Vol. 3, 450–489. Reading, Mass.: Addison-Wesley Publishing Co.

13

"Promotors"—Key Persons for the Development and Marketing of Innovative Industrial Products

HANS GEORG GEMÜNDEN

THE PROBLEM

Innovative products are of strategic importance for both consumer and industrial markets. But, it has been argued that they differ in the way innovative products are developed and diffused. In consumer markets, a *manufacturer-active paradigm* has been very successful. Here the manufacturer initiates and systematically controls the process of idea generation and product development (Brockhoff 1981). However, in many industrial markets a *customer-active paradigm* has been assumed to be more appropriate (von Hippel 1978, 1982a, 1982b). In this case the would-be customer develops the idea for a new product. He selects a supplier he perceives to be capable of making the product, and he takes the initiative to send a request to one or several suppliers: "The role of the manufacturer in this paradigm is to wait for the potential customer to submit a request (. . .) to screen ideas (not needs) for new products; and to select those for development which seem to offer the most promise from the manufacturer's point of view" (von Hippel 1978, p. 40).

Although we share von Hippel's argument that active and challenging customers are very important for the seller's product innovations (Gemünden 1980, 1981; Parkinson 1981), we do not share his opinion that industrial manufacturer's marketing should be passive. Rather, we propose an *interactive paradigm* to explain the development and diffusion of complex innovative capital goods which are sold to organizations. Such innovations are often *systems solutions* with many different hardware and software components tailored to the specific needs of a customer organization. Their transaction is not a simple act, but a long-lasting process during which *both* parties influence each other. They bargain over the terms and they cooperate in the problem-solving process. Especially with important innovative launching customers, long-lasting mutually well-adjusted interorganizational relationships with strong technical, social, legal, and economic bonds exist and are well documented (Ford 1980, 1984; Gemünden 1981, 1985, p. 168; Mattsson 1983). Therefore, an interactive paradigm is clearly needed.

Although it has become a quickly diffusing fashion to propose an "interactive framework" in different variants of "personal" and "organizational", "static" and "dynamic", "dyadic" and "multi-party" approaches, the heavy burden of empirical work has often been laid on future researchers. In particular, no empirical research has been reported which analyzes the role of key decision makers in innovative transaction processes. We have performed such a study (Gemünden 1981). We have analyzed the *problem-solving* and *conflict-handling interaction strategies* of buyer and seller organization, and their influence on the *transactional efficiency* of both parties.

Our model has been tested with a random sample of 1985 interaction processes in which until the 30th of June 1966 an *EDP system* was ordered by a West German organization with no previous EDP equipment. The data were gathered by Witte and his research team (Witte 1969, 1972, 1977 and Gemünden 1981, pp. 77–102), who performed a systematic content analysis of about 147,000 documents filed by the four major EDP manufacturers. The analyzed documents included: ". . . correspondence, offers, system proposals, contracts, estimates of data volume and of the cost effectiveness, complaints, invoices, internal statistics, and internal memos concerning the enterprises in the sample" (Witte 1972, p. 162). It should be noted that this unique database has been a very valuable source for seminal German contributions to organizational decision making. (For a bibliography see Gemünden 1981, appendix).

From our empirical work a *contingency hypothesis* has emerged. There does not exist one efficient way of managing innovative buyer–seller interactions. Rather, both parties must come to an agreement with each other on the *level of aspiration* of their innovative problem solution. They must choose an interactive strategy which *corresponds* to this goal:

• If a solution with a *low* level of aspiration is intended, the *seller* should dominate the interaction process. The problem-solving activity should focus on the *technological* aspects of the innovative problem, and *intensive bargaining* and consultation of competitors should be *avoided*.

• If a solution with a *high* level of aspiration is intended, *both* parties should be involved in an interaction process during which technological *and* organizational problems of innovation using are integratively analyzed. Both parties should consciously *fight out their conflicts in an open manner* by bargaining activities and by consulting competitors. But, they should also *commit* themselves to the results of this process and restrict their conflict-handling activities to the decision stage.

Our results have shown that a *delegation-to-the-seller paradigm* is efficient for a small innovative step, whereas for a big innovative step an *intensive-interaction paradigm* is needed. But, which personal requirements must be met to realize this latter model? *Which roles must the key decision makers play to implement such an interaction strategy successfully?* How can the seller persons recognize if there are key persons in the buying center who can play these roles? How can they counter the high challenge of these ambitious individuals? How can they secure their preference and cooperation?

To answer these questions we need a systematic investigation of the innovator roles:

• a review of the empirical research which has explored the contribution of innovator roles to the success of innovations,
• an analysis of the inter-organizational roles which key individuals play in innovative transaction processes, and
• a systematic empirical investigation which shows how the interplay of the different innovator roles influences the transactional efficiency of buyer and seller organization.

EMPIRICAL RESEARCH ON ROLES AND CHARACTERISTICS OF KEY INDIVIDUALS IN INNOVATIVE DECISION PROCESSES

The "interactive paradigm" has not yet explicitly analyzed the roles of key individuals in transactions of innovative goods. Therefore, we have to report the results from conventional research streams, namely: the research on "buying centers", and the studies on research and (product) development which highlight the roles of "innovators" and "champions". Also reported is Witte's research on the so-called "promotors" which has emerged from his theory of organizing complex innovative decision processes.

Contribution of Buying-Center Studies

The research on *buying centers* offers comprehensive case studies which illustrate e.g. the role of "gatekeepers" (Pettigrew 1972) and the use of "lateral" influence tactics (Strauss 1962). Big surveys have consistently shown that in important complex buying decisions with a high degree of innovation, users, engineers, and top-managers participate with a higher likelihood and presumably exert more influence than in routine buying decisions (Buckner 1967; Scientific American 1969; Spiegel 1972; Erickson and Gross 1980; Spiegel 1982). Medium-scaled studies reveal that the composition of the buying center also depends on the "stage" of the decision process, the novelty and complexity of the buying situation, the type of the transferred commodity, and the organizational and environmental properties of the buying unit. (Given our limited space we reference only the perhaps less known German studies: Kirsch and Kutschker 1978; Klümper 1969; Kutschker and Roth 1975; Roth, Huppertsberg, and Schneider 1974; Schulz 1977; Witte 1970, 1973).

Although many empirical studies have already investigated the buying center, none has explicitly analyzed the role of innovators. Furthermore, an in-depth study and review from Silk and Kalwani (1982) casts severe doubts on the reliability of frequently used influence measures, because they lay the burden totally on the respondents to translate from complex social reality into highly abstract theoretical terms. The buying center studies are too much concerned with departmental memberships and hierarchical ranks. They are caught in the trap of the organizational boundaries of the buying unit. Thus, they take the risk of running down a blind alleyway perpetuating simplifying prejudices of complex buyer–seller relationships. To give us a more comprehensive picture of innovative buying, studies should focus on the *transaction center* and include *inter*-organizational roles. They should systematically analyze the power bases of the key decision makers (Patchen 1974; Brose and Corsten 1981a, 1981b) and describe through which *activities* these individuals exert influence in formal and *informal* communication networks, instead of relying on predetermined stereotypes of vaguely defined "phases" and "functions". (A noteworthy exception is made by Vyas and Woodside 1984, Backhaus ' Günter 1976, Ferguson 1979, Gemünden 1981: Witte 1972 and ·er 1978 critically discuss "phase-theorems".)

Contribution of Selling-Center Studies on Research and Product Development

Studies on selling firms have stressed *personal commitment* to innovation from their beginning. In the early stage, quite heroic one-man theories like Schumpeter's *"dynamic entrepreneur"* (1931) dominated. Later conceptualizations have postulated a *division of labor* between different roles. Recognizing communication problems between inventors and top-managers and potential users, Schon (1963) proposes a second man who promotes the innovation. His *"product champion"* is required to overcome underground resistance to change. He must be strongly committed to the innovation, have considerable power and prestige in the organization, and promote it through a network of good personal contacts, rather than using the formal channels. Schon gives only anecdotal empirical evidence for this hypothesis but a study from Chakrabarti (1974, p. 59) on the success of 45 NASA innovations shows that the presence of a product champion, who plays a dominant role in integrating research-engineering interaction, is very strongly related to success. Further evidence is provided by Globe, Levy, and Schwartz (1973, p.12). However, Schon's "somewhat romantic view that Product Champions, if they have no formal role in a project, can nevertheless overcome barriers in organizations . . ." is *contested* by Jervis (1975, p. 23). His results from SAPPHO research prove that product champions, in successful innovations, always played another more formal role in the innovation process, whereas in unsuccessful cases they normally did not play such a role. This is in line with Chakrabarti's findings of a high correlation between top-management support and product success (1974, p. 62). It is also consistent with Maidique and Zirger (1984, p. 198) who found that senior management's support significantly improved product success, but a "clearly identifiable product champion" showed no significant influence.

The importance of an enthusiastic "top person" or a *"business innovator"* is documented in the Langrish et al. study (1972), by several studies conducted with the SAPPHO approach (Rothwell et al. 1974; Rothwell 1976, 1977; and Szakasitis 1974), and the MIT study from Utterback and his research fellows (1976). The SAPPHO findings indicate that successful business innovators have more power, a higher formal status, more responsibility, more diverse experience and more commitment to the innovation than the managers of failures (Jervis 1975).

A third, but not so well documented figure in this stream of research is the "*technical innovator*", "the 'inventor' or single individual who made the major technical contribution to the development and/or design of the innovation" (Jervis 1975, p. 21). The SAPPHO studies confirm the importance of this promotor in the instruments and textile machinery industries, but not in the chemical industry (ibid., p. 23). Some support is also given in the Langrish et al. study (1972) where an "other person" often described as a "mechanical genius" plays an important role in mechanical engineering, electrical, and crafts industry, but not in chemical industry.

A fourth figure which has been identified by Allen and Cohen (1969) is the "*technological gatekeeper*". He occupies a key position in the communication network of the research laboratory. His formal influence as a superordinate is strengthened by his expertise and his boundary spanning role in external communication (Allen and Cohen 1969; see also Persson 1981, p. 39).

To summarize: *Empirical research on "selling centers" documents that top-managers and experts—who are strongly committed to an innovation—do positively influence the success of research and product development. They occupy central positions in formal and informal networks and use them intensively.* It is important to add that these studies also have proven good external communication and customer involvement during the development process as very critical factors for innovation's success. We therefore suppose that seller's key individuals also play an important role in the *inter*-organizational development and implementation process.

Contribution of Witte's Research Project "Columbus" on the Organization of Innovative Decision Processes

A quite advanced two-power center theory has emerged from Witte and his scholar's research on the organization of innovative decision processes (Witte 1973, 1976, 1977). Their basic assumption is that complex innovations are faced with barriers (Hauschildt 1977; Schulz 1977; Witte 1973). Witte's first barrier, which he names the "*barrier of will*" (Witte 1977, p. 49), arises because innovations alter the status quo and its balance of power. His "*barrier of capability*" (ibid. p. 51) is explained by the very nature of the innovation which is not only unknown as a technological object, but also as a source of new demands for its utilization. To surmount these barriers Witte proposes two corresponding innovators which he calls "promotors" (Witte 1977, p. 67: "I use the German spelling of the word rather than the English term 'promotor' since the latter denotes in a restrictive way a person promoting another person or thing on a professional basis"). The *"promotor by power"* is defined as a person ". . . who actively and intensively promotes an innovation process by means of his hierarchic power" (ibid., p. 54). The *"promotor by know-how . . ."* actively and intensively encourages an innovation process by means of object specific know-how" (ibid., p. 54). Both roles resemble the independently developed "business" and "technical innovator" in the SAPPHO research project. However, for Witte, personal commitment to an innovation is an essential hallmark of any promotor, whereas for the SAPPHO researchers it constitutes a third role which they call "product champion" (Jervis 1975, p. 22. See also his findings on p. 24, showing the personal unions of different roles).

Witte's framework goes one step further than the other conceptualizations by looking at combinations of promotors, which he calls *"promotor structures"*. His preferred model is the *"tandem structure"*, a coalition of a promotor by power and a promotor by know-how (Witte 1977). Using a somewhat harmonizing metaphor, we may compare its functioning with a *bell*: The promotor by power represents the huge body which guards the promotor by know-how against opponents and gives him resources to perform his time-consuming task, the promotor by know-how is the clapper, the dynamic element, which gives the bell its characteristic sound by supplying it with qualified object-specific arguments, which are then disseminated by the promotor by power.

Other promotor structures are supposed to be inferior to this organizational model:

- The Schumpeterian *"personal union of promotor by power and promotor by know-how"* has less management capacity, the creative dialogue between the power-holder and the expert is missing, and there is a permanent conflict between engaging in other important top-management decisions or in interesting technical details of the innovative problem.
- The *"unilateral power structure"* with a solitary promotor by power is unlikely to reach a high degree of innovation since he ". . . neither has the technical insight required for proper evaluation of alternatives nor can he hope to put the innovation into practice by himself" (ibid., p. 56).
- In the *"unilateral know-how structure"* the promotor by know-how is working in isolation. "Although he will be able to plan the innovation project and the method for its implementation, he will be defeated by the barriers of will, or at least be forced

to one compromise after the other until the innovation is reduced to insignificance" (ibid., p. 56).

• The *"structure without promotors"* stands defenseless against both barriers of not wanting and not knowing and is therefore supposed to realize the smallest step of innovation.

In his *empirical test* with a sample of 233 decision processes where computers in Germany were bought or leased for the first time, Witte showed that decision processes with the postulated *tandem-structure* occurred most frequently, and performed the highest number of problem-solving activities in a decision process of medium duration (ibid.). As hypothesized, they reached the highest "degree of innovation", an expert rating considering time of adoption, size of the buyer organization, technology level, and quality of planned usage of the innovation (ibid.).

The personal union structure showed the second highest degree of innovation, but relatively few problem-solving activities during a short decision process. The unilateral power structure also exhibited a short decision process, but more problem-solving activities and the third highest degree of innovation. The unilateral know-how structure needed a very long time and reached only a moderate degree of innovation, although performing many problem-solving activities. The structure without promotors did the smallest innovation step in the shortest time with the least number of problem-solving activities.

Witte's findings were confirmed by a follow-up study by Klümper (1969) and an in-depth study by Kaluza (1979). Other studies that have also claimed to have found evidence for Witte's proposition should be interpreted cautiously, since they did not demonstrate the personal commitment of their "promotors" (Baumberger, Gmür, and Käser 1973; Knopf 1975; Uhlmann 1978; Wintsch 1979). Witte has extended his framework by including opponents by power and by know-how, but this theory has not been tested empirically (Witte 1976, further theoretical advances are reviewed in Corsten 1982; Gaitanides and Wicher 1985; Gemünden 1981; Kaluza 1982; Reber and Strehl 1983; and Schienstock 1975).

To summarize: *Witte's research not only validates the importance of committed top-managers and experts for the success of innovations, it also demonstrates the interplay of these promotors.* Additional findings from Witte and his scholars clearly indicate that salesmen and consultants were markedly influenced by the challenge which buyer's promotors created (Witte 1973, p. 40;

Klein 1974). To explain these influences we will now analyze the inter-organizational roles of the promotors systematically.

THE INTER-ORGANIZATIONAL ROLES OF "PROMOTOR BY POWER" AND "PROMOTOR BY KNOW-HOW"

Our analysis starts with a caveat: the already gathered data did not allow us to identify the promotors in the selling center in the same reliable manner as in the buying center. Therefore, we have already restricted our theoretical analysis on the relationships between buyer organization's promotors and a holistically personalized "seller organization" which has to play several roles. However, the reader should realize that buying and selling centers did differ empirically. The activity on the buyer side was often concentrated on a few key persons who entered very early in the decision process and stayed till the end (Kelly [1974] reports similar results, Rothwell [1977, p. 202] stresses their link function between different stages of the process). In promoting the innovative idea, these individuals often went beyond their formal role. In contrast, the seller persons often entered in the middle of the process, and left it when their special job was done. They had a more formally institutionalized divison of labor, a more equal interpersonal distribution of work-load, and more direct, short-term, financial incentives. But, the selling centers also showed a great variance. Thus, one should not simply postulate a selling center which completely matches the buying center. Rather, we should recognize that the selling centers will vary, depending on the degree of innovation which the trans-action process has for *them*. And we have to wait for empirical research in this rather unexplored area.

The mutual influence of promotors and seller organization's members is based on their motives and power bases. The seller organization's members are interested in the promotors for the following reasons:

- The *promotor by power* is a potential sponsor (Galbraith 1982, p. 10) who also advocates to other top-managers in favor of the innovation. He can protect the salesmen against opponents and avoid wasting their time by effectively coordinating their coop-eration with buyer organization's employees. He can be used to get more support from the own seller organization and he may act as a "troubleshooter" if unexpected problems emerge. Finally, he may argue in favor of an integrated systems solution with a higher sales volume and against proposals from competitors.

- The *promotor by know-how* is an important partner in the problem-solving process. He is required to develop a solution which is suited to the customer-specific needs of innovation usage. He knows underground resistance against change and can defend the seller persons against opponents by know-how. As a member of the buyer organization he may act as a competent and credible opinion leader and reach "influencers" and (potential) users who are not contacted by the sales people. Finally, he may also argue in favor of an integrated systems solution, and against competitor's proposals.

Seller organization's members are also important allies for buyer organization's promotors:

- The *promotor by power* needs problem-solving support. He looks for an experienced and skilled partner who can assist him to control and coordinate the innovative decision process. He searches for a trustworthy partner who can quickly solve unexpected implementation problems, and he wants a highly competent and authorized counterpart who can bargain with him upon the terms of the transfer such as price, delivery date, warranties, and financial conditions.
- The *promotor by know-how* needs relevant, actual technical details to maintain his expert power. He looks for technically competent and credible partners who discuss and constructively criticize his proposals, keep him in touch with specialists in the seller organization and give him further support for his problem-solving activities.

The different inter-organizational influences of promotors become fully evident if we focus on promotor *structures*:

- The *single promotor by know-how* is rather dependent on the seller, because he has no promotor by power who sponsors his work and gives him resources. But, lacking this companion, makes him less interesting for the seller persons. He has to rely on his quite weak referent and expert power, the latter mainly resulting from his knowledge of customer-specific needs and abilities to implement the new system.
- The *single promotor by power* is surely more attractive for the seller, since he alone already offers a good chance for selling an innovative system, which requires high expenditures. But,

lacking a promotor by know-how, he is also very dependent on the seller, because he needs his problem-solving support.

- In contrast to these quite weak positions stands the *tandem structure*. This coalition enables a very *efficient division of labor*: The promotor by know-how screens interesting suppliers, and the promotor by power attracts bids and detailed proposals, which the promotor by know-how in turn carefully evaluates. He can then advise the promotor by power during bargaining sessions and protect him against influence attempts which either exaggerate technical problems or bagatellize them.

From these considerations are derived the following *hypotheses*:

- The level of problem-solving activity of both parties, buyer and seller organization, is highest in interaction processes with a tandem structure. It is lowest in a structure without promotors, and it takes intermediate values in processes with a unilateral promotor structure or a personal union structure.
- The frequency of bargaining activities and consultations of competing seller organizations is highest in interaction processes with a tandem structure. It is lowest in a structure without promotors, and it takes intermediate values in processes with a unilateral promotor structure or a personal union structure.

The *empirical results confirm both hypotheses* (Gemünden 1981). They also show that the ratio of seller-to-buyer activities is lowest when both promotor roles are played, thus emphasizing again the high challenge of the tandem structure for the seller organization. Additional findings show *how* the promotors influence the interaction process: Promotors by power control the decision process by fixing deadlines, whereas promotors by know-how structure the process by decomposing tasks into subtasks of different content (Grün 1973). Promotors by power preferably request formal bids, especially in the unilateral power structure, whereas promotors by know-how pose many written questions about technical details, and the universal promotor searches direct contact in two-sided oral communications (Gemünden 1981).

To summarize: *Promotors are very influential, but they show a janus face. On the one hand they are partners in developing and implementing an innovative solution, advocates who sell a product in the buying organization, fund raisers who look for the money, and allies against opponents to innovation. On the other hand, they demand more problem-solving support, bargain harder, and interact more intensively with competitors. Furthermore, they coordinate these direct and indirect influence modes of coping with buyer-seller conflicts more efficiently, at least in the tandem structure.* This duality makes it interesting to analyze how they influence the transactional efficiency of buyer and seller organization. To do this, we must first operationalize this construct.

THE EFFICIENCY OF BUYER AND SELLER ORGANIZATION: A MEASUREMENT CONCEPT AND ITS APPLICATION

The Measurement Concept

The "transactional efficiency" of buyer and seller organization is very difficult to operationalize, since the goal systems of both parties are complex and include many important intangible elements. Our bibliographic review of 46 empirical investigations consistently documented that diffuse goals and object-specific technical criteria which express the *new* possibilities of the innovation had the highest priorities in buyer's multiple goal systems (see Gemünden [1981] for the detailed references). Research from Hamel (1974) and Hauschildt (1977) revealed a tight link between goal-setting and problem-solving activities: goals are not "given" at the beginning of an innovative decision process, rather they are developed in a continual dialectical process with the offered, searched and developed alternatives. Therefore, our measurement could not rely on a well-defined hierarchy of goals. Our concept had to cover an intangible process of goal-formation and problem solving. A second point was that the concept had to include measures which reflect the efficiency of the *decision process*, and which the involved manager could *influence*. Third, the concept should contain efficiency dimensions which indicate that the buyer-seller relationship will probably develop positively in the *long run*. Although we clearly could not express transactional efficiency in a metric scale of hard dollars, readers should acknowledge that this is the first study which measures the transactional efficiency of buyer and seller organization and relates it to their interaction strategies.

Our measurement concept comprises the following steps:

1. The theoretical derivation starts with the *formal definition* of "efficiency" as a summarizing statement upon the attainment of several efficiency dimensions.
2. To clarify the meaning of efficiency the next step requires a specification of several *theoretical efficiency dimensions*.

3. These dimensions have to be operationalized by *empirical indicators* in the next step.
4. To reduce our many efficiency indicators we used *factor analyses*. The resulting factors can be interpreted as *empirical efficiency dimensions*.
5. Since this step produced several empirical dimensions, a further summarizing step was required. We used *cluster analyses* to find transaction processes with similar efficiency profiles.

Applying the Concept

To obtain our theoretical efficiency dimensions we ask ourselves two questions:

1. Which object is to be evaluated as efficient or inefficient?
2. From whose perspective is something called efficient or inefficient?

The evaluated *objects* are: (a) the activities of the *decision process* from the initiative, when a competent member of the buyer organization decides to allocate resources to the decision problem (Schulz 1977, p. 27), to the final choice of an EDP configuration, when the (first) contract is signed, (b) the *chosen solution*, at the end of the decision process, and (c) the activities of the *implementation process*, reaching from the formal choice to the first installation of an EDP configuration.

The *perspective* of evaluation distinguishes between three categories: the *buyer* organization's point of view, the *seller* organization's way of looking at the problem, and the *common* perspective of buyer and seller organization. With this classification we want to make a pragmatic distinction between efficiency aspects dominantly related to the buyer organization, to the seller organization, and those aspects related to both organizations.

Combining these two criteria gives us nine theoretical efficiency dimensions which are operationalized by 40 efficiency indicators. Figures 1, 2, and 3 show the results of this procedure. (For a detailed description see Gzuk [1975], who operationalized many of the indicators [Gemünden 1978; Gemünden 1981]). These indicators were then factor-analyzed using principal component analysis and varimax rotation. Five factors were extracted, explaining 50.4 percent of the total variance. We interpret the five factors shown in Appendix 4 in the following manner:

1. Our first factor, the *"rationality of the decision process"* includes the efficiency indicators of the decision process. High factor

loadings indicate a wide problem definition, a high willingness to decide, a great degree of transparency, and a wide and deep goal structure. On the other hand, high factor loadings also document high decision expenditures for buyer and seller organization and a long duration of the decision process.

2. Our second factor, the *"stability of choice"* is characterized by high positive loadings on stability and definiteness of choice, and uninterruptedness of the implementation process. The negative loadings show that low stability of choice is empirically equivalent with many major correction steps, and a long duration of the implementation process. The high loading on improvement of volume during the implementation process points out that the seller can only trade up if he is willing to accept a longer implementation stage.

3. High loadings on the third factor document high implementation expenditures. We multiply the factor scores by -1 and name our new factor *"economy of implementation"* since it indicates cost savings, as compared to other transaction processes.

4. Our fourth factor contains the efficiency indicators of our common perspective. High positive factor loadings announce a good bargaining climate, a friendly style of correspondence, high satisfaction with the chosen solution, and a positive evaluation of goal attainment. Negative loadings are symptomatic of many and severe complaints. We therefore label this factor *"harmony of interaction"*.

5. Our last factor shows high loadings for the qualitative flexibility and the sales volume (monetary scale). The substantial positive loadings of quantitative flexibility, technical novelty and sales volume (capacity scale) suggest that a certain "flexibility slack" might have been paid by the buyer organization. This assumption is confirmed by the negative loadings of the adequacy of solution components. Leaving open the question whether some overselling has taken place or not, we can state that higher factor loadings indicate a higher sales volume and label our fifth factor *"sales success"*.

The last step of our procedure is a cluster analysis of the factorial efficiency profiles using Ward's hierarchical method (1963) which produces the following solution:

1. Three clusters which are *inefficient for buyer and seller*. They include solutions with a very low stability of choice, a very bad interaction climate, or very high expenditure for problem-solving and implementation activities.

INDICATOR	MEASUREMENTS
Buyer's Decision Efficiency	
Width of the problem definition	Weighed sum of the problem objects considered by the decision makers
Equality of the problem solving activities	Entropy measure of the distribution of the problem-solving activities over various problem components
Willingness to decide	Weighed sum of problem objects for which choices were made before the final choice
Degree of transparency	Number of alternative EDP systems considered during the decision process, relative to the offer of the EDP market
Degree of rationality	Counts how many kinds of typical rational activities have been performed during the decision process
Thoroughness of decision	General rating from the data collectors reaching from (2) superficial to (6) meticulous
Width of the decision goals	Number of goal sections considered by the decision makers
Deepness of the decision goals	Number of goal attributes considered by the decision makers
Decision expenditure of the buyer organization	Time consume of buyer-organization members during the decision stage
Buyer's Solution Efficiency	
Degree of innovation	Summarizing experts judgement considering chosen hardware, software and utilization of the computer relatively to the firm's historical situation
Technical novelty of the EDP-system	Time lag between the firm's adoption and the adoption pioneers in the firm's branch
Adequacy of solution components	Expert rating of the adequacy of the central unit and the periphery devices
Quantitative flexibility	Number of capacity steps (in a given computer family) without changing the type of the central unit
Qualitative flexibility	Possibilities to connect external devices with the central unit
Buyer's Implementation Efficiency	
Adequacy of means-ends-changes	Ratio of changes related to the EDP hardware and software, to the changes related to the EDP application
Corrections related to reference objects	Number of EDP applications designed during the decision stage minus supplementary EDP applications designed and programmed during the implementation stage
Corrections related to focal objects	Ratio of considerations of the chosen EDP configuration to all EDP configurations considered during the implementation stage
Major correction steps	Weighed sum major corrections steps such as annulations of contracts
Stability of choice	Rating of the data collectors reaching from (2) great and many changes to (5) no changes of the chosen solution
Definiteness of choice	Similarity between the chosen and installed EDP configuration
Uninterruptedness of the implementation	Number of postponements of major implementing activities like installation, programming and training of buyer-organization members
Implementation expenditure of the buyer organization	Time consume of buyer-organization members during the implementation stage

Fig. 1. Buyer's efficiency indicators.

INDICATOR	MEASUREMENT
Seller's Decision Efficiency	
Improvement of volume (DP)	Difference between volume of first offered and contracted volume, expressed in capacity units of the EDP-system
Elimination of competitors (DP)	Intensity of competitors' influence during the decision stage
Decision expenditure of the seller-organization	Time consume of seller-organization members during the decision stage
Expenditure for the training of buying-organization members	Training time given to the buyer-organization members
Seller's Solution Efficiency	
Sales volume (monetary scale)	Measured on a monetary scale with 6 categories
Sales volume (capacity scale)	Measured on a capacity scale with 9 categories
contractual reservations	Rating, reaching from (2) many, and severe reservations to (6) no reservations at all
Seller's Implementation Efficiency	
Improvement of volume (IP)	Difference between size of the contracted and installed EDP-system
Elimination of competitors (IP)	Intensity of competitors' influence during the implementation stage
Implementation expenditure of the seller-organization	Time consume of seller-organization members during the implementation stage

Fig. 2. Seller's efficiency indicators.

2. One *seller-efficient* cluster with a very low expenditure for problem-solving activities, a high stability of choice and average values for the other factors.
3. One *buyer-efficient* cluster: characterized by a low sales volume and average values for the other factors. This means that the buyer has a very good problem-solving-support/price ratio.
4. One cluster which is *efficient for both parties*. This cluster has positive values on all five factors.

THE EMPIRICAL EVIDENCE

To evaluate the efficiency of the different promotor structures we first determine the percentages of *favorable outcomes* per emerged structure. We determine buyer's (seller's) favorable solutions by summing up the relative frequencies of the buyer-efficient (seller-efficient) cluster and the cluster which is efficient for both parties.

INDICATOR	MEASUREMENT
Common Decision Efficiency	
Bargaining climate	Data collector's rating from (2) extreme disputes to (5) kind and open atmosphere
Style of the buyer-seller correspondence	Data collector's rating ranging from (1) extremely cold to (7) extremely warm
Duration of the decision process	Date of the contract event minus date of the start-event
Major interruptions of the decision process	Sum of the three largest interruptions of the decision process
Common Solution Efficiency	
Satisfaction with chosen solution	Data collector's rating of buyer's satisfaction with chosen solution ranging ranging from (1) extremely unsatisfied to (7) extremely satisfied
Common Implementation Efficiency	
Complaint behavior	Number and intensity of buyer's complaints during the implementation stage
Goal attainment evaluation	Positive and negative evaluations of the implemented solution by members of the buyer-organization
Duration of the implementation process	Date of the installation of the EDP-configuration minus date of the contract event

Fig. 3. Common efficiency indicators.

Figure 4 shows the result of this aggregation:

- The *seller* organization's most favorable constellation is the *structure without promotors* followed by the *unilateral know-how structure*. Its third and fourth ranks are occupied by structures where both promotor roles are played, and its worst alternative is clearly the unilateral power structure.
- The *buyer* organization's most favorable solution is the *universal promotor*, followed by the *tandem structure*, and the unilateral know-how structure. Its worst alternative is clearly the structure without promotors. In addition, the single promotor by power also shows an unsatisfactory likelihood of a favorable outcome.

It appears that buyer's efficiency is increased by the engagement of a promotor by know-how, whereas seller's efficiency seems to be influenced negatively by a promotor by power. To explore the reasons for this finding we take a look at figure 5.

We can see that the problem-solving activity of the *promotor by know-how* increases the likelihood of coming to a solution which is an efficient compromise for both parties. His influence decreases the risk for an outcome which is inefficient for both organizations.

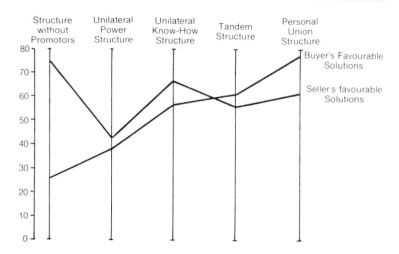

Fig. 4. Aggregated efficiency values for buyer and seller organization for different promotor structures.

In particular, these processes have a higher stability of choice and lower expenditures for implementation. Since buyer-efficient solutions also are more likely with a promotor by know-how, the buyer organization's managers should create a climate and incentive systems which encourage such a promotor.

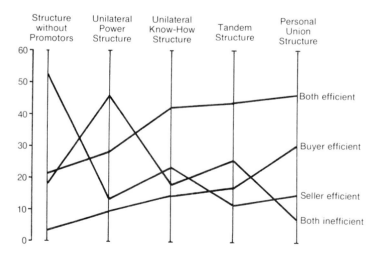

Fig. 5. Efficiency values for buyer and seller organization for different promotor structure.

Remember, however, that the single promotor by know-how took a relatively long time to develop his concept and to overcome resistance to change. His solution only reached a low degree of innovation. If a *big* innovative step is intended, then *both* promotor roles have to be played, i.e. a *tandem structure* or a *universal promotor* is needed. Our detailed findings show that these structures not only realize the greatest innovative step, but they also establish the *best interaction climate*. Since the tandem structure occurs more frequently in large organizations it is characterized by high expenditures for the decision process and a higher sales volume. In contrast, the universal promotor shows a very low sales volume and less decision expenditures, as he typically emerges in small and medium firms.

If we look at the curve with the seller-efficient solutions, we see a marked difference between the *structure without promotors* and the other structures. Without promotors by know-how asking for advice and, without promotors by power demanding problem-solving support, the seller's agents were able to carry through an efficient *hard-selling approach*. The very low expenditures for problem-solving activities for these customers, their small degree of innovation, and the reserved interaction climate in these buyer-seller relationships confirm this hypothesis. We suppose that these customers belong to an "early majority" of adopters who simply follow leading competitors in their industries, not preparing themselves thoroughly for the innovative project. Since their sales volume was not significantly above average, they do however form a big and interesting group for the seller organization.

What makes the *unilateral power structure* so unattractive for both organizations? We had expected that a promotor by power could raise funds and give seller agents chief support in the buying organization. Figure 5 reveals that this structure has a very high occurrence of solutions which are inefficient for both parties. Particularly, these processes often show major corrections of the chosen solution and angry, complaining customers. Symptomatically, competitors are consulted after the contract is signed. The main reason for the failure is the evident disproportion between the high level of aspiration of the *planned* solution and the lack of a promotor by know-how who has enough object-specific insight to recognize the organizational consequences of this concept, and who does the work to implement it. The very low ratio of seller-to-buyer level of problem-solving activity is also symptomatic. The promotor by power probably expected that the seller persons would substitute his lack of expertise. However, the opposite is true: Without a challenging expert partner on the buyer side the seller members did not engage themselves above average. In particular, they strongly neglected the aspects which were related to innovation usage.

SUMMARY OF THE RESULTS, AND DISCUSSION OF THE MANAGERIAL AND THEORETICAL IMPLICATIONS

Summary of the Results

Whereas buying center research has not yet explicitly analyzed the specific roles of innovators, studies on research and product development in selling firms and Witte's research on "promotors" have consistently shown that *committed top managers and experts do positively influence the success of industrial product development and diffusion.* Our systematic investigation of the *inter-organizational roles* played by promotors by power and by promotors by know-how has proven clear-cut differences between the different promotor structures. *Particularly the tandem structure advanced a very efficient division of labor for intra- and inter-organizational tasks.*

The *"transactional efficiency"* of buyer and seller organization, which has been measured for the first time in our study, was strongly influenced by the engagement of the promotors:

- *Promotors by power strive for ambitious solutions, but they often fail without a cooperating promotor by know-how.*
- *Promotors by know-how prepare their proposals thoroughly.* Their solutions need fewer corrections and can be implemented with less expenditures. But, lacking a sponsoring promotor by power, they are more dependent on the seller organization's support and stand defenceless against intra-organizational opponents. Therefore, the single promotor by know-how takes a long time to develop his concept, and, forced to many compromises, realizes only a small innovative step.
- In contrast, *the powerful tandem structure not only designs, but also successfully implements a solution with a high degree of innovation. Particularly, this coalition establishes a very good interaction climate.*
- *The same positive results emerge with a personal union of a promotor by power and a promotor by know-how. However, this universal promotor typically occurs in small and medium-sized organizations.*

- The *structure without promotors* mainly shows seller-efficient solutions. It *favors an efficient hard-selling approach.*

As for problem-solving and conflict management activities, we also propose a *contingency hypothesis* for further research:

- *If a solution with a* low level of aspiration *is intended, then a structure without promotors favors a seller-efficient hard-selling approach. But, by encouraging a* promotor by know-how, *the buyer can strongly improve his efficiency, and the seller can secure a solution which is an efficient compromise for both parties.*
- *If a solution with a* high level of aspiration *is intended, then* both *promotor roles should be played. In small organizations a* universal promotor *may do this, in large organizations a* tandem structure *is required.*

Managerial Implications

First, the "promotor model" is *not* a new proposal. Many managers have already played their promotor roles successfully in the early sixties. If they would not have done so we could not have observed them and could not have offered our findings to the research community. Second, we feel that sellers are quite aware of these promotors and that they have already adjusted their marketing mix to them, e.g. in the communication mix by using different advertising strategies and brochures which are suited to the diverging needs of top-managers, experts and users (Backhaus 1982, p. 169: Strothmann 1979), and in personal selling by matching their teams to buying-center composition. Furthermore, sellers have already established units which systematically acquire and distribute user-developed innovations. Eric von Hippel states that ". . . approximately one-third of all the software IBM leases for usage on large and medium-sized computers is developed by outside users" (Hippel 1982b, p. 118).

But even though our empirical research has only proven facts and relationships that *successful* managers have long recognized, it may still offer valuable systematic insights to future managers who want to learn their lessons without paying dearly for experience. Furthermore, innovative buyer-seller transactions are so complex that each manager can hardly recognize all the problems which we have investigated. It is quite appealing to simplify problems by personalizing them. Managers should be aware of this hazard.

When using our results, they should acknowledge that it is not the persons, but rather the functions which they fulfil that are critical. They should recognize the complex *interplay* of the roles, and the fact that social influence tactics are often carefully hidden (Bonoma 1982 presents illustrative examples). From our measurement procedure of "transactional efficiency" they may derive some stimulating ideas for their difficult task of evaluating costs and benefits of expensive interaction *processes*.

Three hints may be allowed:

1. Buyer organizations should be segmented by *combinations* of promotors, since coalitions of complementary promotors differ markedly from singles (Morris and Freedman 1984).
2. Seller and buyer organizations' managers should also evaluate the *long-range perspective* of their relationship. A minor innovative step may be evaluated as a small success in the short run. However, it is a foot-in-the-door option for the seller and a riskless alternative for the less competent or less willing buyer, who can first get to know the possibilities of the new technology before he (soon) makes the next step. Taking this view, the hard-selling approach in the structure without promotors has to be discounted: It secures an economic success and a relationship with legal and technical bonds, but its often bad interaction climate also favors a bad word-of-mouth and may cause a break of the relationship in the long run (Trawick and Swan 1981).
3. Marquis (1975, p. 15) has made an important distinction between radical breakthrough innovations, complex systems innovations, and "nuts and bolts" innovations. Innovations of the first type change the whole character of an industry and are rare and unpredictable. They have been excluded from our analysis as well as complex systems innovations like weapon systems, communication networks or moon missions, that take many years and many millions of dollars to accomplish. (See Backhaus 1982; Engelhardt and Günter 1981; Günter 1979; Mattsson 1973; and the contrbutions of Engelhardt and Laßmann [1977] for such systems innovations.) But, frontiers are fluent and the remaining category is a very broad one. Therefore, we need further criteria to form homogeneous subgroups. We feel that the standardization/individualization of the hardware, the diversity of the firms in the selling coalition, the internationality of the business, and the strength of competition are important factors which influence the nature of the selling center. They should be acknowledged before transferring our findings.

Theoretical Implications

Our study has shown that research on product development and diffusion in industrial markets cannot be restricted to "buying centers" or to "selling centers", rather it should focus on *transaction centers*. It should consider the *level of aspiration of the innovative solution and efficiency for both parties* if it wants to derive managerial implications, and not simply describe how many persons, hierarchical ranks, and departments have been involved. Furthermore, it should include "promotors" on the seller side and opponents by power and by know-how on both sides. It could measure barriers to change explicitly, control for organizational and for environmental properties of both organizations, and for informal and formal networks within and between them. It could control for properties of the innovation and for the antecedents of the buyer-seller relationship, and it can be extended from a dyadic to a multi-organizational approach. Measurement of power bases, influence tactics, and motives can surely be improved by using recently developed instruments in the sociology, social psychology, and marketing science, and, last but not least, a triangulation of Witte's mirror image content analysis might also be helpful. However, proposals for empirical research are easily articulated, but hard to put into action. Clearly, an interaction approach is also needed to attack this problem. I am ready to do it.

Appendix A: Matrix of factor loadings of the efficiency indicators.

Efficiency Indicator	Factor I	Factor II	Factor III	Factor IV	Factor V	Comm.
Width of problem definition	.79					.62
Equality of problem definition	.56					.32
Willingness to decide	.79		.15	−.17		.67
Degree of Transparency	.60		.20	.15	.42	.61
Degree of Rationality	.79					.63
Thoroughness of the decision	.37			.14		.17
Width of the decisison goals	.75			−.11		.58
Deepness of the decision goals	.77					.60
Decision expenditure (BO)	.78		.44			.79
Degree of innovation	.18	−.23			.11	.11
Technical novelty of EDP-System	−.16	−.15	.27		.47	.34
Adequacy of solution components			.18	−.27	.11 .37	.26
Quantitative flexibility		−.28			.45	.28

Appendix A—continued

Efficiency Indicator	Factor I	Factor II	Factor III	Factor IV	Factor V	Comm.
Qualitative flexibility		−.21			.74	.60
Adequacy of means-ends-changes		.10	−.49	.13	−.19	.30
Corrections of reference objects	−.24	−.25	.36			.26
Corrections of focal objects		−.53	.16	−.19	.45	.55
Major corrections steps		−.87	.10	−.13		.79
Stability of choice	−.18	.67		.14		.50
Definitness of choice		.81		.14	−.16	.72
Uninterruptedness of implement	−.16	.52	−.41			.74
Implementation expenditure (BO)	.18	−.12	.98			.84
Improvement of volume (DP)				−.38	.14	.17
Elimination of competitors (DP)	−.62		−.31	.20		.53
Decision expenditure (SO)	.76		.38			.73
Expenditure for training	.24	−.13	.66		.12	.52
Sales volume (monetary scale)	.24	−.12	.11	−.20	.70	.61
Sales volume (capacity scale)	.29	.17		−.29	.50	.45
Contractural reservations	−.23	.13	−.17	.26	.25	.22
Improvement of volume (IP)		−.60			.14	.39
Elimination of competitors (IP)	−.23		−.74			.60
Implementation expenditure (SO)	.22	−.22	.82			.78
Bargaining climate				.69		.48
Style of correspondence				.67		.46
Duration of decision process	.68	−.13	−.12	.17	.38	.67
Interruption of decision process	.37	−.21	−.22	.28	.41	.47
Satisfaction with solution				.59		.37
Complaint behavior	.11	−.10	.21	−.71		.57
Goal attainment evaluation		.30	−.10	.61	−.10	.48
Duration of implementation proc.	−.10	−.72	.28		.23	.66
Explained total variance	.17	.10	.10	.07	.07	

REFERENCES

Allen, T. J. & Cohen, S. I. 1969: Information flow in research and development laboratories. *Administrative Science Quarterly*, *14*, 12–19.
Backhaus, K. 1982: *Investitionsgüter-Marketing*. Munchen: Verlag Franz Vahlen.

Backhaus, K. & Günter, B. 1976: A phase-differentiated interaction approach to industrial marketing decisions. *Industrial Marketing Management*, *5*, 255–70.

Baumberger, J., Gmür, U. M. & Käser, H. 1973: *Ausbreitung und Übernahme von Neuerungen, Ein Beitrag zur Diffusionsforschung*, 2 vols. Bern: Verlag Paul Haupt.

Bonoma, T. V. 1982: Major sales: Who really does the buying? *Harvard Business Review*, *60* (May/June), 111–19.

Brockhoff, K. 1981: *Produktpolitik*. Stuttgart, New York: Gustav Fischer UTB.

Brose, P. & Gorsten, H. 1981a: Interaktionstabellen—ein Ansatz zur Anwendbarkeit des Promotorenmodells. *Zeitschrift für Organisation*, *8*, 447–52.

Brose, P. & Corsten, H. 1981b: Anwendungsorientierte Weiterentwicklung des Promotoren-Ansatzes. *Die Unternehmung*, *13*, 89–104.

Buckner, H. 1967: *How British industry buys*. London: Hutchinson & Company.

Chakrabarti, A. K. 1974: The role of champions in product innovation. *California Management Review*, *17* (Winter), 58–62.

Corsten, H. 1982: *Der nationale Technologietransfer. Formen—Elemente—Gestaltungsmöglichkeiten—Probleme*. Berlin: Verlag Erich Schmidt.

Engelhardt, W. H. & Günter, B. 1981: *Investitionsgüter-Marketing. Anlagen, Einzelaggregate, Teile, Roh- und Einsatzstoffe, Energie-träger*. Stuttgart, Berlin, Köln, Mainz: Verlag W. Kolhammer.

Engelhardt, W. H. & Laßmann, G. (Eds.) 1977: *Anlagen-Marketing. Zeitschrift für betriebswirtschaftliche Forschung, Sonderheft 7/77*. Opladen: Westdeutscher Verlag.

Erickson, R. A. & Gross A. C. 1980: Generalizing industrial buying: A longitudinal study. *Industrial Marketing Management*, *9*, 253–65.

Ferguson, W. 1979: An evaluation of the buygrid analytic framework. *Industrial Marketing Management*, *8*, 40–4.

Ford, D. 1980: The development of buyer–seller relationships in industrial markets. *European Journal of Marketing*, *14*, 339–53.

Ford, D. 1984: Buyer/seller relationships in international industrial markets. *Industrial Marketing Management*, *13*, 101–12.

Gaitanides, M. & Wicher, H. 1985: Venture Management-Strategien und Strukturen der Unternehmungsentwicklung. *Die Betriebswirtschaft*, *45*, 414–26.

Galbraith, J. R. 1982: Designing the innovation organization. *Organizational Dynamics* (Winter), 5–25.

Gemünden, H. G. 1978: The effectiveness of buyer-seller transactions. A measurement concept and its application. Arbeitspapier, Lehrstuhl für Bankbetriebslehre, University of Saarbrücken.

Gemünden, H. G. 1980: Effiziente Interaktionsstrategien im Investitionsgütermarketing. *Marketing-ZEP*, *2*, 21–32.

Gemünden, H. G. 1981: *Innovationsmarketing—Interaktionsbeziehungen zwischen Hersteller und Verwender innovativer Investitionsgüter*. Tübingen: J. C. B. Mohr (Paul Siebeck).

Gemünden, H. G. 1985: Coping with inter-organizational conflicts—Efficient interaction strategies for buyer and seller organization. *Journal of Business Research*, *13*, 167–81.

Globe, S., Levy, G. W. & Schwartz C. M. 1973: Key factors and key events in the innovation process. *Research Management*, *8* (July), 8–15.

Grün, O. 1973: *Das Lernverhalten in Entscheidungsprozessen der Unternehmung*. Tübingen: J. C. B. Mohr (Paul Siebeck).

Günter, B. 1979: *Das Marketing von Großanlagen, Strategieprobleme des Systems Selling*. Berlin: Duncker u. Humblot.

Gzuk, R. 1975: *Messung der Effizienz von Entscheidungen. Beitrag zu einer Methodologie der Erfolgsfeststellung betriebswirtschaftlicher Entscheidungen*. Tübingen: J. C. B. Mohr (Paul Siebeck).

Hamel, W. 1974: *Zieländerungen im Entscheidungsprozeß*. Tübingen: J. C. B. Mohr (Paul Siebeck).

Hauschildt, J. 1977: *Entscheidungsziele. Zielbildung in innovativen Entschei-dungsprozessen: theoretische Ansätze und empirische Prüfung*. Tübingen: J. C. B. Mohr (Paul Siebeck).

Hippel, E. von 1978: Successful industrial products from customer ideas. Presentation of a new customer-active paradigm with evidence and implications. *Journal of Marketing*, *42*, 39–49.

Hippel, E. von 1982a: Appropriability of innovation benefit as a predictor of the source of innovation. *Research Policy*, *11*, 95–115.

Hippel, E. von 1982b: Get new products from customers. *Harvard Business Review*, *60* (March/April), 117–22.

Jervis, P. 1975: Innovation and technology transfer—the roles and characteristics of individuals. *IEEE Transactions on Engineering Management*, *22*, 19–27.

Kaluza, B. 1979: *Entscheidungsprozesse und empirische Zielforschung in Versicherungsunternehmen*. Karlsruhe: Verlag Versicherungswirtschaft e.V.

Kaluza, B. 1982: Das Promotoren-Modell. WiSt—*Wirtschaftswissenschaftliches Studium*, *11*, 408–12.

Kelly, P. J. 1974: Functions performed in industrial purchasing decisions with implications for marketing strategy. *Journal of Business Research*, *2*, 421–33.

Kirsch, W. & Kutschker, M. 1978: *Das Marketing von Investitionsgütern. Theoretische und empirische Perspektiven eines Interaktionsansatzes*. Wiesbaden: Gabler-Verlag.

Klein, H. 1974: Die Konsultation externer Berater. *Entscheidung unter Außeneinfluß*. H. Klein & J. Knorpp (Eds.), Tübingen: J. C. B. Mohr (Paul Siebeck).

Klümper, P. 1969: Die Organisation von Entscheidungsprozessen von Industrieanlagen. Dissertation, University of Mannheim.

Knopf, R. 1975: Dimensionen des Erfolgs von Reorganisationsprozessen. Dissertation, University of Mannheim.

Kutschker, M. & Roth, K. 1975: Das Informationsverhalten vor industriellen Beschaffungsentscheidungen. Ein empirischer Bericht aus dem Sonderforschungsbereich 24 für sozial- und wirtschaftspsychologische Entscheidungsforschung der Universität Mannheim, University of Mannheim.

Langrish, J., Gibbons, M., Evans, W. G. & Jevons, F. R. 1972: *Wealth from knowledge. Studies of innovation in industry*. London: Macmillan.

Maidique, M. A. & Zirger, B. J. 1984: A study of success and failure in product innovation: The case of the U.S. electronics industry. *IEEE Transactions on Engineering Management*, *31*, 192–203.

Marquis, D. G. 1976: The anatomy of successful innovations. In R. R. Rothberg (Ed.), *Corporate strategy and product innovation*. New York: The Free Press 14–25.

Mattsson, L.-G. 1973: Systems selling as a strategy on industrial markets. *Industrial Marketing Management*, *3*, 107–20.

Mattsson, L.-G. 1983: An application of a network approach to marketing—defending and changing market positions. Paper presented to the Scandinavian–German Symposium on Empirical Marketing Research at the University of Kiel.

Morris, M. H. & Freedman, S. M. 1984: Coalitions in organizational buying. *Industrial Marketing Management*, *13*, 123–32.

Parkinson, S. T. 1981: Successful new product development—an international comparative study. *R&D Management*, *11*, 79–85.

Patchen, M. 1974: The locus and basis of influence on organizational decisions. *Organizational Behavior and Human Performance*, *11*, 195–221.

Persson, O. 1981: Critical comments on the gatekeeper concept in science and technology. *R&D Management*, *11*, 37–40.

Pettigrew, A. M. 1972: Information control as power resource. *Sociology*, *6*, 187–204.

Reber, G. & Strehl F. 1983: Zur organisatorischen Gestaltung von Produktinnovationen. *Zeitschrift für Organisation*, *52*, 262–6.

Roth, K., Huppertsberg, B. & Schneider, J. 1974: Empirische Untersuchung zum Investitionsgütermarketing in der BRD—Verwenderbericht. Veröffentlichung aus dem Sonderforschungsbereich 24 für sozial- und wirtschaftspsychologische Entscheidungsforschung der Universität Mannheim, University of Mannheim.

Rothwell, R. 1976: Innovation in textile machinery. Some significant factors in success and failure. Science Policy Research Unit, Occasional Papers Series, No. 2, University of Sussex.

Rothwell, R. 1977: The characteristics of successful innovators and technically progressive firms. *R&D Management*, 7, 191–206.

Rothwell, R., Freeman, C., Horsley, A., Jervis, V. T. P., Robertson, A. B. & Townsend, J. 1974: SAPPHO undated—Project SAPPHO phase II. *Research Policy*, 3, 258–91.

Schienstock, G. 1975: *Organisation innovativer Rollenkomplexe*. Meisenheim an Glan: Anton Hain.

Schon, D. A. 1963: Champions for radical new inventions. *Harvard Business Review, 41* (March/April), 77–86.

Schulz, D. H. 1977: *Die Initiative zu Entscheidungen*. Tübingen: J. C. B. Mohr (Paul Siebeck).

Schumpeter, J. 1981: *Theorie der wirtschaftlichen Entwicklung. Eine Untersuchung über Unternehmergewinn, Kapital, Kredit, Zins-und den Konjunkturzyklus*, 3rd ed., München und Leipzig: Duncker u. Humblot.

Scientific American 1969: *How industry buys 1970*. New York: *Scientific American*.

Silk, A. J. & Kalwani, U. 1982: Measuring influence in organizational purchase decisions. *Journal of Marketing Research, 19*, 165–81.

Spiegel Verlag 1972: *Entscheidungsprozesse und Informations-verhalten in der Industrie*, Hamburg: Spiegel Verlag.

Spiegel Verlag 1982: *Der Entscheidungsprozeß bei Investitions-gütern: Beschaffung, Entscheidungskompetenzen, Informationsverhalten*. Hamburg: Spiegel Verlag.

Strauss, G. 1962: Tactics of lateral relationship: The purchasing agent. *Administrative Science Quarterly, 7*, 161–86.

Strothmann, K.-H. 1979: *Investitionsgütermarketing*. München Verlag Moderne Industrie.

Szakasatis, G. D. 1974: The adoption of the SAPPHO method in the Hungarian electronics industry. *Research Policy, 3*, 18–28.

Trawick, F. I. & Swan, J. E. 1981: A model of industrial satisfaction/complaining behavior. *Industrial Marketing Management, 10*, 23–30.

Uhlmann, L. 1978: *Der Innovationsprozeß in westeuropäischen Industrieländern*, Vol. 2. Der Ablauf industrieller Innovationsprozesse, Berlin und München: Duncker u. Humblot.

Utterback. J. M., Allen, T. J., Hollomon, H. J. & Sirbu, M. A. 1976: The process of innovation in five industries in Europe and Japan. *IEEE Transactions on Engineering Management*, EM-23, 1, 3–9.

Vyas, N. & Woodside, A. G. 1984: An inductive model of industrial supplier choice processes. *Journal of Marketing, 48* (Winter), 30–45.

Wagner, G. R. 1978: Die zeitliche Desaggregation von Beschaffungsentscheidungen aus der Sicht des Investitionsgütermarketing. *Zeitschrift für betriebswirtschaftliche Forschung, 30*, 266–89.

Ward, J. H. 1963: Hierarchical grouping to optimize an objective function. *Journal of the American Statistical Association, 58*, 236–44.

Wintsch, E. 1979: Die Analyse des Entscheidungsprozesses beim Kauf von Computern als Grundlage für die Marktbearbeitungsmaßnahmen der Hersteller. 2 Vols. Dissertation, University of St. Gallen.

Witte, E. 1969: Organization of management decision processes. *The German Economic Review, 7*, 256–62.

Witte, E. 1970: Informationsaktivität und Entscheidungsobjekte. Eine Tangente zu Parkinsons "Gesetz der Trivialität" *Hamburger Jahrbuch für Wirtschafts- und Gesellschaftspolitik, 15*, 163–75.

Witte, E. 1972: Field research on complex decision-making processes—the phase theorem. In *International Studies of Management and Organization*, 2, 156–82.

Witte, E. 1973: *Organisation für Innovationsentscheidungen. Das Promotoren-Modell.* Göttingen: Verlag Otto Schwarz und Co.

Wittte, E. 1976: Kraft und Gegenkraft im Entscheidungsprozeß. *Zeitschrift für Betriebswirtschaft*, 46, 319–26.

Witte, E. 1977: Power and innovation: a two-center theory. *International Studies of Management and Organization*, 7 (Spring), 47–70.

14

On Social Factors in Administrative Innovation

ARTHUR L. STINCHCOMBE

The basic postulate of this presentation (from the author's *Stratification and Organization*) is that resistance to administrative innovations is mostly due to (more or less) rational anticipation that concrete interests (career prospects, risks of failure, chance for exceptional success, etc.) are endangered. Consequently what we want to do is to classify administrative innovations simultaneously by the interests they advance (generally collective interests or goals of some corporate group) and interests they damage (generally interests "vested in", i.e. protected by the normative structure embedded in, the old régime). In general the idea is that in order to advance some corporate interest in higher efficiency, one has to sacrifice the career or authority interests of some people who are advantaged by the old régime. If we can connect together sources of higher efficiency with *types of damage* to "vested interests", we will have the basis for a differentiated theory of resistance to administrative innovation.

The general reasons that this is a special problem are that: (1) a general way to reward administrative responsibility is by the promotions—and administrative reorganization reorganizes promotion chances; (2) a sign of one's future career chances, and a symbol that can be used in interpersonal prestige seeking, is the weight of the decisions one is responsible for—and administrative reorganization redistributes authority; (3) administration is a matter of the relations between information and decisions, but the right to take a decision is a property right, because different decisions give profits to different people—in order to reorganize information processing for decisions, therefore one often must reorganize property rights, and such reorganization often involves therefore top levels of the organization and property, rather than technical, considerations.

So in order to invent a classification of administrative innovations that will lead to productive analyses, we have to locate connections between sources of efficiency and impact on promotions, authority and responsibility, or property rights.

ELEMENTARY ADMINISTRATIVE REORGANIZATIONS

The least problematic type of innovation is clearly one which merely reorganizes communication for greater effectiveness within a fixed status structure. Even such trivial rearrangements may create tensions or resistance. For example, reorganizing a flow of information that must be acted on in such a way that a low status person transmits information that has the character of a command to a higher status person creates tension. An example given by William Foote Whyte is the client's order for food in a restaurant, transmitted by a waitress to a male cook. In one case he solved the tension by having the waitress put the order on a spindle, which the cook removed at his own discretion (Whyte 1948).

A second general problem with even minor reorganization for greater efficiency of communication is that, as Crozier (1964) has analyzed at length, uncertainty creates interpersonal power. The informal organization of power is therefore often destroyed if, for example, anyone can get the information from a computer through a terminal instead of having to ask the person in charge of that information.

Generally speaking these status consequences of elementary administrative reorganization are invasions of status rights that were not legitimate in the first place. Consequently the opposition to them usually does not have any effect politically, but rather works by bad temper, tears of frustration, unexplained sulking, and the like.

REORGANIZATION OF SYMBOLS OF STATUS

The chief case of reorganization of symbols of status is the allocation of space. Space, and the characteristics of offices, symbolize status and authority, and allocate rights to include or exclude people in an activity or a bit of information. That is, they function both as symbols of status and also as "property", as the right to control one's own fate and arrange one's work to suit oneself. At the time I was at Johns Hopkins University, one right which was maintained in the office of the President of the University rather than delegated was the allocation of space. Ordinarily when space is reorganized people invent all sort of reasons why their future plans involve the use of all the unused space, why secretaries need private offices, and the like. When low status people like students are originators of demands on higher status people like secretaries, or even worse professors, you generally find strong arguments, first, in favor of installing counters, windows, doors, or other barriers to communication, second to restrict the hours in which such demands can be initiated, and third to put the lowest status person at the window or counter. Any plan to reorganize the space so as to provide freer access for lower status people also sets off a burst of rationalizations for the restrictions. As long as one maintains the restriction one has to then provide for a caste system so that high status people can originate demands in spite of the barriers.

Privileges that are explicitly allocated by the organization on the basis of rank are also hard to reorganize. I recall a professor of sociology at Berkeley who seriously maintained that letting students park in a parking structure previously reserved for faculty and staff was "unconstitutional".

Because these symbols are recognized by the organization as having a serious connection to authority and rank, their reorganization tends to produce serious political efforts with legitimate arguments (if not always true or organizationally rational arguments) to hinder the innovation. It takes a serious mobilization of political power in an organization to reallocate space, or to change the rules by which parking privileges are given out. Authority over space allocations should always be concentrated at the highest practical level.

DECENTRALIZATION OF AUTHORITY

Reorganization of authority systems presents very different problems when authority is to be decentralized, that is, when lower status people are to be given more discretion or looser controls, than when authority is to be centralized. For example, the problem of allowing students a choice between courses is a lot different than the problem of imposing a new course requirement. The main problems with decentralization are: (a) the chance that lower participants have different utility functions, so that decentralization implies a change in policy, (b) the invention of new, more abstract flows of control information so that details can be delegated without delegating overall control of policy, and (c) motivation of the intermediate authorities to conform to the decentralization.

(a) A common feature of debates about decentralization is the "myth of the happy-go-lucky slave". The basic notion is that the underclass do not know what is good for themselves, will not take the time and energy to investigate things thoroughly, or are inherently incompetent to maximize what needs to be maximized. For example, the myth would say that if allowed to choose courses for themselves students would take only easy courses or only interesting courses or only politicized courses, certainly never statistics, that they would not listen to advice of the faculty about what they should take, and that they do not know enough about the structure of the field as a whole to choose wisely. That is, the arguments urge that if one decentralizes decisions then the utility function that will be maximized will be altered. The difference between advising students that they will probably need and want to have had statistics, and requiring statistics, is that it is the students utilities that are maximized in the first case, the faculty's idea of student utilities in the other. Conservative or "radical planning" arguments generally oppose "classical liberal" arguments (say John Stuart Mill) by some such assertion of the ignorance or evil qualities of the underclass.

Sometimes the debate on this question takes the form of asking whose utilities ought to be maximized. In such cases the success of the decentralization often depends on the constellation of political power. A well organized and powerful underclass like the proletariat often gets liberties that a poorly organized and weak underclass like students cannot get. Sometimes the debate takes the form of investigating the factual pattern of preferences of the underclass, for example, by experimentation and new monitoring. Sometimes the debate takes the form of proposals to add new substitute controls for the ones that are being taken off, such as personal advising of students in place of requirements.

(b) A common reason for decentralizing is simply that the flow of tactical decisions to the top of the hierarchy prevents that top from paying enough attention to strategic considerations. When the decentralization takes place there is a technical problem of inventing a flow of more abstract information, so that strategic decisions are not also decentralized. The crucial feature of this process from a social point of view is that lower levels of the organization must perform this abstraction process (if the higher level performs the abstraction, then they will be as burdened with details about tactics as before). But lower level people have many other motives in talking to their superiors than conveying exactly correct information. Particularly when the abstracted information is used to evaluate the subordinate unit (e.g. cost accounts or performance statistics), there are reasons to expect new flows of information to lead to new ways to evade accurate performance measurement.

Quite often the solution to this problem is to create a police force with a separate hierarchy, and particularly a separate career structure. One of the reasons the chief cost accountant is often of very high rank, and often has his people (or her people, if chief accountants are ever women—I do not know of any female chief accountants) detached in various departments, so that the chief source of performance statistics will not be corrupted near its source. Obviously setting up a new policing apparatus, and solving all the problems of mutual trust and distrust that have to be dealt with, can create political and trade union conflicts and is generally a tension-filled process.

(c) Quite often decentralization decisions are taken by central authorities, who want to redistribute authority from the middle levels to lower levels. For example, usually the rules enforced by lifeguards in swimming pools are more restrictive than the rules set down by the directorate in charge of the athletic facility as a whole—this generalization is based only on seventeen pools in four countries, but seventeen out of seventeen is quite a statistic. Suppose for example that a pool is open for adults from 20.00 to 22.00 hours, and the lifeguards rule that no one can enter after 21.00 hours (because that makes it harder to persuade them to leave at 22.00 hours). If the central directorate did not want to make such a restriction, they would have to persuade the lifeguards to let people in after 21.00 hours, unless of course they wanted to come down to the pool to let people in themselves. The central directorate may have as much trust in the utility function of the lowest level as of the intermediate level, and it already has a readymade police force for new control information in the former intermediate level, but the intermediate level has no real motive for destroying its own authority.

In particular intermediate level authorities have all kinds of ways for using the power of interpretation of the rules, and the power to explore in networks of mutual trust at the intermediate level to find solutions for problems of lower participants, so as to recreate the authority relations. If we study the telephone traffic of a department in an American university, most of the calls will be to one or another minor official elsewhere in the bureaucracy, often asking for an exception to the rule or asking for the way to formulate a particular decision so that it is in conformity with the rules. Since everything sensible is always a violation of the rules—

I think it was the railroad magnate Hill in the U.S. who said, "you can't build a railroad within the law"—this means that rule manipulators have power. For example, I once had a Dean confess to me that he could not permit me to carry out a collective bargain I had made with my office staff—which was formally within my powers as a Department chairman—because his administrative assistant did not agree that it was a good idea. The reason his administrative assistant knew about it was because my administrative assistant opposed it, and told him.

IMPOSITION OF HIERARCHICAL RELATIONS

The distinctive problem I want to point to here is subordinating people to new authorities. That is, decentralization involves freeing subordinates. What I want to analyse under (4) is creating new bonds on subordinates. This may take the form of (a) governing a new area of subordinate behavior, by extending the jurisdiction of the superior, (b) adding new, more complex, control information or sanctions so that a finer tuning of control becomes possible, (c) adding an intermediate level in a hierarchy or other special control structures such as an inspector's department, or (d) increasing the punitiveness of sanctions for violating orders or regulations. We will deal with the case of abolishing an independent jurisdiction in the next section.

The general problem here is the reaction of subordinates to the reduction of their discretion. Some of the best studies of this type of innovation are those by James Q. Wilson (1968) of reform police administrations. The usual reason why one has reform administrations in American municipal police departments is that lower level police discretion is being used to allow prostitution or gambling to go on, or to allow violation of building codes, evasion of special taxes, special patrolling of private functions at public expence, in return for bribes. All the devices mentioned above are used by reform police administrations, because the general problem is that *any* discretion in a police department with a culture of corruption can be turned to corrupt purposes. The lower policemen enter into a sort of orginality competition with the police chief, to see whether they can invent evasions as fast as the chief can invent new controls.

What is distinctive of these innovations is that, in the Dahrendorf sense, they are class conflicts. They pit people who are dominated against people who dominant them, who now propose to dominate them in a new way. Ordinarily trade unions and other class conflict organizations will get involved in such conflicts, if these organizations exist. Passive resistance, "conscientious withdrawal of efficiency", grievances, and other informal equivalents of open class conflicts are likely when there are no trade unions. As Gouldner (1954) found in his study of a wildcat strike, if the formal union apparatus cannot legitimate conflict under these circumstances (the withdrawal of the "indulgency pattern" is a clear case of increasing hierarchical control, which involved all four processes above), the conflict is likely to break out of the collective bargaining framework.

DESTRUCTION OF CAREER CHANCES

A common problem in administrative reform is the "Buy or make?" decision. Formally one should go on the market to buy a component or service whenever there are economies of scale, accidental variations between organizations in abilities of personnel, inefficiencies in one's own organization, or other competitive advantages in competing firms. In fact we hardly ever find the destruction of a department in one organization in order to buy the same service in another, but if no such department exists we do find organizations failing to create one. The difference is that it is much harder to destroy the jobs and career chances of people already in the organization than of those not yet hired.

The capacity to destroy subunits, or to drastically reorganize career chances, can generally be legitimated only by reference to outside contingencies. That is, unless an organization as a whole is going bankrupt, it is very hard to legitimate internal reorganization to save money. Consequently business organizations are much more likely than other kinds to carry out "buy or make" administrative reforms, and to do those only in hard times. Ordinarily a competitive disadvantage of a section of an organization is regarded as temporary and remediable, and as a good investment of slack resources from the rest of the organization.

One circumstance in which this is not so true is when a merger takes place. What happens in a merger is that the "constitution" of the organization is being rewritten, and consequently legitimacy is reconcentrated toward the very top of the organization's political system. Drastic pruning of relatively unprofitable parts of an organization is therefore one of the implicit profits of mergers. Succession crises and other occasions for regarding the "constitution" as problematic also make the destruction of career chances possible. Machiavelli somewhere urges in a rather similar context that the

Prince should appoint a prime minister to carry through such punitive policies, then "discover" what the prime minister has been doing and execute him.

REORGANIZATION OF PROPERTY RIGHTS

Administrative reorganization may involve the relocation of decision powers which carry with them rights to benefits. For example, long-term contracts for ships for dry bulk goods or oil redistributes the risks of rising or falling rates for rental of ships, and also enables both the shipping company and the client to design the ship for the particular trade envisaged. So the administrative advantages of long-term contracts carry with them redistribution of risks of profit and loss, i.e. contingent redistributions of property rights. If the price of ships goes down the long-term contract is a valuable property of the shipowner; if the price of ships goes up, the long-term contract is a valuable property of the client, say the oil company. When an administrative reform requires a redistribution of property rights, the owners of those rights are likely to be concerned.

The general solution here is exchange, that if one requires property rights for an administrative reform, one has to buy them. Such administrative reforms therefore tend to be easier in the private sector (as one President of the University of Chicago once observed, "there is not much market for a used university campus"), but to require the attention of top authorities in the organization. The previous comment on mergers legitimating destroying parts of the organization is a particular application of the general relation between the nonroutine exchange of property and centralization of decisions.

Quite often informal property rights within organizations are dealt with in the same way. For example an implicit property right to the promotion chances that inhere in a certain position of authority may be bought off by "kicking him upstairs", by paying the promotion cost in order to get the authority back to reallocate a new way.

GENERALIZATIONS ABOUT ADMINISTRATIVE REFORM

The classification above is roughly in the order of seriousness of the opposition to administrative reoganization, and consequently in the order of the concentrations of political power required to carry them through. We can therefore state some tentative generalizations about the relations between political structures or organizations and the probability of different sorts of reform. Some of these have been suggested above.

- The more difficult the administrative reorganization (the higher its number in the classification above), the more it must be legitimated by pressures from outside the organization, such as bankruptcy, legislative requirement, merger or ownership change, etc.
- The more difficult a reform, the more concentration of power toward the center of the organization will facilitate it, and so the more it will be found only in highly hierarchical organizations like armies or business firms, and not in less hierarchical ones like universities or trade unions.
- The more difficult a reform, the more it matters whether the central authorities in the organization favor that reform, the more the initiative for such reform will come from the top, and in other ways the higher the hierarchical level that will be involved in the reform.
- The more difficult a reform, the higher the cost in political turmoil, strikes, personnel loss, and other social disruption one will have to be willing to pay, and perhaps the higher the money cost for buying off the opposition.
- The more difficult the reform, the more selection there will be for what Selznick (1957) calls "institutional leadership", and the more likely it is that routine bureaucratic leadership will fail—serious administrative reforms are entrepreneurial, *not* administrative, matters.

REFERENCES

Crozier, M. 1964: *The bureaucratic phenomenon.* London: Tavistock Publications.
Gouldner, A. W. 1954: *Patterns of industrial bureaucracy.* Glencoe, Ill.; Free Press.
Selznick, P. 1957: *Leadership in administration.* Evanston, Ill.: Row, Peterson.
Whyte, W. Foote 1948: *Human relations in the restaurant industry.* New York: McGraw-Hill.
Wilson, J. Q. 1968: *Varieties of police behavior.* Cambridge, Mass.: Harvard University Press.

Part III
Knowledge, Innovation, and Growth

15

Lead Users: A Source of Novel Product Concepts

ERIC VON HIPPEL

Accurate understanding of user need has been shown near-essential to the development of commercially successful new products (Rothwell et al. 1974). Unfortunately, current market research analyses are typically not reliable in the instance of very novel products or in product categories characterized by rapid change, such as "high technology" products. In this paper (*Management Science*, Vol. 32, No. 7 (1986)) I explore the problem and propose a solution: marketing research analyses which focus on what I term the "lead users" of a product or process.

Lead users are users whose present strong needs will become general in a marketplace months or years in the future. Since lead users are familiar with conditions which lie in the future for most others, they can serve as a need-forecasting laboratory for marketing research. Moreover, since lead users often attempt to fill the need they experience, they can provide new product concept and design data as well. How lead users can be systematically identified, and how their perceptions and preferences incorporated into industrial and consumer marketing research analyses of emerging needs for new products, processes and services is examined below.

MARKETING RESEARCH CONSTRAINED BY USER EXPERIENCE

Users selected to provide input data to consumer and industrial market analyses have an important limitation: Their insights into new product (and process and service) needs and potential solutions are constrained by their own real-world experiences. Users steeped in the present are thus unlikely to generate novel product concepts which conflict with the familiar.

The notion that familiarity with existing product attributes and uses interferes with an individual's ability to conceive of novel attributes and uses is strongly supported by research into problem solving (Table 1). We see that experimental subjects familiar with a complicated problem-solving strategy are unlikely to devise a simpler one when this is appropriate (Luchins 1942). Also, and germane to our present discussion, we see that subjects who use an object or see it used in a familiar way are strongly blocked from using that object in a novel way (Duncker 1945; Birch and Rabinowitz 1951; Adamson 1952). Furthermore, the more recently objects or problem-solving strategies have been used in a familiar way, the more difficult subjects find it to employ them in a novel way (Adamson and Taylor 1954). Finally, we see that the same effect is displayed in the real world, where the success of a research group in solving a new problem is shown to depend on whether solutions it has used in the past will fit that new problem (Allen and Marquis 1964). These studies thus suggest that typical users of existing products—the type of user-evaluators customarily chosen in market research—are poorly situated with regard to the difficult problem-solving tasks associated with assessing unfamiliar product and process needs.

As illustration, consider the difficult problem-solving steps which potential users must go through when asked to evaluate their need for a proposed new product. Since individual industrial and consumer products are only components in larger usage patterns which may involve many products, and since a change in one component can change perceptions of and needs for some or all other products in that pattern, users must first identify their existing multiproduct usage patterns in which the new product might play a role. Then they must evaluate the new product's potential contribution to these. (For example, a change in the operating characteristics of a computer may allow a user to solve new problem types if he makes related changes in software and perhaps in other, related products and practices. Similarly, a consumer's switch to microwave cooking may well induce related changes in food recipes, kitchen practices, and kitchen utensils.) Next, users must invent or select the new (to them) usage patterns which the proposed new product makes possible for the first time, and evaluate the utility of the product in these. Finally, since substitutes exist for many multiproduct usage patterns (e.g. many forms of problem analysis are available in addition to the novel ones made possible by a new computer) the user must estimate how the new possibilities presented by the proposed new product will compete (or fail to compete) with existing options. This problem-solving task is clearly a very difficult one, particularly for typical users of existing products whose familiarity with existing products and users interferes with their ability to conceive of novel products and uses when invited to do so.

The constraint of users to the familiar pertains even in the instance of sophisticated consumer marketing research techniques such as multiattribute mapping of product perceptions and preferences (Silk and Urban 1978; Roberts and Urban 1985). "Multiattribute" (multidimensional) marketing research methods, for example, describe a consumer's perception of new and existing products in terms of a number of attributes (dimensions). If and as a complete list of attributes is available for a given product category, a consumer's perception of any particular product in the category can be expressed in terms of the amount of each attribute the consumer perceives it to contain, and the difference between any two products in the category can be expressed as the difference in their attribute profiles. Similarly, consumer preferences for existing and proposed products in a category can in principle be built up from consumer perceptions of the importance and desirability of each of the component product attributes.

Although these methods frame user perceptions and preferences in terms of attributes, they do not offer a means of going beyond the experience of the users interviewed. First, for reasons discussed above, user subjects are not well positioned to accurately evaluate novel product attributes or "amounts" of familiar product attributes which lie outside the range of their real-world experience. Second, and more specific to these techniques, there is no mechanism to induce users to identify all product attributes potentially relevant to a product category, especially attributes which are currently not present in any extant category member. To illustrate this point, consider two types of such methods, similarity-dissimilarity ranking and focus groups.

In similarity-dissimilarity ranking, data regarding the perceptual dimensions by which consumers characterize a product category are generated by inviting a sample of consumers to compare products in that category and assess their similarity-dissimilarity. In some variants of the methods, the consumer specifies the ways in which the products are similar or different. In others, the consumer simply provides similarity and difference rankings, and the market analyst determines—via his personal knowledge of the product type in question, its function, the marketplace, the consumer, etc.—the important perceptual dimensions which "must" be motivating the consumer rankings obtained.

The similarity-dissimilarity method clearly depends heavily on an analyst's qualitative ability to interpret the data and correctly identify all the critical dimensions. Moreover, by its nature, this method can only explore perceptions derived from attributes which exist in or are associated with the products being compared. Thus, if a group of consumer evaluators is invited to compare a set of cameras and none has a particular feature—say, instant developing—then the possible utility of this feature would not be incorporated in the perceptual dimensions generated. That is, the method would have been blind to the possible value of instant developing prior to Edwin Land's invention of the Polaroid camera.

In focus group methods, market analysts assemble a group of consumers familiar with a product category for a qualitative discussion of perhaps two hours' duration. The topic for the focus group, which is set by an analyst, may be relatively narrow (e.g. "35 mm amateur cameras") or somewhat broader (e.g. "the photographic experience as you see it"). The ensuing discussion is recorded, transcribed, and later reviewed by the analyst whose task it is to identify the important product attributes which have implicitly or explicitly surfaced during the conversation. Clearly, as with similarity-dissimilarity ranking, the utility of information derived from focus group methods depends heavily on the analyst's ability to accurately and completely abstract from the interview data the attributes which consumers feel important in products.

In principle, however, the focus group technique need not be limited to only identifying attributes already present in existing products, even if the discussion is nominally focused on these. For example, a topic which extends the boundaries of discussion beyond a given product to a larger framework could identify attributes not present in any extant product in a category under study. If discussion of the broad topic mentioned earlier, "the photographic experience as you see it", brought out consumer dissatisfaction with the time lag between picture taking and receipt of the finished photograph, the analyst would be in possession of information which could induce him to identify an attribute not present in any camera prior to Land's invention, instant film development, as a novel and potentially important attribute.

But how likely is it that an analyst will take this creative step? And, more generally, how likely is it that either method discussed above, similarity-dissimilarity ranking or focus groups, will be used to identify attributes not present in extant products of the type being studied, much less a complete list of all relevant attributes? Neither method contains an effective mechanism to encourage this outcome, and discussions with practitioners indicate that in present-day practice, identification of any novel attribute is unlikely.

Finally, both of these methods conventionally focus on familiar product categories. This restriction, necessary to limit the number of attributes which "completely describe" a product type to a manageable number, also tends to limit consumer perceptions to attributes which fit products within the frame of existing product categories. Modes of transportation, for example, logically shade off into communication products as partial substitutes ("I can drive over to talk to him—or I can phone"), into housing and entertainment products ("We can buy a summer house—or go camping in my recreational vehicle"), indeed, into many other of life's activities. But since a complete description of life cannot be compressed into 25 attribute scales, the analysis is constrained to a narrower—usually conventional and familiar—product category or topic. This has the effect of rendering any promising and novel cross-category new product attributes less visible to the methods I have discussed.

In sum, then, we see that marketing researchers face serious difficulties if they attempt to determine new product needs falling outside of the real-world experience of the users they analyze.

LEAD USERS' EXPERIENCE IS NEEDED FOR MARKETING RESEARCH IN FAST-MOVING FIELDS

In many product categories, the constraint of users to the familiar does not lessen the ability of marketing research to evaluate needs for new products by analyzing typical users. In the relatively slow-moving world of many consumer products, new cereals and new car methods do not often differ radically from their immediate predecessors. Therefore, even the "new" is reasonably familiar, and the typical user can thus play a valuable role in the development of new products.

In contrast, in high technology industries, the world moves so rapidly that the related real-world experience of ordinary users is often rendered obsolete by the time a product is developed or during the time of its projected commercial lifetime. For such industries I propose that "lead users" who *do* have a real-life experience with novel product or process concepts of interest are essential to accurate marketing research. Although the insights of lead users are as constrained to the familiar as those of other users,

Table 1. The effect of prior experience on users' ability to generate or evaluate novel product possibilites.

Study	Nature of research	Impact of prior experience on ability to solve problems
Luchins (1942)	Two groups of subjects ($n =$) were given a series of problems involving water jars, e.g.: 'If you have jars of capacity A, B and C how can you pour water from one to the other so as to arrive at amount D?' Subject group 1 was given 5 problems solvable by formula, $B - A - 2C = D$. Next, both groups were given problems solvable by that formula *or* by a simpler one (e.g. $B - C = D$).	81% of experimental subjects who had previously learned a complex solution to a problem type applied it to cases where a simple solution would do. No control group subjects did so ($p =$ NA[a]).
Duncker (1945)	The ability to use familiar objects in an unfamiliar way was tested by creating 5 problems which could only be solved by that means. (For example, one problem could be solved only if subjects bent a paper clip provided them and used it as a hook.) Subjects were divided into two groups. One group of problem solvers saw the crucial objects being used in a familiar way (e.g. the paper clip holding papers), the other did not (e.g. the paper clip was simply lying on a table unused).	Subjects were much more likely to solve problems requiring the use of familiar objects (e.g. paper clips) in unfamiliar ways (e.g. bent into hooks) if they had been shown the familiar use just prior to their problem-solving attempt. Duncker called this effect "functional fixedness" ($n = 14$; $p =$ NA[a]).
Birch and Rabinowitz (1951)	Replication of Duncker, above.	Duncker's findings confirmed ($n = 25$; $p < 0.05$).
Adamson (1952)	Replication of Duncker, above.	Duncker's findings confirmed ($n = 57$; $p < 0.01$).
Adamson and Taylor (1954)	The variation of "functional fixedness" with time was observed by the following procedure. First, subjects were allowed to use a familiar object in a familiar way. Next, varying amounts of time were allowed to elapse before	If a subject uses an object in a familiar way, he is partially blocked from using it in a novel way. ($n = 32$; $p < 0.02$). This blocking effect decreases over time (see graph).

subjects were invited to solve a problem by using the object in an *un*-familiar way.

Functional fixedness as a function of log time

Allen and Marquis (1964)

Government agencies often buy R&D services via a "Request for Proposal" (RFP) which states the problem to be solved. Interested bidders respond with Proposals which outline their planned solutions to the problem and its component tasks. In this research, relative success of eight bidders' approaches to the component tasks contained in 2 RFPs was judged by the Agency buying the research ($n = 26$). Success was then compared to prior research experience of bidding laboratories.

Bidders were significantly more likely to propose a successful task approach if they had prior experience with that approach only, rather than prior experience with inappropriate approaches only.

aThis relatively early study showed a strong effect but did not provide a significance calculation—or present data in a form which would allow one to be determined without ambiguity.

lead users are familiar with conditions which lie in the future for most—and so are in a position to provide accurate data on needs related to such future conditions.

I define "lead users" of a novel or enhanced product, process or service as those displaying two characteristics with respect to it:

• Lead users face needs that will be general in a marketplace—but face them months or years before the bulk of that marketplace encounters them, *and*

• Lead users are positioned to benefit significantly by obtaining a solution to those needs.

These two lead user characteristics are shown schematically in Figure 1. Two specific examples of lead users: Firms who today need and could obtain significant benefit from a type of office automation which the general market will need tomorrow are lead users of office automation; a semiconductor producer with a current strong need for a process innovation which many semiconductor producers will need in two-years' time is a lead user with respect to that process.

Users whose present need foreshadow general demand exist because important new technologies, products, tastes, and other

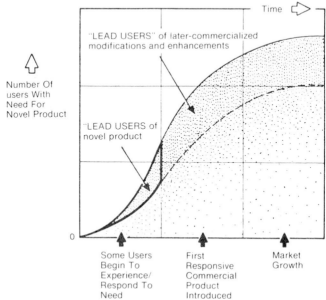

Fig. 1. A schematic of lead users' position in the life cycle of a novel product, process or service. Lead users (1) encounter the need early and (2) expect high benefit from a responsive solution. (Higher expected benefit indicated by deeper shading.)

factors related to new product opportunities typically diffuse through a society, often over many years, rather than impact all members simultaneously (Rogers and Shoemaker 1971). Thus, when Mansfield (1968) explored the rate of diffusion of twelve very important industrial goods innovating into major firms in the bituminous coal, iron and steel, brewing, and railroad industries, he found that in 75 percent of the cases it took over 20 years for complete diffusion of these innovations to major firms. Accordingly, some users of these innovations could be found far in advance of the general market.

Users of new products and processes have been shown to differ on the level of benefit they can obtain from these (Mansfield 1968). The greater the benefit a given user can obtain from a needed novel product or process, the greater his effort to obtain a solution will be. (This link between innovation activity and expectation of economic benefit was first empirically established by Jacob Schmookler [1966], who conducted a careful study of the correlation between changes in sales volumes of some capital goods and appropriately lagged changes in rates of patent applications in categories related to those goods.) I therefore reason that users able to obtain the highest net benefit from the solution to a given new product (or process or service) need will be the ones how have devoted the most resources to understanding it. And it follows that this subset of users should have the richest real-world understanding of the need to share with inquiring market researchers.

UTILIZING LEAD USERS IN MARKETING RESEARCH

How then can lead users be incorporated into marketing research? I suggest a four-step process:

(1) Identify an important market or technical trend;
(2) Identify lead users who lead that trend in terms of (a) experience and (b) intensity of need;
(3) Analyze lead user need data;
(4) Project lead user data onto the general market of interest.

I consider how each of these steps might be approached below with regard to industrial and consumer products.

Identifying an Important Trend

Lead users are defined as being in advance of the market with respect to a given important dimension which is changing over time. Therefore, before one can identify lead users in a given product category of interest, one must identify the underlying trend on which these users have a leading position.

Identification of important trends affecting promising markets is already commonly performed by many firms as a necessary component of their corporate strategy. Methods used range from the intuitive judgments of experts, perhaps formalized in a technique such as the "Delphi" method, to simple trend extrapolation to more complex correlation or econometric models. (See Chambers, Mullick, and Smith [1971], for a useful practitioner's overview. See Martino [1972] for the special case of forecasting trends in technology.) Despite the existence of formal trend assessment methods, however, trend identification and assessment remains something of an art. Thus, analysts typically must judge which of many important trends in a market they will focus on, or must combine several into a suitable index variable.

In the case of industrial goods, trend identification and assessment can often be both informal and accurate. Since potential buyers typically measure the value of proposed new industrial products in economic terms, important underlying trends related to product value are often inescapably clear to those in the industry. For example, it is clear to those in the semiconductor and computer fields that computer memory and microprocessor chips are getting more capable and less expensive for a given capability every year. It is also clear that, for technical reasons, this trend is also likely to continue for a number of years. Finally, it is clear that this trend has very important cost/performance implications for firms which incorporate these semiconductors in computers or myriad other increasingly "intelligent" products.

In the case of consumer goods, accurate trend identification is often more difficult because there is often no underlying stable basis for comparison such as that played by economic value for industrial goods. Therefore, while consumer perceptions of trends and their subjective assessment of the importance of these can be determined straightforwardly by survey at any given point in time, these perceptions may not be consistent over time. (For example, we cannot expect to predict the trend in consumer interest in auto fuel economy as a function of fuel cost as accurately as we could predict industrial buyer interest in fuel economy on that basis.)

In sum, reliable methods for formal prediction of trends over time which will have an important effect on a given product area are not yet well developed. In some product areas, however, notably in industrial goods, the needed data on important trends are clear to those with expertise, and in these instances the poor state of formal methods is not an impediment to incorporating analyses of lead users into marketing research.

Identifying Lead Users

Once a firm has identified one or more significant trends which appear associated with promising new product opportunities, the market researcher can begin to search for lead users, users (1) who are at the leading edge of each identified trend in terms of related new product and process needs and (2) who expect to obtain a relatively high net benefit from solutions to those needs. Let us consider practical means for identifying lead users in the instance of industrial goods and consumer goods in turn.

The first task, identifying users at the leading edge of a given trend, is usually straightforward in the case of industrial goods because a given firm's position on a range of trends is usually well known to industry experts. Thus, in many instances, industrial good manufacturers have only a few or a few score major potential customers for a given product type and often know the characteristics of each user quite well. As illustration, recall the important trend toward cheaper, more capable computer memory and microprocessor semiconductor chips. Manufacturers of semiconductor process equipment would recognize that cheaper, more capable semiconductors are achieved in major part by the packing of circuit elements ever more densely on chip surfaces by semiconductor makers (equipment users), a trend involving significant new user needs for semiconductor process equipment. They would also know that many users at the leading edge of the need for density trend are makers of VLSI memories, with makers of other types of semiconductors such as linear ICs lying further back on the curve. Therefore, VLSI memory manufacturers can be flagged as potential lead users of process equipment with regard to this trend.

The second task is identifying the subset of those user firms positioned at the forefront of the trend under study who are also able to obtain relatively high "net benefit" from adopting a solution to trend-related needs. In the case of industrial goods, net benefit is typically measured in economic terms. And when this is so, the net benefit (B) which a user firm expects to obtain from a solution to a given need can be stated as: $B = (V)(R) - C - D$ where (V) is the dollar "volume" of product sales or processing activity to which the user plans to apply his solution; (R) is the increased rate of profit per dollar of this volume resulting from application of that solution; (C) is the user's anticipated costs in developing and/or adopting the solution; and (D) is the net benefit which the user would have obtained from old practices, equipment, etc., displaced by the novel solution. (Industrial firms typically make the calculation described above when they assess the return on investment they may anticipate from investing in a new product, process or service. I will not describe specific methods of making such calculations here. General methods can be found in accounting texts. A discussion of net benefit calculations useful in the specific instance of innovations will be found in von Hippen [1982].)

An additional, very practical method for identifying lead users involves identifying those users who are actively innovating to solve problems present at the leading edge of a trend. Thus, the semiconductor process equipment makers mentioned in our earlier example could seek those few VLSI memory manufacturers (equipment users) who are actively developing processes for the manufacturer of denser chips. A user conducting such R&D is probably a lead user because innovation is expensive, and the user engaged in it surely expects to reap high net benefit from a problem solution. Identifying lead users by seeking innovating users can be very economical, because the identity of users conducting R&D on a given problem area is often common knowledge to industry participants.

In the case of consumer goods, lead users with respect to specified trends can readily be identified by appropriately designed surveys. For example, if the trend toward increasing consumption of "health foods" is selected, a survey of consumer food preferences could identify those on the leading edge of that trend. The lead users among this group could then be identified by additional questions concerning the value respondents place on improvements in the healthfulness of food. (Such a screening question might be: "How much extra would you be willing to pay for X food free of Y additive?") Those found to place a significantly higher value than most on such improvements, e.g. x standard deviations above the mean, are the users who anticipate obtaining the highest net benefit from a solution to the need. They are therefore lead users with respect to this trend.

Finally, three important complexities with regard to identifying lead users should be noted. First, key lead users should not necessarily be sought within the usual customer base of the manufacturer performing the market research. They may be customers of a competitor—or totally outside of the industry he serves. For example, if a manufacturer of composite materials used in autos identifies an important trend toward lighter, higher strength materials, he may find the lead users at the front of this trend are aerospace firms rather than auto firms, because aerospace firms may be willing to pay more than auto firms for improvements on these attributes. Often, consumer products manufacturers will find valuable lead users among users of analogues industrial goods, because the benefit which an industrial user can expect from a given advance often far outstrips that which an individual consumer could expect. Thus, a manufacturer seeking to develop centralized controllers for home heating, lighting and security systems might well seek lead users among firms seeking to use controllers of similar function in commercial buildings. (Note that one must always be aware of both similarities and differences between the lead users one is assessing and the user population one intends to serve. This point will be developed below in the section entitled "Projecting Lead User Data onto the General Market of Interest".)

The second complexity with respect to identifying lead users is that one need not be restricted to identifying lead users who can illumine the *entire* novel product, process or service which one wishes to develop. One may also seek out those who are lead users with respect to only a few of its attributes—or indeed of a single attribute, defined as narrowly as one likes. Thus, to elaborate on the example begun above, a manufacturer of centralized controllers of home systems might seek lead users with respect to the energy management aspects of such a controller among firms using controllers of analogues function in commercial or industrial applications where a great deal of energy is used and/or energy costs are high. At the same time, he might seek lead users with respect to the security system attributes of such a home controller system among a totally different set of users—perhaps individuals or firms who feel at high risk for burglary and have very valuable goods to protect.

The third and final complexity regarding the identification of lead users which I will mention has the following source: Users driven by expectations of high net benefit to develop a solution to a need might well have solved *their* problem and no longer feel that need. Therefore a survey seeking to identify lead users on the basis of high unmet need might not identify these particular users. This can be a significant loss, since lead users who perceive that they have successfully developed a responsive solution to the need at issue clearly have valuable data for market researchers. In practice, however, I find that in the instance of industrial products, users at the leading edge of an important (moving) trend have to innovate again and again to maintain a level of satisfaction with their current practice, and so will seldom express expectations of low net benefit from additional improvements over their present practice. In the instance of consumer products, the problem can be addressed via additional survey questions specifically inquiring about possible consumer-developed solutions to the needs under study.

Analyzing Lead User Data

Data derived from lead users and their real-life experience with novel attributes and/or product concepts of commercial interest can be incorporated in market research analyses using standard market research methods. However, the analyst might wish to be on the lookout for somewhat more user-developed product solutions and more substantive need statements in lead user data than he is used to finding in analyses of other user populations. Recall that, since problem-solving activity has been shown to be motivated by expectations of economic benefit, and since lead users have been defined in part as users positioned to obtain high net benefit from a solution to their needs, it is reasonable that lead users may have made some investment in solving the need at issue. Sometimes lead user problem-solving activity takes the form of applying existing commercial products or components in ways not anticipated by their manufacturers. Sometimes lead users may have developed complete new products responsive to their need.

Product development by users receiving relatively high returns from such activity has been empirically documented (von Hippel 1982). In some product areas (e.g. semiconductor process machinery [von Hippel 1977] and scientific instruments [von Hippel 1976]), moreover, users have developed *most* of the commercially successful product innovations. To illustrate, the results of several studies of the functional locus of innovation are summarized in Table 2. (The absence of actual product development by users in a given area [e.g. engineering plastics (Berger 1975) and conductor attachment equipment (VanderWerf 1982)] does not mean that lead users with their rich insights are absent here; it simply means that the distribution of economic benefits flowing from product development in that area makes product development by manufacturers or other nonuser groups so attractive that nonusers preempt user product development activity [von Hippen 1982].)

Users develop both industrial and consumer products. An example of each will help convey the flavor:

Table 2. Data regarding the role of users in product development.

Study	Nature of innovations and sample selection criteria	n	Innovative product developed by:[a] User	Mfg.	Other
Knight (1963)	Computer innovations 1944–1962: —systems reaching new performance high	143	25%	75%	
	—systems with radical structural innovations	18	33%	67%	
Enos (1962)	Major petroleum processing innovations	7	43%	14%	43%[b]
Freeman (1968)	Chemical processes and process equipment available for license, 1967	810	70%	30%	
Berger (1975)	All engineering polymers developed in U.S. after 1955 with > 10^6 pounds produced in 1975	6	0%	100%	
Boyden (1976)	Chemical additives for plastics—all platicizers and UV stabilizers developed post World War II for use with 4 major polymers	16	0%	100%	
Lionetta (1977)	All pultrusion processing machinery innovations first introduced commercially 1940–1976 which offered users a major increment in functional utility[c]	13	85%	15%	
von Hippel (1976)	Scientific instrument innovations: —first of type (e.g. first NMR)	4	100%	0%	
	—major functional improvements	44	82%	18%	
	—minor functional improvements	63	70%	30%	
von Hippel (1977)	Semiconductor and electronic subassembly manufacturing equipment: —first of type used in commercial production	7	100%	0%	
	—major functional improvements	22	63%	21%	16%[d]
	—minor functional improvements	20	59%	29%	12%[d]
VanderWerf (1982)	Wirestripping and connector attachment equipment	20	11%	33%	56%[e]

[a]NA data excluded from percentage computations.
[b]Attributed to independent inventors/invention development companies.
[c]Figures shown are based on reanalysis of Lionetta's (1977) data.
[d]Attributed to joint user-manufacturer innovation projects.
[e]Attributed to connector suppliers.

IBM designed and built the first printed circuit card component insertion machine of the X-Y Table type to be used in commercial production. (IBM needed the machine to insert components into printed circuit cards which were in turn incorporated into computers.) After building and testing the design in-house, IBM sent engineering drawings of their design to a local machine builder along with an order for eight units. The machine builder completed this and subsequent orders satisfactorily and, two years later, applied to IBM for permission to build essentially the same machine for sale on the open market. IBM agreed and the machine builder became the first commercial manufacturer of X-Y Table component insertion machines. (The above episode marked that firm's first entry into the component insertion equipment business. They are a major factor in the business today.) (von Hippel 1977)

In the early 1970's, store owners and salesmen in southern California began to notice that youngsters were fixing up their bikes to look like motorcycles, complete with imitation tailpipes and "chopper-type" handlebars. Sporting crash helmets and Honda motorcycle T-shirts, the youngsters raced their fancy 20-inchers on dirt tracks.

Obviously on to a good thing, the manufacturers came out with a whole new line of "motorcross" models. By 1974 the motorcycle-style units accounted for 8 percent of all 20-inch bicycles shipped. Two years later half the 3.7 million new juvenile bikes sold were of the motorcross model . . . (*New York Times* 1978)

Of course, completely developed new products are not the only useful "solution data" available from lead users. All need statements implicitly or explicitly contain more or less information about possible solutions to the need at issue. Consider the following sequence of need statements which deal with a consumer product:

*I am unhappy . . .

*about my children's clothes . . .

*which are often not fully clean even when just washed.

*I find that X type stains on Y type clothes are especially hard to remove.

*If I mix my powdered detergent into a paste and apply it to the stain before washing, I find it helps get things clean.

Each succeeding statement clearly provides a valuable increment of data useful for defining a new product need and devising a responsive solution. On the basis of the last statement we see that liquid detergent could be invented. We also are able to learn that the user is approaching the problem as a "stain removal" problem rather than "keep the kids away from X staining agent" problem. And, probably, the user is ranking this choice after having experimented with both approaches. In essence, such *experience* with the need/problem is what makes lead user's data so valuable.

Projecting Lead User Data onto the General Market of Interest

The needs of today's lead users are typically not precisely the same as the needs of the users who will make up a major share of tomorrow's predicted market. Indeed, the literature on diffusion suggests that, in general, the early adopters of a novel product or practice differ in significant ways from the bulk of the users who follow them (Rogers and Shoemaker 1971). Thus, analysts will need to assess how lead user data apply to the more typical user in a target market rather than simply assume such data straightforwardly transferable.

In the instance of industrial goods, the translation problem is typically not serious. As we pointed out earlier, industrial products are typically evaluated on economic grounds whereby users calculate the relative costs and benefits of the proposed product. When an objective economic analysis is possible, all users—not just lead users—will make similar calculations and thus provide a common basis for market projections.

In the instance of consumer goods, and in the instance of industrial goods for which the costs and benefits of the proposed product for the user do not form the basis for product preferences, a test of the applicability of lead user needs and concepts to the future general market is not so simple. One approach involves prototyping the novel product and asking a sample of typical users to use it. Such users would then be in a position to provide accurate product evaluation data to market research (a) if presenting the user with the product created conditions for him similar to the conditions a future user would face, and (b) if the user were given enough time to fully explore the new product and fully adapt his usage patterns to it. If a new product were being tested in this manner in a field where little else was expected to change by the time of product introduction—say, a new detergent for the home laundry—conditions (a) and (b) could probably be effectively met. However, in rapidly moving fields in which the proposed new product will interact with many other not-yet-developed products in unforeseen ways, new approaches may be needed.

SUMMARY

In this paper I have defined lead users, and explored the valuable insights they can offer regarding needs—and, often, prototype solutions—for novel products, processes and services. I have also presented four general steps by which one may identify and analyze lead users in any given instance. Practitioners may wish to use the method in its present early form, while researchers may wish to explore, test and refine it. Both are possible, and we make suggestions regarding each below.

I suggest that interested practitioners have no hestitation about experimenting with the general methodology described here. Lead users are often accessible enough to allow successful identification of interesting data regarding desirable new products and/or product modifications with little effort—given that the practitioner has a good knowledge of the customers and application area he is analyzing. As evidence, during the past two years Professor Glen Urban and I have helped approximately 100 MIT Master's students to undertake short projects involving the identification of lead users in areas they were familiar with. With very little coaching, almost all have succeeded. (Examples: lead users of sports equipment have been identified and studied in sports ranging from rock climbing to trail biking to street hockey. Other projects have dealt with lead users of various types of industrial process equipment and various types of computer hardware and software.) Therefore, I urge practitioners to "learn by doing", and conduct a rough initial test for their own interest and satisfaction. If the results are positive, I hope they will be motivated to do still more.

Researchers who wish to systematically explore the value of lead user methods will find many possible approaches. I propose that initial empirical studies of the value of lead user data be focused on industrial goods rather than consumer goods. (As was noted earlier, lead users of industrial goods can typically be identified more reliably than lead users of most consumer goods given today's state of the art.) The value of lead user data under real-world conditions can be assessed via a longitudinal study design which tests the predictive accuracy of data collected earlier from lead users against the actual future general market as it evolves. A less ambitious effort could focus on industrial products and compare the economic performance of novel product concepts proposed by lead versus typical users.

Researchers who wish to improve lead users methods will find much needs to be done in both industrial and consumer goods arenas. For example, the means for identifying lead users and analyzing their needs obviously must be improved and extended. Also, organizational schemes for routinely acquiring lead user data (special interface groups, special incentives which will induce lead users to interact on a continuing basis, etc.) need to be developed and tested. Valuable research on all these topics appears to be exciting and well within reach.*

*I gratefully acknowledge the helpful comments and suggestions provided by Professor Glen L. Urban of MIT's Sloan School of Management and Professor John H. Roberts of the Australian Graduate School of Management.

REFERENCES

Adamson, Robert E. 1952: Functional fixedness as related to problem solving: a repetition of three experiments. *J. Experimental Psychology*, *44*(4) (October), 288–91.

Adamson, Robert E. & Taylor, Donald W. 1954: Functional fixedness as related to elapsed time and to set. *J. Experimental Psychology*, *47*(2) (February), 122–6.

Allen, T. J. & Marquis, D. G. 1964: Positive and negative biasing sets: the effects of prior experience on research performance. *IEEE Trans. Engineering Management*, EM-11(4) (December), 158–61.

Berger, Alan J. 1975: Factors influencing the locus of innovation activity leading to scientific instrument and plastics innovations. S.M. thesis. Cambridge, MA: MIT Sloan School of Management.

Birch, Herbert G. & Rabinowitz, Herbert J. 1951: The negative effect of previous experience on productive thinking. *J. Experimental Psychology*, *41*(2) (February 1951), 121–5.

Boyden, Julian W. 1976: A study of the innovative process in the plastics additives industry. S.M. thesis. Cambridge MA: MIT Sloan School of Management.

Chambers, John C, Mullick, Satinder K. & Smith, Donald D. 1971: How to choose the right forecasting technique. *Harvard Business Rev.* (July–August), 45–74, provides a useful practitioner's overview.

Duncker, K. 1945: On problem-solving. Trans. Lynne S. Lees, *Psychological Monographs*, *58*(5), (Whole No. 270).

Enos, John Lawrence 1962: *Petroleum progress and profits: a history of process innovation*. Cambridge: MIT Press.

Freeman, C. 1968: Chemical process plant: Innovation and the world market. *National Institute Economic Rev.* 45 (August), 29–57.

Knight, K. E. 1963: A study of technological innovation: the evolution of digital computers. Ph.D. dissertation, Carnegie Institute of Technology. Data shown in Table 2 of this paper obtained from Knight's Appendix B, parts 2 and 3.

Lionetta, William G., Jr., 1977: Sources of innovation within the pultrusion industry. S.M. thesis. Cambridge, MA: MIT Sloan School of Management.

Luchins, Abraham S. 1942: Mechanization in problem solving: the effect of einstellung. *Psychological Monographs*, *54*(6), (Whole No. 248).

Mansfield, Edwin 1968: *Industrial research and technological innovation: an econometric analysis*. New York: W. W. Norton, 134–235.

Martino, Joseph P. 1972: *Technological forecasting for decision-making*. New York: American Elsevier.

New York Times, 29 January 1978, F3.

Roberts, John H. & Urban, Glen L. 1985: New consumer durable brand choice: modeling multiattribute utility, risk, and dynamics. Working Paper WP 1636–85. Cambridge, MA: MIT Sloan School of Management.

Rogers, Everett with Shoemaker, F. Floyd. 1974: *Communication of innovations: a cross-cultural approach,* 2nd ed. New York: The Free Press.

Rothwell, R., Freeman, C. et al. 1974: SAPPHO updated—project SAPPHO phase II. *Research Policy, 3*, 258–91. See also Achilladelis, B., Robertson, A. B. & Jervis, P. 1971: *Project SAPPHO: A study of success and failure in industrial innovation*, 2 vols. London: Center for the Study of Industrial Innovation.

Schmookler, Jacob 1966: *Invention and economic growth*. Cambridge, MA: Harvard University Press.

Silk, Alvin J. & Urban, Glen L. 1978: Pre-test-market evaluation of new packaged goods: A model and measurement methodology. *J. Marketing Res. 15* (May), 189; Also: Shocker, Allan D. & Srinivasan, V. 1979: Multiattribute approaches for product concept evaluation and generation: a critical review. *J. Marketing Res. 16* (May), 159–80.

Vander Werf, Pieter 1982: Parts suppliers and innovators in wire termination equipment. Working Paper WP 1289–82. Cambridge, MA: MIT Sloan School of Management.

von Hippel, Eric 1976: The dominant role of users in the scientific instrument innovation process. *Research Policy, 5*, 212–39.

von Hippel, Eric 1977: The dominant role of users in semiconductor and electronic subassembly process innovation. *IEEE Trans. Engineering Management*, EM-24(2) (May), 60–71.

von Hippel, Eric 1982: *The sources of innovation*. New York: Oxford University Press. Also, Eric von Hippel 1982: Appropriability of innovation benefit as a predictor of the source of innovation. *Research Policy, 11*, 95–115.

16

The New Product Learning Cycle

MODESTO A. MAIDIQUE AND BILLIE JO ZIRGER

Many factors influence product success. That much is generally agreed upon by researchers in the field. The product, the firm's organizational linkages, the competitive environment, and the market can all play important roles. On the other hand, the results of research on new product success and failure (Cooper 1983) is reminiscent of George Orwell's *Animal Farm* in that some factors seem to be "more equal than others". But, exactly which set of factors predominates seems to be, at least in part, a function of both the methodology and the specific population studied by the researcher (Maidique and Zirger 1983).

The Stanford Innovation Project (SINPRO)

In a survey of 158 products in the electronics industry, half successes and half failures, we developed our own list of major determinants of new product success. The eight principal factors we identified are listed below roughly in the order of their statistical significance. Products are likely to be successful if:

(1) The developing organization, through in-depth understanding of the customers and the marketplace, introduces a product with a high performance to cost ratio.
(2) The create, make, and market functions are well coordinated and interfaced.
(3) The product provides a high contribution margin to the firm.
(4) The new product benefits significantly from the existing technological and marketing strengths of the developing business units.
(5) The developing organization is proficient in marketing and commits a significant amount of its resources to selling and promoting the product.
(6) The R&D process is well-planned and coordinated.
(7) There is a high level of management support for the product from the product conception stage to its launch into the market.

(8) The product is an early market entrant.

The study that led to these conclusions consisted of two exploratory surveys described in detail elsewhere (Maidique and Zirger 1984). The first survey was open ended and was divided into two sections. In the first part we asked the respondent to select a pair of innovations, one success and one failure. Successes and failures were differentiated by financial criteria. The second section of the original survey asked each respondent to list in his own words the factors which he believed contributed to the product's outcome. Seventy-nine senior managers of high technology companies completed this questionnaire.

The follow-up survey was structured into 60 variables derived from three sources: (1) analysis of the results of the first survey, (2) review of the open literature and; (3) the authors' own extensive experience in high technology product development. Each respondent, on the basis of the original two innovations identified in survey 1, was asked to determine for each variable whether it impacted the outcome of the success, failure, neither or both. Survey 2 was completed by 59 of the original 79 managers.

The results from these two initial surveys were reported earlier (Maidique and Zirger 1984). To summarize, we conducted several statistical analyses for each variable and innovation type including determination of means and standard deviations, binomial significance, and clustering. Appendix A shows the binomial significance for the 37 variables which differentiated between success and failure. Combining our statistical results with the content analysis of the initial survey, we derived the eight propositions listed earlier.

Using these eight factors as a starting point, we then developed a block diagram of the new product development process that focuses on the product characteristics and the functional interrelationships and competences that are most influencial in determining new product success or failure (Figure 1). In our view the innovation process is a constant struggle between the forces of change and the status quo. Differences in perceptions between the innovator and the customer and also between the groups that make the building blocks of the innovation process—engineering, marketing and manufacturing—all conspire to shunt new product development or to deflect it from the path of success. Effective management attempts to integrate these constituencies and to allocate resources in a way that makes the new possible. These ideas are the basis of a model of the new product development

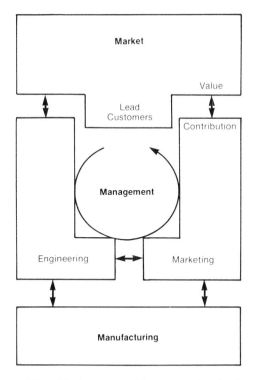

Fig. 1. Diagram of the critical elements of the new product development process.

process that we describe more fully and validate empirically in a forthcoming paper (Zirger and Maidique 1986).

The eight propositions resulting from our analysis were the objective "truths" that resulted from statistical analysis of our large sample of new product successes and failures. Though coincident in their salient aspects with the work of others (Cooper 1983), these results, however, did not fully satisfy us. Had we missed important variables in our structured surveys? Had our respondents understood our questions? Had we failed to detect significant relations between some of the variables we identified—or between these and some yet undiscovered factors? How valid were our final generalizations? And most important, what were the underlying conceptual messages in this list of factors? In short, we were concerned that perhaps our statistical analysis might have blurred important ideas.

Reflecting on his research on the individual psyche, Carl Jung (1987) once put it in this way:

The statistical method shows the facts in the light of the average, but does not give a picture of their empirical reality. While reflecting an indisputable aspect of reality, it can falsify the actual truth in a most misleading way . . .. The distinctive thing about real facts, however, is their individuality. Not to put too fine a point on it, one could say that the real picture consists of nothing but exceptions to the rule, and that, in consequence, reality has predominately the characteristic of irregularity. (p. 17)

Such irregularities have caused one of the most experienced researchers in the field to wonder out loud if any fundamental commonalities exist at all in new product successes. "Perhaps", Cooper observed (1979), "the problem is so complex, and case so unique, that attempts to develop generalized solutions are in vain".

METHODOLOGY

To address the concerns noted above, we prepared individual in-depth case studies for 40 of the original 158 products to search for methodological flaws or significant irregularities that might challenge the results of our statistical analysis (Table 1). The case studies were prepared under the supervision of the authors by 45 graduate assistants. Seventeen West Coast electronics firms which had participated in the 1982 Standford-AEA Executive Institute and in our original two surveys served as sites for the 20 case studies. This subset of the original product pairs served as the subject of analysis for the case studies. Two or more project assistants interviewed managers and technologists and prepared written reports that included interview transcripts or summaries, background information on the firm, the competitive environment, the product development process, the characteristics of each of the two products, validation of the original survey 2 and a critical review of the factors that contributed to success or failure in each case. Overall, 101 managers and technologists were interviewed in 148 hours of interviews.

Most of the companies supplied the research terms with detailed financial, marketing and design information regarding each one of the products, including in some cases internal memoranda that traced the products' development histories. Because of the confidentiality of this data, we must not identify any of the firms, much as we would like to thank them for their contributions to the project. In some cases, to illustrate a point, we have chosen to use examples from the public domain, or from published cases we or others have written about, and we may mention a company by name; however, the companies that collaborated with the project are either left anonymous or given fictitious names which, when first introduced, are placed in quotation marks.

This paper reports how these case studies and the associated interview transcripts enriched our earlier conclusions. In "Defining 'User Needs' and 'Product Value'" we begin to clarify the terms that we had employed in our survey, specifically "user needs" and "product value". In "How Should Product Success be Measured" we explore the meaning of success. The case studies led us to expand our concept of success and failure beyond the one-dimensional confines of financial return. Indeed, success and failure often appear to be close partners, not adversaries, in organizational and business development. Finally in "The New Product Learning Cycle" we postulate an evolutionary model of new product development, which we believe leads to a better understanding of the relationship between success and failure. For many of the propositions we present here, we lack the analytical support that underlies the eight factors identified in our original research. Nonetheless, we feel that these findings, which we hope will help to illuminate further research—including our own—are as important as our statistical results.

DEFINING "USER NEEDS" AND "PRODUCT VALUE"

The detailed case studies largely reinforced the principal findings of the overall study (Maidique and Zirger 1984). But the case studies also enriched some of the findings from structured questionnaires by providing fresh insights on several of the key variables. In this paper, we focus on the most important and perhaps the least specific variable, "understanding of the market" and "user needs" which is believed to result in products with "high value".

One of the principal findings of our large sample survey was that "user needs" and "customer and market understanding" are of central importance in predicting new product success or failure, a result that parallels the findings of the pioneering SAPPHO pairwise comparison study (Rothwell et al. 1974; Freeman 1974). This result, however, does little to illuminate how a firm goes about achieving such understanding. What's more, citing user needs ex post facto as a key explicatory variable in product success can be simply disregarded as tautological. Of course, it can be argued the company "understood" user needs if the product was successful.

Expanding on such criticisms, Mowery and Rosenberg (1979) have pointed out that the term "user need" is any event vague and lacks the precision with which economists define related market variables such as demand. What seems to be important, however, is to determine whether there are identifiable ex post ante actions that organizations take that develop and refine the firm's understanding of the customer's needs.

In most of the instances in which interviewees indicated a product had succeeded because of "better understanding of customer needs", they were able to support this view by citing specific actions or events. Both the experiential background of the management and developing team as well as actions taken during the development and launch process were viewed as important.

One line of argument went thus: we understood customer needs because the managers, engineers and marketing people associated with the product were people with long-term experience in the technology and/or market. In such a situation, some executives argue, very little market research is required because the company's management has been close to the customer and to the dynamics of his changing requirements all along. As the group vice-president of a major instrument manufacturer explained, "We were able to set the right design objectives, particularly cost goals, because we knew the business, *we could manage by the gut* (authors' emphasis)".

This approach was evident in other firms also. When "Perfecto", a leading U.S. process equipment manufacturer, induced by a request from one of its European customers, commissioned a domestic market survey to assess potential demand for a new product that combined the functions of two of its existing products, the result was almost unanimously negative. Because of a quirk in the process flow in U.S. plants (which differed from European plants), domestic customers did not immediately see significant value in the integrated product. Notwithstanding the market survey data, Perfecto executives continued to believe that the product would prove to be highly cost effective for their worldwide customers. Buoyed by enthusiasm in Europe and a feeling of deep understanding of his customers that was the result of 13 years of experience in the numerically controlled process equipment market, Perfecto's president gave the project the go ahead. His experience and self-confidence paid off. There was ultimately a significant demand for the new machine on both sides of the Atlantic.

These experience-based explanations, however, are only partially useful blueprints for action. The argument simply says that experienced people do better at new product development than the inexperienced, a hypothesis confirmed by our earlier research and that of others (Cooper and Bruno 1977).

Most of our informants, however, characterized the capture of "user needs" in action-oriented terms. For the successful product in the dyad, they described the company as having more openly, frequently, carefully and continuously solicited and obtained customer reaction before, after and during the initiation of the development and launch process. In some cases, the attempt to get customer reaction went to an extreme. "Electrotest", a test equipment manufacturer conducted design reviews for a successful new product at their lead customers' plants. In general the successful products were the result of ideas which originated with the customers, filtered by experienced managers. In one case customers were reported to have "demanded" that an instrument manufacturer develop a new logic tester. As a rule, the development process for the successful products was characterized by frequent and in-depth customer interaction at all levels and throughout the development and launch process. While we did not find (and did not look for), what von Hippen discovered in his careful research on electronic instruments, that users had in many cases already developed the company's next product, it was clear that, more so than any other constituency, they could point out the ideas that would result in future product successes (von Hippel 1976).

But when listening to customers, it's not enough to simply put in time. It is of paramount importance to listen to potential users without preconceptions or hidden agendas. Some companies become enamored with a new product concept and fail to test the idea against the reality of the marketplace. Not surprisingly, they find later that either the benefit to the customer was more obvious to the firm than to the customer himself or the product benefits were so specific that the market was limited to the original customer. For these reasons, the president of an automated test equipment manufacturer provided the following admonition, "When listening to customers, clear your mind of what you'd like to hear—Zen iistening".

Unless this careful listening cascades throughout the company's organization and is continually "market"-checked, new products will not have the value to the customer that results in a significant commercial success. A predominant characteristic of the 20 successful industrial products that we examined in our case studies is that they resulted in almost immediate economic benefits to their users, not simply in terms of reduced direct manufacturing or operating costs. The successful products seemed to respond to the utility function of potential customers, which included such considerations as quality, service, reliability, ease of use and compatibility.

Low cost or extraordinary technical performance, per se, did not result in commercial success. Unsuccessful products were often technological marvels that received technical excellence awards and were written up in prestigious journals. But typically such extraordinary technical performance comes at a high price and is often not necessary. "Very high performance, at a very high price. This is the story behind virtually every one of our new product failures", is how a general manager at "International Instruments", an instrument manufacturer with a reputation for technical excellence, described the majority of his new product disappointments.

In contrast to this phenotype, new product successes tend to have a dramatic impact on the customer's profit-and-loss account directly or indirectly. "Miltec", an electronic systems manufacturer, reported that its successful electronic counter saved their users 70 percent in labor costs and down-time. "Informatics", a computer peripheral manufacturer, developed a very successful magnetic head that was not only IBM compatible but 20 percent cheaper and it offered a three times greater performance advantage. An integrated satellite navigation receiver developed by "Marine Technology", a communications firm, so drastically reduced on-board downtime in merchant marine ships in comparison to the older modular models that the company was overwhelmed by orders. The first 300 units paid for the $2.5 million R&D investment; overall 7,000 units were sold. By comparison, the unsuccessful products provided little economic benefit. Not only were they usually high-priced but often they were plagued by quality and reliability problems, both of which translate into additional costs for the user.

HOW SHOULD PRODUCT SUCCESS BE MEASURED?

Our original surveys used an unidimensional success taxonomy. Success was defined along a simple financial axis. Successful products produced a high return while unsuccessful products resulted in less than break-even returns. Using this measure, our population of successes and failures combined to form a clearly bimodal distribution (Figure 2) that reinforced our assumption that we were dealing with two distinct classes of phenomena. While obtaining and plotting this type of data went beyond what most prior success-failure researchers had deemed necessary to provide, our detailed case studies lead us to conclude that this may not have been enough.

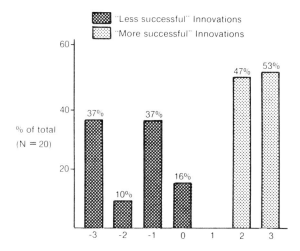

Fig. 2. Distribution of successes and failures by degree of success/failure for case studies.

Success is defined as the achievement of something desired, planned, or attempted. While financial return is one of the most easily quantifiable industrial performance yardsticks, it is far from the only important one. New product "failures" can result in other important byproducts: organizational, technological and market development. Some of the new product failures that we studied led to dead ends and resulted in very limited organizational growth. On the other hand, many others—the majority—were important milestones in the development of the innovating firm. Some were the clear basis for major successes that followed shortly thereafter.

International Instruments, a large electronics firm, developed a new instrument based on a new semiconductor technology (diode arrays) that the firm had not yet used in one of its commercial products. The instrument, though technically excellent, was developed for a new market where the company did not have its traditionally keen sense of what value meant to the customer. Few units were sold and the product was classified as a failure. On the other hand, the experience gained with the diode array technology became the basis for enhancement of other product families based on this newly gained technical knowledge. Secondly, the organization learned about the characteristics of the new market through the diode array product, and, armed with new insights, a redesigned product was developed which was a commercial success. Was the diode array product really a failure, its developers asked?

In this and other cases we observed, the failure contributed naturally to the subsequent successes by augmenting the organization's knowledge of new markets or technologies or by building the strength of the organization itself. An example from the public domain illustrates this point. After Apple Computer has been buffeted by the manufacturing and reliability problems that plagued the Apple III launch, which caused Apple to lose its lead in the personal computer market and to yield a large slice of the market to IBM. Apple's chairman summed up the experienced thus (Maidique et al. 1985). "There is no question that the Apple III was our most maturing experience. Luckily, it happened when we were years ahead of the competition. It was a perfect time to learn". As demonstrated by the manufacturing quality of the Apple IIe and IIc machines. Apple, that is the Apple II division,-learned a great deal from the Apple III mishaps. Indeed, Sahal has pointed out that success in the development of new technologies is a matter of learning (Sahal 1981). "There are few innovations", he points out, "without a history of lost labor. What eventually makes most techniques possible is the object lesson learned from past failures". In his classic study of technological failures. Whyte argues that most advances in engineering have been accomplished by turning failure into success (1975). To Whyte engineering development is a process of learning from past failures.

Few would think of the Boeing Company and its suppliers as a good illustration of Sahal's and Whyte's arguments. Rosenberg, however, has pointed out that early 707s, for many years considered the safest of airplanes, went into unexplainable dives from high altitude flights. The fan-jet turbine blades used in the jumbo jet par excellence, Boeing's famed 747, failed frequently under stress in the 1969–1970 period (Rosenberg 1982). Despite these object lessons, or perhaps because of them, Boeing makes more than half of the jet-powered commercial airlines sold outside of the Soviet bloc. According to the executive vice-president of the Boeing Commercial Airplane Company, himself a preeminent jet aircraft designer, "We are good partly because we build so many airplanes. We learn from our mistakes, and each of our airplanes absorbs everything we have learned from earlier models and from other airplanes" (Newhouse 1982).

Learning by Doing, Using and Failing

It has long been recognized that there is a strong learning curve associated with manufacturing activity. Arrow (1962) characterized the learning that comes from developing increasing skill in manufacturing as "learning by doing". Learning by doing results in lower labor costs. The concept of improvement by learning from experience has been subsequently elaborated by the Boston Consulting Group and others to include improvements in production process, management systems, distribution, sales, advertising, worker training, and motivation. This enhanced learning process, which has been shown for many products to reduce full costs by a predictable percentage every time volume doubles, is called the experience curve (Henderson 1968).

Rosenberg, based on his study of the aircraft industry, has proposed a different kind of learning process, "learning by using" (Arrow 1962, pp. 120–40). Rosenberg distinguishes between learning that is "internal" and "external" to the production process. Internal learning results from experience with manufacturing the product, "learning by doing"; external learning is the result of what happens when users have the opportunity to use the product for extended periods of time. Under such circumstances, two types of useful knowledge may be derived by the developing organization. One kind of learning (embodied) results in design modifications that improve performance, usability, or reliability, a second kind of learning (disembodied) results in improved operation of the original or the subsequently modified product.

In our study we found another type of external learning, "a learning by failing", which resulted in the development of new market approaches, new product concepts, and new technological alternatives based on the failure of one or more earlier attempts (Figure 3). When a product succeeds, user experience acts as a feedback signal to the alert manufacturer that can be converted into design or operating improvements (learning by using). For products that generate negligible sales volume, little learning by using takes place. On the other hand, products that fail act as important probes into user space that can capture important information about what it would take to make a brand new effort successful, which sometimes makes them the catalyst for major reorientations. In this sense, a new product is the ultimate market study. For truly new products it may be the only effective means of sensing market attitudes. According to one of our respondents, a vice-president of engineering of "California Computer", a com-

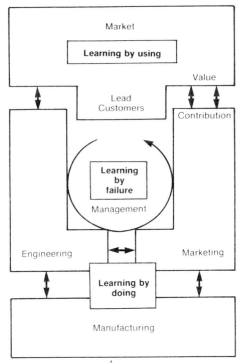

Fig. 3. A model of internal and external learning.

puter peripherals manufacturer, "No one really knows if a truly new product is worth anything until it has been in the market and its potential has been assessed".

Another dimension of "learning by failing" relates to organizational development. A failure helps to identify weak links in the organization and to inoculate strong parts of the organization against the same failure pattern. The aftermath of the Apple III resulted in numerous terminations at Apple Computer, from the president to the project manager of the Apple III project. Those remaining, aided by new personnel, accounted for the well-implemented Apple III redesign and reintroduction program and the highly successful Apple IIe follow-on product (Maidique et al. 1983).

When the carryover of learning from one product to another is recognized, it becomes clear that the full measure of a product's impact can only be determined by viewing it in the context of both the products that followed. While useful information can be obtained by focusing on individual products or pairs of products, *the product family is a far superior unit of analysis from which to derive prescriptions for practicing managers.* The product family incorporates the interrelationship between products, the learning from failures as well as from successes. Thus, it is to product families, including false starts, not to individual products, that financial measures of success should be more appropriately applied.

Consider a triplet of communications products developed over a 10-year period by an electronics system manufacturer. For several years Marine Technology had developed and marketed commercial and military navigation systems. These systems were composed of separate components manufactured by others, such as a receiver, teletype, and a minicomputer, none of which was specifically designed for the harsh marine environment. Additionally, this multicomponent approach, though technically satisfactory, took up a great deal of space, which is at a premium on the bridge of a ship. Each of Marine Technology's new product generations attempted to further reduce the number of components in the system. By 1975 a bulky HP minicomputer was the only outboard component.

The need for a compact, rugged, integrated navigation system has thus been abundantly clear to Marine Technology engineers and sales people. Therefore, when microprocessors became available in the early 1970s, it was not surprising that Marine Technology's general manager initiated a program to develop a new lightweight integrated navigation system specially designed for the marine environment. The product was developed by a closely knit design team that spent six to eight months working with potential customers, and later market testing prototypes. Two years later the company introduced the MT-1, the world's first microprocessor-based integrated navigation system. The product was an instant success. Over 7,000 units were sold at a price of $25,000 per unit. At this price, margins exceeded 50 percent.

Shortly after the success of the MT-1 was established, engineering proposed a new product (the MT-2) to Marine Technology's newly appointed president. The MT-2 was to be about one-sixth the volume of the MT-1 and substantially cheaper in price. The president was so impressed with a model of the proposed product that he directed a team, staffed in part by the original MT-1 design team members, to proceed in a top secret effort to develop the MT-2. The team worked in isolation, only a handful of upper management and marketing people were aware of the project. Three and a half years and $3.5M later, the team had been able—by sacrificing some features—to shrink the product as promised to one-sixth the size of the MT-1 and to reduce the price to about $10,000. But almost simultaneously with the completion of the MT-2's development, a competitor had introduced an equivalent product for $6,000. Furthermore, the product's small size was not considered a major advantage. Key customers indicated the previous product was "compact enough". The company attempted to eliminate some addition features to tailor the product to the consumer navigation market, but it found that it was far too expensive for this market, yet performance and quality were too low for its traditional commercial and military markets. The product was an abject economic failure. Most of the inventory had to be sold below the cost.

A third product in the line, the MT-3, however, capitalized on the lessons of the MT-2 failure. The new MT-3 was directed specifically at the consumer market. Price, not size, was the key goal. Within two years the MT-3 was introduced at a price of $3,000. Like the MT-1, it was a major commercial success for the company. Over 1,500 have been sold and the company had a backlog of 600 orders in 1982 when the case histories were completed.

At the outset of this abbreviated product family vignette, we said the family consisted of three products. In a strict sense, this is correct, but in reality there were four products, starting with what we will call the MT-0, the archaic modular system. The MT-0 was instrumental to the success of the MT-1. Through the experience with customers that it provided, it served to communicate to the company that size and reliability improvements would be highly valued by customers in the commercial and military markets in which the company operated. With the appearance of microprocessor technology, what remained was a technical challenge, usually a smaller barrier to success than deciphering how to tailor a new technology to the wants of the relevant set of customers, as the company found out through the MT-2.

The success of the MT-1 was misread by the company to mean, "the smaller, the more successful", rather than, "the better we understand what is important to the customer the more successful". The company had implicitly made an inappropriate trade-off between performance, size, and cost. They acted as if they had the secret to success—compact size—and by shutting off its design team from its new as well as its old customers ensured that they would not learn from them the real secrets to success in the continually evolving market environment. It remained for the failure of the MT-2 to bring home to management that, by virtue of its new design, the company was now appealing to a new customer group that had different values from its traditional commercial and military customer. Equipped with this new learning, the company was now able to develop the successful MT-3.

THE NEW PRODUCT LEARNING CYCLE

There are several lessons to be learned from the history of Marine Technology's interrelated succession of products. First, their experience clearly illustrates the importance of precursors and follow-on products in assessing product success. To what extent, for instance, was the MT-2 truly a failure, or, alternatively, how necessary was such a product to pave the way for the successful MT-3? With hindsight one can always argue that the company should have been able to go directly to the MT-3, but wasn't to some extent the learning experience of the MT-2 necessary? Secondly, the story illustrates once again the importance of in-depth customer understanding as well as continuous interaction with potential customers throughout the development process even at the risk of revealing some proprietary information. Whatever learning might have been possible before entering the consumer market was shunted aside by the company's secretive practices.

The product evolution pattern of "Computronics", a startup computer systems manufacturer, reinforces these findings. One of the Computronics' founders had developed a new product idea for turnkey computer inventory control system for jobbers (small distributors) in one of the basic industries. From his experience as a jobber in this industry, he knew that it was virgin territory for a well-conceived and supported computerized system. During the development process, the company enlisted the support of the relevant industry association. Association members offered product suggestions, criticized product development, and ultimately the association endorsed the product for use by its members. The first Compu-100 system was shipped in 1973. Ten years later, largely on the strength of this product and its accessories, corporate sales had doubled several times and reached nearly $100 million, and the company dominated the jobber market.

As the company's market share increased, however, management recognized that new markets would have to be addressed if rapid growth was to continue. In early 1977, the company decided to take what seemed a very logical step to develop a system that would address the needs of the large wholesale distributors in the industry. Based on its earlier successes, the company planned to take this closely related market by storm. After a few visits to warehousing distributors, the product specifications were established and development began under the leadership of a new division established to serve the high end of the business. Since no one at Computronics had first-hand experience with this higher level segment of the distribution system, a software package was purchased from a small software company, but it took a crash program and several programmers a year to rewrite the package so that it was compatible with Computronics hardware. After testing the Compu-200 at two sites, the company hired a team of additional sales representatives and prepared for a national roll-out. Ten million dollars of sales were projected for the first year.

First year revenues, however, were minimal. Even three years after the launch the product had yet to achieve the first year's target revenue. What had happened? The new market would appear at first glance to be a perfect fit with Computronics' skills and experience, yet a closer examination revealed considerable differences in the new customer environment which, nonetheless, were brushed away in a cavalier manner by Computronics' management, who were basking in the glow of the Compu-100's success.

As organizations in such a euphoric state often do, Computronics grossly underestimated the task at hand. The large market for warehousing distributor inventory systems was attractive to major competitors such as IBM and DEC. But only a cursory study of the new customers and their buying habits was carried out. The tacit assumption was that large warehousing distributors were simply grownup jobbers. Yet these new customers were now much more sophisticated, had used data processing equipment for other functions, and generally required and developed their own specialized software. Increased competition and radically greater customer sophistication combined to require that Computronics be represented by a highly experienced and competent sales force. But because of the hurry to launch the product, Computronics skipped the customary training for sales representatives and launched the field sales force into a new area for the firm: the complex long-term business of selling large items ($480,000 each) to a technically knowledgeable customer. By believing that repeating past practices would reproduce past successes, Computronics had turned success into failure.

"Every victory", Carl Jung (1970) once wrote, "contains the germ of a future defeat". Starbuck and his colleagues (1978) have observed that successful organizations accumulate surplus resources that allow them to loosen their connections to their environments and to achieve greater autonomy, but they explain, "this autonomy reduces the sensitivity of organizations to changing environmental conditions . . . organizations become less able to perceive what is happening so they fantasize about their environments and it may happen that reality intrudes only occasionally and marginally on these fantasies". Fantasies create a myth of invincibility, yet an old Chinese proverb says, "There is no greater disaster than taking on an enemy lightly" (Tzu 1983).

Marine Technology's management fantasized that they had the secret of success: smaller is better. Computronics had a fantasy that similarly extrapolated their past victories: new markets will be like old markets. This is a pattern that repeats itself over and over in business. We have already alluded to one of the best publicized contemporary examples of this phenomenon: Apple Computer's Apple III on the heels of its colossally successful Apple II precursor. Even IBM is not exempt from this cycle. After taking one-third of the market with its PC (personal computer) despite its late entry, a senior executive at the IBM PC Division stated, "We can do anything" (Business Week 1984). "Anything" did not, as it happens, include the follow-on product to the PC, the PCjr, which—in contrast to the original PC—was not adequately test marketed to determine how consumers would react to its design features and ultimately had to be dropped from the IBM product line. IBM and Computronics, however, have both, at least temporarily, become more humble. Both are soliciting customer inputs so that they can redesign their disappointments.

The flow from success to failure, and back to success again, at Marine Technology illustrated a rhythm that we were to encounter repeatedly in our investigations. In the simplest terms, failure is the ultimate teacher. From its lessons the persistent build their successes. Success, on the other hand, often breeds complacency. Moreover, success seems to create a tendency to ignore the basics, to believe that heroics are a substitute for sound business practice. As the general manager of "Automatrix", a test equipment manufacturer, pointed out, "It's hard, very hard, to learn from your successes". Ironically, success can breed failure for firms that continue to view the future through the prism of present victories, especially in a dynamic industry environment.

These observations have led us to propose a model of new product success and failure in which successes and failures alternate with an irregular rhythm. This is not to say that for every success there must be a complementing failure. Most industrial products— about three out of five—succeed despite popular myths to the contrary (Cooper 1983). For some highly successful companies, as many as three out of four new products may be commercially successful. Most companies continuously learn by using, through their successful new products, and—as in the case of Marine Technology—they continuously develop improved designs. This is what most new product efforts are about—minor variations on existing themes.

But continued variations on a theme do not always lead to a major successes. In time, further variations are no longer profitable, and the company usually decides to depart from the original theme by adopting a new technology—microprocessors, lasers, optics—or to attack a new market—consumer, industrial, government—or, alternatively, organizational changes, defections, or promotions destroy part of the memory of the organizations so that the old now seems like the new. Changes in any of these three dimensions can result in an economic failure or, in our terms, new learning about a technology, a market, or about the strengths and weaknesses of a newly formed group as shown in Figure 4, an extension of the familiar product—customer matrix originally proposed by Ansoff (1965). This recurrent cycle of success and failure is shown graphically in Figure 5.

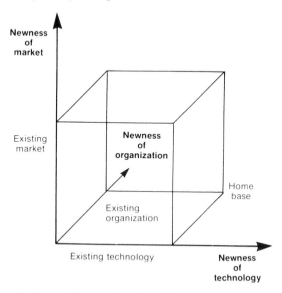

Fig. 4. Learning by moving away from home base.

In the model, a sequence of successes is followed by either a major organizational change, changes in product design,

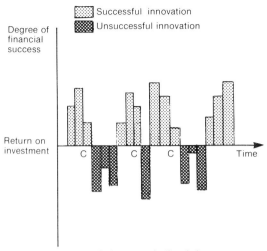

C = market, technological or organizational change

Fig. 5. A typical new product evolution pattern.

technology, or market directions that prompt an economic failure, which in turn spurs a new learning pattern. The model assumes a competitive marketplace, however, and is less likely to be applicable to a monopolistic situation in which a single firm dictates the relationship between customer and supplier. A second caveat is that while the pattern is roughly depicted as regular, in general it will be irregular, but the cycle of oscillation between economic success and failure, we believe, will still hold.

Success as a Stochastic Process

Success in new products is never assured. Too many uncontrolled external variables influence the outcome. Occasional or even frequent failure is a way of life for product innovations. As Addison reminds us in his Cato, "Tis not in mortals to command success". But while it is not possible to assure the outcome of any one product trial, it is possible to increase the likelihood that a product or group of products will be successful. Addison goes on to add, "We'll do more, Sempronius, *we'll deserve it*" (authors' emphasis). The eight factors that we identified at the outset of this article, and the cyclic model we propose, are an attempt to help managers conceptualize the new product development process so as to improve the proportion of economic successes. But on the other hand, it would be a mistake to attempt to increase the number of successes by reducing new product risk to zero by cautious, deliberate management. In the process, rewards may be also reduced to the same level.

Failure, as we have tried to argue, is part and parcel of the learning process that ultimately results in success. Sahal (1981) sums up the process thus, "What eventually makes the development of new techniques possible is the object lesson learned from past failures . . . profit by example". The important thing is to have a balance between successes and failures that results in attractive returns. Here a lesson from experienced venture capitalists, masters at the success forecasting game, is useful. As part of another research project, the authors have interviewed some of the nation's most successful and experienced venture capital investors whose portfolios have generally shown gains of 25–35 percent over the past 10 years. Given a large pot of opportunities, experienced venture capitalists believe they can select a group that will on the average yield an excellent return, yet few professionals are so sanguine that they believe that they can with certainty foretell any one success; they've seen too many of their dreams fail to meet expectations.

A new product developed by "Electrosystems", a military electronics company, seemed to fit a venture capitalist's dream. The company's new product, a phase-locked loop, had its origin with one of their key customers, the requisite technology was within their area of expertise, and a powerful executive championed the product throughout its development. A large market was anticipated. Thus far, a good bet, one might conclude. What's more, the resulting product was a high quality instrument. Yet the product brought little in the way of revenues to the firm for an alternative technology that solved the same problem in a cheaper way was simultaneously developed and introduced by a competitor.

Five years later, again spurred by a customer requirement, Electrosystems developed an electronic counter as part of a well-funded and visibly championed development program. This time, however, the product saved the customer 70 percent of his labor costs, considerably reduced his downtime, and there was no alternative technology on the horizon. This product was very successful. The point here is not that the product that ultimately produced a cost advantage to the customer was more successful. That much is self-evident. The point is that at the outset both products looked like they would provide important advantages to the customer. After all, they both originated with customers. Both projects were well-managed and funded and technologically successful. Yet one met with unpredictable external competition that blunted its potential contribution. The company did not simply fail and then succeed. It succeeded because it pursued both seemingly attractive opportunities. In other words, success generally requires not one but several, sometimes numerous, well-managed trials. This realization prompted one of our wisest inter-viewees, the Chief Engineer of "Metalex", an instrument manu-facturer, to sit back and say, "I've found the more diligent you are, the more luck you have".

This is the way both venture capitalists and many experienced high-technology product developers view the new product process. Venture capitalists who have compiled statistics on the process have found that only 60 percent of new ventures result in com-mercial success, the rest are a partial or complete loss. (Not surprisingly, this is about the same batting average that Cooper found in his study of industrial products.) About 40–50 percent of new venture-capital backed ventures produce reasonable returns, and only 10–15 percent result in outstanding investments. But it can be easily computed that such a combination of investments can produce a 25–30 percent return or more as a portfolio.

New Research Directions

Our research on new product success and failure has led us to reconsider our unit of analysis. Choosing the new product as the basic unit of analysis has many advantages. New products are clearly identifiable entities. This facilitates gathering research data. New products have individualized sales forecasts and return on investment criteria, and managements generally know whether these criteria are satisfied. "Successes" can be culled from "failures".

Our results, however, indicate that if financial measures of success are to be applied as criteria, a more appropriate unit of analysis is the project family. Before an individual product is classifed as a failure, its contribution to organizational growth, market development, or technological advance must be gauged. New products strongly influence the performance of their successors, and in turn are a function of the victories and defeats of their predecessors. Before the laurels are handed over to a winning team, an examination should be made of the market, technological, and organizational base from which the team launched its victory (Figure 4).

One of the IBM's most notable product disasters was the Stretch computer. IBM set out to develop the world's most advanced computer, and, after spending $20 million in the 1960s for development, only a few units were sold.

On the heels of the Stretch fiasco came one of the most successful products of all time, the IBM 360 series. But when IBM set out to distribute kudos, it recognized that much of the technology in the IBM 360 was derived from work done on the Stretch computer by Stephen Dunwell, once the scapegoat for the Stretch "setback". Subsequently Mr. Dunwell was made an IBM fellow, a very prestigious position at IBM that carries many unique peaks (Fortune 1966). As Newton once said, "If I have seen far it is because I stood on the shoulders of giants" (Bartlett 1968).

We were able to gain insight into this familial product interrelationship because our success-failure dyads were often members of the same product family. But even though they were interrelated, they represented only a truncated segment of a product family. Nonetheless, in some sites, for example Marine Technology, we were able to collect data on three or four members of a product family. On the other hand, our efforts, to date, fall far short of a systematic study of product families. This is the central task of the next stage of our research.

Our limited results, however, bring into question research that focuses on the product as the unit of analysis, including our own. Consider one of our principal research findings, which is also buttressed by the findings of several prior investigators: successful products benefit from existing strengths of the developing business unit. The implication of this finding is that organizations should be wary of exploring new territories. In contrast to this result, our observations would lead us to argue just the opposite, that firms should continuously explore new territories even if the risk of failure is magnified. The payoff is the learning that will come from the "failures" which will pave the way for future successes.

Careful validation of the cyclic model of product development proposed here could have other important consequences for our understanding of technology-based firms. If indeed the pattern proposed in Figure 5 is generalizable to firms that are continuously attempting to adapt to new markets and technologies, then there are important implications for management practice.

First, the model implies that new product development success pivots on the effectiveness of intra and inter company learning. This conclusion puts a premium on devising a managerial style and structure that serves to catalyze internal and external communication. Second, by implicitly taking a long-term view of the product development process, the model emphasizes the importance of long-term relationships with employees, customers, and suppliers. Out of such a view comes a high level of understanding, and therefore of tolerance for failure to achieve commercial success at any one given point in the product line trajectory. Firms need to learn that product development is a journey, not a destination. These preliminary findings are compatible with an exploratory study of new product development in five large successful Japanese companies completed by Imai and his colleagues (Imai, et al. 1982). One of the principal findings of their research was that the firms studied were characterized by an almost "fanatical devotion towards learning—both within organizational membership (sic) and with outside members of the interorganizational network". This learning, according to the authors, played a key role in facilitating successful new product development. It appears that when successful at new product development, small and large U.S. companies operate in a very similar manner to the best-managed Japanese firms.

Many key questions, however, remain to be settled. Is there an optimal balance between successes and failures? Are Japanese firms susceptible to the same oscillating pattern between success and failure as American firms? How does this balance change across industries? How can tolerance for failure be communicated without distorting the ultimate need for economic success? How can a firm learn from the failures of others? Are there characterisic success-failure patterns for a group of firms competing in the same industry? These and other related questions will occupy us in the next phase of our research.

Appendix A. Significant variables from survey 2 grouped by index variable.

Successful innovations were:	No. of observations	Cumulative binomial	Significance rating
(1) Better matched with user needs			
better matched to customer needs	44	8.53 E-09	+ + +
developed by teams which more			
fully understood user needs	44	1.27 E-05	+ + +
accepted more quickly by users	49	7.01 E-04	– – –
(2) Planned more effectively and efficiently			
forecast more accurately (market)	43	1.25 E-07	+ + +
developed with a clearer market strategy	45	1.24 E-04	+ + +
formalized on paper sooner	45	3.30 E-03	+ + +
developed with less variance between			
actual and budgeted expenses	46	2.70 E-02	– –
expected initially to be more			
commercially successful	42	8.21 E-02	+
(3) Higher in benefit-to-cost			
priced with higher profit margins	51	6.06 E-08	+ + +
allowed greater pricing flexibility	52	1.02 E-06	+ + +
more significant with respect to			
benefit-to-cost ratio	43	6.86 E-03	+ + +
(4) Developed by better-coupled organizations			
developed by better-coupled functional divisions	39	1.68 E-07	+ + +
(5) More efficiently developed			
less plagued by after-sales problems	35	5.84 E-05	– – –
developed with fewer personnel changes on			
the project team	28	6.27 E-03	– – –
impacted by fewer changes during production	41	1.38 E-02	– –
developed with a more experienced project team	39	2.66 E-02	+ +
changed less after production commenced	47	7.19 E-02	–
developed on a more compressed time schedule	39	9.98 E-02	+
(6) More actively marketed and sold			
more actively publicized and advertised	39	4.64 E-03	+ + +
promoted by a larger sales force	28	6.27 E-02	+ + +
coupled with a marketing effort to educate users	37	1.00 E-02	+ +
(7) Closer to the firm's areas of expertise			
aided more by in-house basic research	25	7.32 E-03	+ + +
required fewer new marketing channels	25	7.32 E-03	– – –
closer to the main business area of firm	30	8.06 E-03	+ + +
more influenced by corporate reputation	29	3.07 E-02	+ +
less dependent on existing products in the market	36	6.62 E-02	–
required less diversification from traditional markets	24	7.58 E-02	–
(8) Introduced to the market earlier			
than competition			
in the market longer before competing products introduced	44	1.13 E-02	+ +
first-to-the-market type products	39	1.19 E-02	+ +
more offensive innovations	46	5.19 E-02	+
generally not second-to-the-market	36	6.62 E-02	–
(9) Supported more by management			
supported more by senior management			
potentially more impactful on the careers of	31	1.66 E-03	+ + +
the project team members	32	5.51 E-02	+
developed with a more senior project leader	39	9.98 E-02	+
(10) Technically superior			
closer to the state-of-the-art technology	36	3.26 E-02	+ +
more difficult for competition to copy	45	3.62 E-02	+ +
more radical with respect to world technology	42	8.21 E-02	+

REFERENCES

Ansoff, H. I. 1965: *Corporate Strategy.* New York: McGraw-Hill, pp. 131–133.

Arrow, K. 1962: The Economic Implications of Learning by Doing, *Review of Economic Studies,* June.

Bartlett, J. 1968: *Bartlett's Familiar Quotations,* 13th ed. New York: Little, Brown and Company.

Bartlett, J. 1968: *Bartlett's Familiar Quotations,* 14th ed. New York: Little, Brown and Company.

Business Week 1984: How IBM Made Junior and Underachiever, June 25, p. 106.

Cooper, A. C. and A. V. Bruno 1977: Success among High-Technology Firms, *Business Horizons* 20 (2), pp. 16–22.

Cooper, R. C. 1983: A Process Model for Industrial New Product Development, *IEEE Transactions on Engineering Management* EM-30 (1), pp. 2–11.

Cooper, R. C. 1979: The Dimensions of Industrial New Product Success and Failure. *Journal of Marketing* 43, p. 102.

Cooper, R. G. 1983: Most Products *Do* Succeed, *Research Management,* Nov.–Dec., pp. 20–25.

Freeman, C. 1974: *The Economics of Industrial Innovation.* Harmondsworth: Penguin Books, pp. 161–197.

Henderson, B. 1968: *Perspectives on Experience.* Boston Consulting Group.

Hippel, E. A. von 1976: Users as Innovators, *Technology Review* No. 5, pp. 212–239.

Imai, K., I. Nonaka and Takeuchi, H. 1982: *Managing the New Product Development Process: How Japanese Companies Learn and Unlearn.* Tokyo: Institute of Business Research, Hitotsubashi University, Kunitachi, pp. 1–60.

Jung, C. G. 1957: *The Undiscovered Self.* New American Library, p. 17.

Jung, C. G. 1970: *Psychological Reflections.* Princeton, N.J.: Princeton University Press, p. 188.

Maidique, M. A. and B. J. Zirger 1984: A Study of Success and Failure in Product Innovation: The Case of the U.S. Electronics Industry, *IEEE Transactions on Engineering Management* EM-31 (4), pp. 192–203.

Mowery, D. and N. Rosenberg 1979; The Influence of Market Demand upon Innovation: A Critical Review of Some Recent Empirical Studies, *Research Policy* 8, pp. 101–153.

Newhouse, J. 1982: *The Sporty Game* New York: Alfred A. Knopf, p. 7.

Rosenberg, N. 1982: *Inside the Black Box, Technology and Economics.* Cambridge: Cambridge University Press pp. 124–126.

Rothwell, R., C. Freeman, A. Horley, V. I. P. Jervis, Z. B. Robertson and J. Townsend 1974: SAPPHO Updated-Project SAPPHO, Phase II, *Research Policy* 3, pp. 258–291.

Sahal, D. 1981: *Patterns of Technological Innovation.* Reading, MA: Addison-Wesley, p. 306.

Starbuck, W., Greve, A. and Hedberg B. L. 1978: Responding to Crisis, *Journal of Business Administration* No. 9, p. 111–137.

Tzu, Lao 1983: *Tao Te Ching.* New York: Penguin Boks, p. 131.

Whyte, R. R. 1975: *Engineering Progress Through Trouble.* London: Institution of Mechanical Engineers.

Wise, T. 1966: IBM's $5B Gamble, *Fortune,* September

Wise, T. A Rocky Road to the Market Place, *Fortune,* October.

Zirger, B. J. and Maidique, M. A. (forthcoming) Empirical Testing of a Conceptual Model of Successful New Product Development. To be submitted to *Management Service.*

17

Evaluating Research—ROI is not Enough

GEORGE F. MECHLIN AND DANIEL BERG

Technological innovation in the United States has been a dominant factor in this country's history of economic leadership. A 1977 Department of Commerce report notes, for example, that "technological innovation was responsible for 45 percent of the nation's economic growth from 1929 to 1969" (*Business Week* 1978, p. 47). In recent years, however, informed observers have expressed increasing alarm over a lagging U.S. commitment to scientific endeavor.

The facts are unmistakable. In his 1979 commencement address at the University of Pittsburgh, Richard C. Atkinson, then director of the National Science Foundation, identified these central trends of the decade ending in 1978:

- R&D as a fraction of GNP had decreased 20 percent.
- Basic research as a fraction of GNP had decreased 24 percent.
- The proportion of scientists and engineers in the labor force who are engaged in R&D had dropped 13 percent.
- Industrial investment in basic research as a fraction of net sales had declined 32 percent.

A comparison of the patents issued by the U.S. Patent Office tells much the same story. Michael Boretsky of the Department of Commerce has noted that 80 percent of the patents issued in 1965 originated in this country. By 1977, that figure had slipped to 63 percent (Boretsky 1979).

More to the point, in 1956 industry performed 38 percent of the basic research in the United States; in 1976, just 16 percent (Business Week 1978). True, some portion of this shift is the result merely of cosmetic alterations in the description of industrial research projects as either "basic" or "applied". (According to the National Science Foundation, the term *basic* applies properly to original investigations for the advancement of scientific knowledge not having specific commercial objectives.) But since basic research, however defined, is responsible for the vast majority of major applied technological breakthroughs, this slippage is legitimate cause for alarm.

Why the decline? Why the abandonment of an apparently successful industrial strategy? The usual answers given—an inhibiting regulatory environment, unfavorable tax and accounting policies, and the like—are true as far as they go, but they do not go far enough. It would seem that so fundamental a movement away from basic research must also have something to do with the techniques managers use to determine its value. And it does. Something seems to be telling them that it is not worth doing today. And there is.

THE LIMITATIONS OF ROI

For business, a commitment to research is first and foremost an investment. Over the past two decades, the most popular techniques for evaluating investments have led managers to place more and more emphasis on short-term profitability—especially in cases where short-term profitability is the basis for executive incentive programs. "Return on investment" has become the catch phrase of the new managerial class, and inevitably the principles of ROI have come to dominate industrial budgeting for research every bit as much as they have budgeting for other purposes.

We do not deny the proven utility of ROI techniques. Nor do we challenge for a moment the fact that investments in research must be evaluated and must in some sense pay off. We merely wish to point out that there are many considerations to which the principles of ROI, for all their seeming accuracy, are simply not responsive. However much research managers may gain by using ROI, the plain fact is that they stand to lose something too. It is no accident that laboratory staffers often come to interpret *ROI* as "restraint on innovation." (Hayes and Abernathy 1980, p. 67).

If industry is going to reverse its shift away from basic research, it will need new tools for both structuring and evaluating its research efforts. Hence, we need a better understanding of the limitations of using ROI in industrial research and, equally important, a better sense of where and with what it needs to be supplemented. First, the limitations.

Time Span

Though ROI calculations account for only present activities, most technological breakthroughs take years to produce results. A study of eight cases of technological innovation (out of which ten new products have emerged) shows that each took an average of 19.2 years to move successfully from laboratory to marketplace (Globe 1973).

At Westinghouse, for example, the development of a super-conducting generator has been in the works for more than 30 years. The underlying scientific principle—that resistance to the flow of an electric current through metals virtually disappears as the temperature of the metals approaches absolute zero—has been known since 1911. Even so, Westinghouse did not begin to study the phenomenon until 1947, when it first acquired a low-temperature liquefier. From the outset, company researchers were well aware that many years would pass before they could send the technology to the marketplace.

Today Westinghouse is finally building an experimental 300 megavolt-ampere superconducting generator, which it hopes to complete by 1983 and have running at a commercial power station by the mid-1980s. Actual commercial production of the generators is not expected until 1988 or later, yet Westinghouse spent approximately $3 million on research into superconductivity during the 1950s, another $10 million during the 1960s, and an additional $10 million during the past decade.

Had simple ROI considerations been applied to the research, it surely would have died somewhere along the way, for there is what Robert C. Dean, Jr. has called a "temporal mismatch" between the natural pace of innovation and management's desire for immediate results (Dean 1974, p. 12).

Unpredictability of Results

Another limitation of ROI has to do with the unpredictable nature of industrial research. A project may be a resounding technological success—it may even reveal profound new relationships among natural phenomena—but still fall far short of commercial success. The danger of using ROI is greatest in basic research, since its discoveries may have no obvious commercial application at the time they are made. Even short-term, low-risk projects may succeed technically but fail commercially, for there can be no hard-and-fast protection against a competitor's new product, a shift in consumer preferences, or a sudden change in the sponsoring company's priorities.

Still, even negative results from a current project may add substantially to overall knowledge and produce unexpected dividends in the future. Recently Westinghouse invested significant time and money in a project to develop a coal-based fuel cell as a source of power. Eventually, having concluded that the fuel cell would not be a readily marketable product, Westinghouse discontinued the project—an apparent failure. Yet, within a matter of months, the supply of oil and gas for electric power generation became an urgent national problem and coal gasification a very important technology.

Under these changed circumstances, its work on the fuel cell, so recently put aside as commercially inappropriate, proved to be the key to the economical conversion of coal into gas. Lacking a way to measure the economic value of negative results, conventional ROI thinking had shown the project to be a clear failure.

Imprecision of Measurement

Managers can rarely attribute corporate profits and losses to particular research efforts. It is not uncommon for a single development in a research laboratory to benefit many divisions of a company, yet typical ROI calculations are really not able to express the true value to the company of that development.

Just a few months ago at the Westinghouse R&D center outside Pittsburgh, one of the research people was studying the technology of how water flows through porous geological formations. The goal he had in mind was to help perfect solution mining techniques for in situ uranium mining.

As it turned out, the knowledge he gained had an important application to the work of a real estate development group, which needed to be able to evaluate for regulatory authorities the environmental impact of draining a lake. In addition, another group has applied his results to its work on the problems of heat flow through porous metals in the development of high-temperature turbine blades. Still another group has found his results useful in its work on a below-ground water-coolings system for heat pump technology.

Take another example. A chemist recently made a staff presentation on a polyurethane material originally developed for use on a missile launcher. That same technology was subsequently applied, quite successfully, to a new process for encapsulating things to be used underwater. Nobody had the remotest idea, when the project was first undertaken, that a second application for it would exist. But investing in making people knowledgeable in the first area turned out to be of value in the second and cost only a minimal amount.

No ROI system we know of can measure the real value of these single research efforts as a whole. Nor can it measure the true closeness of "fit" between any piece of current research and long-range corporate objectives. The time frame is simply too short and the results too unpredictable.

CENTRALIZED RESEARCH

The problems with ROI as a fair measurement of industrial research become even worse when a multi-division company combines many of its research activities into a central laboratory. Although the responsiveness of research to corporate objectives significantly improves, the primary advantages of centralizing research—the improvements in performance it makes possible—involve intangible, "overhead" benefits of the sort that ROI techniques are especially ill-suited to handle but that greatly facilitate the exercise of strategic control.

Technology Transfer

Consider, for example, the value a centralized research laboratory has as a communications nexus for technical functions throughout a company. Working with the technical staffs of various divisions, laboratory personnel are in an excellent position to know that work in one area—say, solving problems of stress analysis for a turbine blade—may have important applications elsewhere—in work on fans, for example, or in processing equipment or air-conditioning machinery. Such knowledge helps avoid both obvious duplication in research and seat-of-the-pants "fixes" for complex problems. At the same time, it allows the laboratory to define and initiate research beneficial to several divisions but out of their financial reach individually.

An executive of a Pittsburgh company with three physically separate R&D labs, each reporting to a different division, recently asked how to coordinate research activity. The answer was simple: put all research personnel together in the same physical place and let the coordination happen over lunch, in the library, in the hallways. Until these people can rub shoulders daily, until they can work informally on each other's projects, the degree of communication needed for an effective coordination of effort simply will not exist.

The great virtue of a centralized laboratory is to make such communication possible. It may be supplemented, of course, with the establishment of research committees whose members come from different divisions, functions, or major projects. It may be supplemented with guest lectures, task forces, newsletters, seminars, symposia, and reports. But the surest way to facilitate the transfer of new ideas is to put researchers together in the same place.

How can ROI formulas place a fair value on this exchange of information? How can they determine whether the funds spent on centralizing operations earn an adequate return? For that matter, how can they measure in the present the money to be saved by avoiding technical problems in the future?

For example, having encountered certain stress problems in turbines, central laboratory personnel are in an especially good position to recognize comparable design problems in other rotating equipment—before that equipment is manufactured or sold. Obviously, preventing such problems at the design stage is far less costly than attempting to solve them in the customer's plant, yet conventional budgeting techniques have no real way to account for the avoidance of a future loss of profit.

Such techniques are simply not designed to address many of the ways in which technology transfer is of value to a company. A centralized research staff, standing a little apart from the commercial and financial deadlines of operating divisions, provides a kind of strategic overhead service to its parent organization.

It is well situated to be professionally objective in its judgment of new technology within or outside the company, to give unbiased advice to line managers, to act as an independent troubleshooter for divisional problems, to provide the corporate linkage with government research projects, and to evaluate the long-range implications of that research for the technological progress of the industry. These are all things ROI is not suited to measure.

Use of Facilities and Personnel

Very rarely is it possible or desirable for a diversified corporation to perform all its R&D on a centralized basis. Responsibility for engineering, including product development, is often left to its various business units. Those units, in turn, are vulnerable to the same fluctuations in the business cycle that afflict every organization, and their peaks and valleys of activity inevitably produce high and low degrees of demand for R&D effort.

If the parent company does not maintain real flexibility in its centralized research operation, its business units will meet cyclical fluctuations in demand with excessively wasteful personnel policies.

At the peak of the cycle, the units will experience all the costs and delays of recruiting and training skilled personnel, at the bottom, they will lay off talented people in ways guaranteed to demoralize those who remain. Even at the best of times, they are not likely to provide a sufficient level of activity to occupy genuinely specialized researchers at their full capability.

This argument applies as well to the use of capital resources—expensive equipment and buildings. In the long run, if there is no centralized operation able to pick up or reduce the slack as needed, it is the parent company that will suffer most. One of the most valuable services provided by a company's research establishment is precisely this ability to manage slack effectively. It is a service, however, which finds no place in conventional ROI calculations.

Research as Overhead

The same is true of many of the other overhead functions performed by a centralized research operation. Customers, for instance, often have technical problems that the supplier's sales and marketing representatives cannot solve. Quite understandably, the customer's technical staff prefers to raise those problems firsthand with their counterparts in the supplier's organization. Working with research personnel from a central laboratory has a special appeal for them because the role of a central laboratory is neither to make a sale nor to defend a product but dispassionately to solve technical problems.

What do the categories of ROI have to say about the value of this customer service function? Or about that of a central laboratory as a device for the recruitment of managers? In many a corporation, key management positions are filled by people with technical backgrounds who first entered the organization through its research operation. However useful this recruitment function, no ROI system can measure the function's true value or attach to it the blessing of a price.

An investment in research as a means of customer service or of executive recruitment—as in say, legal services or stockholder relations—represents a portion of corporate overhead on which it is close to impossible to calculate an accurate return. As a result, it is all too easy to allow the categories of ROI to become a hidden determinant, and not a tool, of corporate policy. Put simply, managers are not apt to rally to projects with invisible or unmeasurable returns.

NEW PARAMETERS OF MEASUREMENT

Yet managers must measure the value of research, intangibles and all. But how? Though no fully appropriate method is yet available, the main parameters of such a method are already clear. For some things—for example, particularly well-defined, short-term development projects—ROI techniques do work reasonably well. For most, they must be supplemented with others. It is to these others that we now turn our attention.

Periodic Reviews

The evaluation of a research project's worth must take place periodically during the project's life. Before it is formally begun and then on an annual basis, research managers must ask, "Is it worth doing?" as well as "Is it in accord with corporate goals?" and must be able to answer with a firm yes. They must ask these questions with special emphasis before deciding to commercialize the project because commercialization usually involves an investment several orders of magnitude greater than typical research costs. Competent research on an unsound or undesirable project is still competent research. Poor project selection is—and must be treated as—a failure of management, not of research itself.

Discretionary Choice

The group within a company responsible for the final commercial product must have some discretion over project choice and must contribute to the costs of the research effort. This arrangement materially sharpens the quality of judgments along the way about the project's ultimate worth at the same time that it improves the chances that the final evaluation of the project will contain no major surprises. Improving judgment in this fashion is not without its cost, for discretion of this sort does tend to bias project choice toward low-risk and relatively short-term efforts.

As a practical matter, to ensure that the research laboratory has an appropriate mix of short- and long-term projects, it is important that the researchers themselves select a certain fraction of the projects on which they work. In making these selections, they should look to areas in which a scientific discovery has either just been made or is reasonably likely to be made. They should also seek out those areas where the development of a new technology would be useful whether embodied in a new product or in improved manufacturing processes. By contrast, they should avoid those projects that merely seek to improve existing products. And if they do choose to work on significantly longer-range projects, a council of senior operating managers should have the opportunity to review them.

Product Life Cycles

Given the length of time some projects take to pay a return and given the temporal mix of projects in any laboratory's portfolio, we have found it helpful to tote up on an annual basis the financial return on all the laboratory's projects to date, especially those that have left the laboratory. Since the cash flow related to any project varies in a rather predictable way with its stage of development or commercialization, it is only fair to consider together projects at every stage of evolution.

Product Line Contribution

Another good method for determining in retrospect the contribution of a company's research effort is to observe the growth in product lines for which that effort has been responsible. In addition, where licensing a company's technology is an important part of strategic policy, it is useful—and fairly simple—to establish the contribution of the central laboratory to the stream of license revenue.

Self-developed Products

When a company has done much of its own product development, that portion of total corporate ROI flowing from operations based on self-developed products offers still another reasonable approximation of ROI derived from research. If corporate returns in these areas are lower than expected, additional analysis will show whether the fault—say, excessive warranty costs—lies with inadequate laboratory work. Though still inexact, this approach does take fully into account the lasting financial effects of those hard-to-quantify aspects of research effort.

BALANCED JUDGMENT

There can be no shortcut, no easy formula, for assessing the value of corporate research. The complex expectations of its sponsors affect the measures those sponsors use to determine whether their expectations have been met. And the choice of methods affects in turn, what kind of expectations they can have in the first place.

Every accounting procedure, no matter how carefully employed, becomes a hidden determinant—and not just a measurement—of corporate policy.

What each company needs, therefore, is a reasonable, deliberately thought-through set of expectations for its research activity. What we must not have is a return to the euphoric illusions of the post-World War II era when there seemed no limit to the social and financial benefits of industrial research. Companies must use a technique like ROI only to the extent that it conveys significant information and never allow themselves to become the captives of their own accounting procedures. Finally, they must supplement ROI calculations with a host of more subjective judgments. With the evaluation of research—as with so much else managers do—there is just no substitute for disciplined professional judgment.

REFERENCES

Boretsky, Michael 1979: Technology, Technology Transfers and National Security. Speech to the U.S. Army War College, October 12, 1978 (updated February 26, 1979.)

Business Week 1978: Vanishing Innovation. July 3, p. 47.

Dean, Robert C. Jr. 1974: The Temporal Mismatch—Innovation's Pace vs. Management's Time Horizon. *Research Management* (May), p.12.

Globe, Samuel 1973: *Interaction of Science and Technology in the Innovative Process: Some Case Studies* (Columbus, Ohio: Battelle Memorial Institute), now available as Report PB228508/AS from the National Technical Information Service, Department of Commerce, Springfield, Va. 22161.

Hayes, Robert H. & Abernathy, William J. 1980: Managing Our Way to Economic Decline. *Harvard Business Review* July–August, p. 67.

18

The Role of the Research University in the Spin-off of High-Technology Companies

EVERETT M. ROGERS

The purpose of this article is to summarize what is known about the process that leads to the spin-off of private forms from research universities. Our focus here is (1) upon the United States, and (2) upon high-technology firms in the microelectronics industry, a sector in which a high rate of technological innovation is occurring and which represents the main arena for close university–industry relations in the 1980s.

THE INFORMATION SOCIETY

In recent years, the United States, Japan, and most Western European nations have passed through an important transition in the make-up of their workforce, the basis of their economy, and in the very nature of their society. Information has become the vital element in the new society that has emerged, and so these nations are called "Information Societies".

An *Information Society* is a nation in which a majority of the labor force is composed of information workers, and in which information is the most important element (Rogers 1986). Thus the Information Society represents a sharp change from the Industrial Society in which a majority of the workforce was employed in manufacturing occupations, such as auto-assembly and steel-production, and where the key element was energy. In contrast, *information workers* are individuals whose main activity is producing, processing, or distributing information, and producing information technology. Typical information worker occupations are teachers, scientists, newspaper reporters, computer programmers, consultants, secretaries, and managers. These individuals will teach, sell advice, give orders, and otherwise deal in information. The main activity is not to raise food, put together nuts and bolts, or to deal with physical objects.

Information is patterned energy that affects the probabilities available to an individual making a decision (Rogers and Kincaid 1981). Information lacks a physical presence of its own; it can only be expressed in a material form (such as ink on paper) or in an energy form (like electrical impulses). Information can often be substituted for other resources, such as money and/or energy. Information behaves somewhat oddly as an economic resource in the sense that one can sell it (or give it away) and still have it.

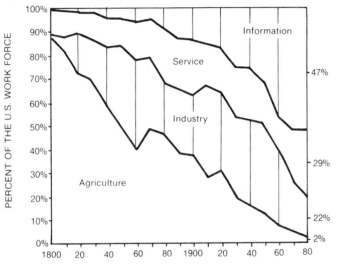

Fig. 1. The United States became an Industrial Society in about 1900, and an Information Society in about 1950.

Applications of the steam engine to manufacturing and transportation, beginning around 1750 in England, set off the Industrial Revolution that began the transition from an Agricultural Society to an Industrial Society. The Agricultural Society had been dominant for about 10,000 years until this point, and most Third World nations are still Agricultural Societies today. The Industrial Revolution spread throughout most of Europe, to North America, and later to Japan. Figure 1 shows that the United States began to industrialize in the mid-1880s, from 1900 to 1955, the largest part of the workforce was employed in industrial jobs. Then, in 1955, a historical discontinuity happened in the United States when industrial employment began to decrease and information workers became most numerous. Today they are a majority, representing about 55 percent of the workforce. While the United States led other nations in becoming an Information Society, Canada, England, Sweden, France, and other European countries are not far behind.

Certain of the important characteristics of the Agricultural Society, the Industrial Society, and the Information Society, are compared in Table 1.

*Table 1.*Comparison of the agricultural society, industrial society, and the information society.

Key characteristics	Agricultural society	Industrial society	Information society
1. Time period	10,000 years (and continues today in most Third World countries)	200 years (began in about 1760 in England)	? years (began in about 1955 in the U.S.)
2. Key element basic resource	Food	Energy	Information
3. Main type of employment	Farmers	Factory workers	Information workers
4. Key social institutions	Farm	Steel factory	Research university
5. Basic technology	Manual labor	Steam engine	Computer and electronics
6. Nature of mass communication	One way print media	One way electronic media (radio, film, television)	Interactive media that are de-massified in nature

THE RESEARCH UNIVERSITY AS THE KEY INSTITUTION IN THE INFORMATION SOCIETY

Fundamental to the growth of the Information Society is the rise of knowledge industries that produce and distribute information, rather than material products, or goods and services. The research university (1) produces information as the result of the research that it conducts, especially basic research, and (2) produces information-producers (individuals with graduate degrees, who are trained to conduct research). This information-producing role is particularly characteristic of the 50 or so leading research universities in the United States. A *research university* is an institution of higher learning whose main function is to perform research and to provide graduate training.

The research university fulfils a role in the Information Society analogous to that of the factory in the Industrial Society. It is the key institution around which growth occurs, and it determines the direction of that growth. Each of the several major high-technology regions in the United States is centered around a research university: Silicon Valley and Stanford University, Route 128 and MIT, and Research Triangle and the three main North Carolina universities (Duke, North Carolina State, and the University of North Carolina). The research university is especially important to its nearby high-technology firms when they are relatively new.

A *high-technology industry* is one in which the basic technology underlying the industry changes very rapidly. A high-tech industry is characterized by (1) highly-educated employees, many of whom are scientists and engineers, (2) a rapid rate of technological innovation, (3) a high ratio of R&D expenditures to sales (typically about $1:10$), and (4) a worldwide market for its products (Rogers and Larsen 1984). The main high-technology industries today are electronics, aerospace, pharmaceuticals, instrumentation, and biotechnology. Microelectronics, the sub-industry of electronics centered on semiconductor chips and their applications (such as in computers), is usually considered the highest of high technology because the underlying technology is changing more rapidly than in other high-technology industries.

Microelectronics technology, applied in the form of computers (especially microcomputers) and telecommunications, is driving nations like the United States into becoming Information Societies. That is why the role of the research university is so important in understanding the emergence of the Information Society. Research universities today are helping to redraw the economic map of the United States, by creating clusters of high-technology industrial firms around certain university campuses.

THE TREND TO CLOSER UNIVERSITY–INDUSTRY RELATIONSHIPS

What caused the trend in recent years to closer university–industry relationships, especially in the conduct of research?

During the 1980s the federal government cut back severely on its funding of university research (except for military research). Consequently, universities looked to private industry for research funds. The National Science Foundation estimates that industry funding of university research increased fourfold in the past decade, to about $300 million. During the 1980s many state and local governments launched initiatives to encourage the development of high-technology industry, in order to create new jobs and to fuel economic growth. Fearful of Japanese competition, U.S. micro-electronics firms formed university–industry collaborative research centers, and invested considerable resources in funding these centers.

Largest of the new R&D centers is the Microelectronics and Computer Technology Corporation (MCC), which located on the campus of the University of Texas at Austin in 1983. Fifty-six universities in 27 states competed with Austin for the MCC, with state and local governments offering a variety of incentives. Three hundred Texas leaders in state and local governments, universities, and private companies put together a multi-million dollar package to win the MCC.

Arizona Governor Bruce Babbitt, whose state was a finalist in the selection process, remarked (1984): "Some 60 mayors and 27 governors complained about the unfair advantage of Texas oil money, and promised their constituents a better showing next time." Certainly the 1983 MCC decision heightened awareness among state and local officials about the importance of high-technology development, and created a fuller realization of the role of research universities in attracting high-technology firms.

What did the University of Texas, the state of Texas, and the city of Austin get in return for their efforts to attract the MCC? The MCC is supported at $75 million per year by a consortium of 21 U.S. firms that are the giants of the microelectronics industry, plus government research grants (mainly from the U.S. Department of Defense). The MCC presently has a research staff of about 400. During its first year of operation, the MCC created a boomtown mentality in Austin. Fourteen, high-technology firms moved all or part of their operations employing 6,100 people to Austin during 1983, while in 1982 only four companies with 900 jobs move to Austin. The average selling price of a new single-family home rose 20 percent to $106,157 during 1983.

But the main benefits to Austin of getting the MCC will appear years from now, when a high-technology complex in micro-electronics develops around Austin. It is possible that this complex may eventually rival or surpass California's Silicon Valley as a center for the production of information technology. In this sense, the MCC decision may have settled the location of the future capital of the Information Society.

The MCC is only one of several new university–industry research centers in microelectronics.

- The Center for Integrated Systems (CIS) at Stanford University was founded in 1981 and is supported by $15 million from 20 U.S. microelectronics firms, and $15 million from the U.S. Department of Defense.
- Arizona State University launched its Center for Solid-State Electronics Research as part of its Excellence in Engineering Program in 1981, to encourage the development of high-technology industry. Funding for the first five years consists of $20 million from the state of Arizona, $10 million from private firms, and $3 million from the federal government.
- The Microsystems Industrial Group at MIT is sponsored by about a dozen companies, many of them on Route 128.
- The Microelectronics Center of North Carolina (MCNC), a research and training facility in Research Triangle Park, was launched in 1980 with $24 million from the state legislature.

Today there are about 25 university–industry microelectronics research centers at U.S. universities. Other such collaborative R&D centers have been founded for robotics, biotechnology, and other high-technology fields. In addition, many other technology transfer mechanisms are used by research universities:

- Research parks, like the Research Triangle Park in North Carolina and the Stanford Research Park (which was the first of its kind, and still is the most successful).
- Industrial liaison programs by universities, which provide a means for private firms to get an early look at research results and to identify promising students in order to hire them as future employees, in exchange for paying an annual membership fee to the university.
- Faculty consulting by universities professors. President Carl Taylor Compton of MIT in the 1930s not only allowed his faculty to consult for pay one day a week, but strongly encouraged them to do so. After World War II, the MIT faculty consulting policy spread to Stanford University, and, in recent years, to a number of other research universities that wish to foster technology transfer. Typically, a university allows its professors to work for pay one day per week.
- Especially in recent years, professors have formed companies of their own, often launching their start-up around a technological idea that they bring with them from their university laboratory. In some cases, the faculty entrepreneur who founds a high-technology company then cuts off his/her academic ties.

WHAT GOVERNMENT INITIATIVES PROMOTE HIGH-TECHNOLOGY INDUSTRY?

Figure 2 shows how federal, state, and local governments directly encourage high-technology development, and how they indirectly seek this goal through facilitating the role of local research universities. Although state and local governments are far more active than the federal government in promoting high-technology development, a variety of federal policies and programs encourage high-technology industry to cooperate with research universities. The Economic Recovery Tax Act of 1981 provides a 25 percent tax credit for increased corporate R&D expenditures over a base year; up to 65 percent of research contracted to universities and to certain other institutions is covered by this Act. Further, federal policy allows firms to deduct from their taxes as charitable contributions part of the cost of equipment donated to universities. Changes in federal tax policies in recent years also have aided the expansion of venture capital; as a result, it has been easier to obtain financing to launch start-up firms (Rogers and Larsen 1984).

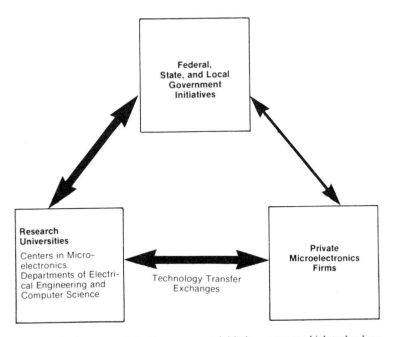

Fig. 2. Federal, state and local government initiatives promote high-technology development through technology transfer exchanges between research universities and microelectronics firms.

The recent easing by the federal government of anti-trust restrictions on collaborative R&D activities has facilitated closer university–industry relationships. The MCC in Austin could not have been founded had it not been for a favorable opinion by the U.S. Department of Justice on December 28, 1982. In 1984, Federal legislation was passed to remove certain anti-trust barriers to collaborative R&D.

State governments have initiated programs to coordinate the activities of state and local governments, private industry, and universities to facilitate technology transfer between the research university and its surrounding high-technology firms. State and local laws are factors affecting technology transfer and the development of high-technology industry. For example, absence of a state income tax in Texas is cited by some firms as an important reason why they moved to Austin.

BENEFITS AND COSTS OF UNIVERSITY–INDUSTRY RELATIONSHIPS

Clearly the university–industry collaborative research centers in microelectronics represent a new and important force on the university campus, and one that has already generated a great deal of policy controversy. "Few subjects have received as much attention recently in university circles as the prospect of closer relations with American business" (Rosenzweig 1982).

Some observers see this recent development in a very positive light. Governor Bruce Babbitt (1984) of Arizona refers to: "a new awareness that the fruits of university research and development activity have little economic value unless they are systematically harvested in the marketplace". The university obviously benefits from the research funds that it receives, and the professors may gain useful experience which they can incorporate in their courses for the benefit of their students.

In general, both the research universities and private industry are pleased with their new, closer relationships. Microelectronics companies feel that their membership fees paid to the collaborative university–industry research centers are one means of dealing with Japanese competition (for example, the MCC justifies its existence publicly as the U.S. response to the Japanese Fifth-Generation Computer Project). In addition, semiconductor companies in Silicon Valley report that they participate in the collaborative centers as a means to identify future employees and faculty consultants from among university students and professors (Larsen 1984).

But there is also a variety of problems connected with the industry-sponsored research centers on university campuses. For instance, the priorities accorded by a university to certain disciplines and to certain research problems may be affected by the priorities of private firms that donate research funds, thus causing a feeling of relative deprivation on the part of departments and professors not so favored. The growing emphasis on technology may come at the expense of arts and humanities. University scientists fear that the price of industry collaboration will be a shift from basic research to more product-oriented development.

Problems of inequality also can occur between universities, as well as within a university. When Stanford, Texas, Arizona State, and MIT create university–industry collaborative research centers in microelectronics, are other universities adversely affected? Hancock (1983) noted that in the new era of university–industry relations: "One trend is apparent; corporate money goes to the academic haves, not the have-nots".

According to Culliton (1983): "Concern about the propriety of university–industry relations has been a central theme on the country's research campuses for the past couple of years". Perhaps one turning point in the recognition of this problem occurred in 1982 when the presidents of five universities met with 11 corporate leaders at Pajaro Dunes (California) to explore such questions as:

How can universities preserve open communication and independence in the direction of basic research while also meeting obligations to industry? Is it acceptable for one corporation to dominate research in an entire [academic] department? Are there adverse consequences in terms of collaboration among faculty in various departments if one group must worry about protecting corporate right to licenses? Will extensive corporate ties erode public confidence in university faculty as disinterested seekers of truth? (*Science* 1982).

Three main points of potential conflict between the university and private firms exist in the present era of closer relationships:

1. Restrictions on the communication of research results, where company secrecy policies may conflict with the scientific desire for free communication.

2. The relatively short-term research orientation of the private firms versus the longer-term orientation of university scientists toward basic research.
3. The agenda of priorities for university research may be affected by corporate sponsorship, with emphasis upon scientific fields with a direct potential for commerical payoff.

Clearly the new relationships between industry and the university, often fostered by government initiatives and intended to encourage high-technology development, mark a very important change in the role of the U.S. university.

LESSONS LEARNED ABOUT PROMOTING HIGH-TECHNOLOGY INDUSTRY

The experience of recent years indicates that state and local governments often are engaging in a futile activity when they try to attract high-technology firms away from other locales in the United States. *It is more effective for a state or city to grow its own high-technology industry, than to try to steal it from another city or state.* One seed for starting local entrepreneurial activities is to invest in improving a nearby research university, especially in such academic departments as electrical engineering, computer science, and molecular biology. Investment in improving a local university is likely to pay off, eventually, in technology transfer to private firms and, later, to generating an entrepreneurial head of steam in starting-up high-tech firms.

A considerable time lag is usually involved from the improvement of a research university, to the rise of a local high-technology industry. Evidence on this point is provided by the case of the Research Triangle in North Carolina. Governor Luther H. Hodges had the original vision of a North Carolina high-technology center back in the 1950s. Hodges followed the Stanford University model of establishing a university research park in order to create a high-technology complex. Only after 20 years of concerted efforts did many high-tech firms begin to move to Research Triangle, with the key turning point occurring in 1965 when IBM located an R&D unit there. Total employment in Research Triangle Park's 40 firms is now over 20,000 with an annual payroll of $500 million (Rogers and Larsen 1984, p. 241). But there is not yet much entrepreneurial activity in starting up new firms. Several decades were also necessary for the rise of Silicon Valley in Northern California and for the beginnings of Route 128 around Boston. So high-technology industry does not usually get underway overnight (on the other hand, the present Austin take-off is occurring very rapidly).

The presence of an outstanding research university in a locale does not necessarily cause the development of a high-technology center. Evidence of this point is provided by such excellent universities as Harvard, Columbia, Chicago, Berkeley, and Cal Tech, none of whom have played an important role in technology transfer to local firms. Obviously, other factors than just the presence of a local research university are involved in launching a high-technology center: a high quality-of-life, models of the entrepreneurial spirit, and the presence of venture capital. Even when a research university is present, it must have policies that encourage faculty to assist local firms, or else not much technology transfer will occur. Such favorable policies exist at Stanford, MIT, Texas, Arizona State, the North Carolina universities, and at other universities.

I have implied throughout this essay that the research university and private firms each have an important role to play in the rise of a high-technology community: "Universities are not good incubators because they are too far removed from the marketplace (Miller and Côté 1985). Obviously then, the increasing degree of collaboration between industry and the research university augurs well for high-tech spin-offs. However, *not much technology is transferred from university to industry (or vice versa) through the mechanism of a formal agreement such as a collaborative R&D center, unless the firms are immediate neighbors to the university.* "Physical proximity facilitates the absorption of new technologies. Technological diffusion is still a geographical phenomenon" (Miller and Côté 1985).

Striking evidence for the importance of proximity in university–industry technology transfer is provided by Eveland's (1985) communication network analysis of nine collaborative R&D centers that were assisted by the National Science Foundation. Figure 3 shows the communication networks among the 47 industry representatives of 14 private firms belonging to one of the university-based R&D centers, plus the five university administrators, 14 faculty, and 22 students affiliated with this center. Very little communication occurs between university personnel and their industry counterparts, at least on a weekly or monthly basis (once-a-year contacts are not shown here, because not much technology-exchange could happen on such a limited basis). Eveland's (1985, p. 46) network analysis shows that the professors are fairly interconnected, and that students constitute the next layer of the onion, with each student communicating mainly to his/her professor. The next layer out in the sociogram (Figure 3) consists of industry

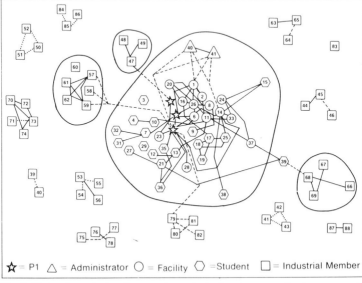

——— Weekly direct contact

- - - - Monthly U/I or interfirm contact

············ Less frequent interfirm contact

★ = P1   △ = Administrator   ◯ = Facility   ◯ =Student   ☐ = Industrial Member

Fig. 3. Network analysis of research project-related communication among 88 individuals in a collaborative industry–university research center. *Source:* Eveland (1985, p. 46).

representatives, about 70 percent are isolates and only six of the 47 communicate with university people on a monthly basis.

The R&D center depicted in Figure 3 conducts basic research in a very traditional scientific discipline, and this mission may be one reason for the lack of much industry–university communication. Also, the network data were gathered in the second year of the R&D center's operation, and perhaps effective industry–university communication requires longer to begin. Nevertheless, Eveland's network data convey an important cautionary lesson for those who think that industry–university technology-exchange will occur readily and directly as a result of forming a collaborative R&D center.

The research university in the United States is changing its role from mainly that of conducting basic research to also taking a more active role in transferring technology to private firms. The most advanced case of this trend is happening in microelectronics, the engineering/scientific field that is driving the Information Society.

Whether the trend to closer university–industry relationships will result in greater benefits than costs to the participants, and to society, remains to be seen.

REFERENCES

Babhitt, Bruce 1984: The states and the reindustrialization of America. *Issues in Science and Technology*, *1*, 84–93.

Culliton, Barbara J. 1983: Academe and industry debate partnership. *Science*, *219*, 150–1.

Eveland, J. D. 1985: Communication networks in university–industry cooperative research centers. Washington, D.C.: National Science Foundation, Division of Industrial Science and Technological Innovation, Productivity Improvement Section, Report.

Hancock, Elise 1983: Academe meets industry: charting the bottom line. *Alumni Magazine Consortium*, *7*, 1–9.

Larsen, Judith K. 1984: Policy alternatives and the semiconductor industry. Los Altos, California: Cognos Associates, Report to the National Science Foundation.

Miller, Roger & Côté, Marcel 1985: Growing the next Silicon Valley. *Harvard Business Review*, *63*, 114–123.

Rogers, Everett M. 1983: *Diffusion of innovations*. New York: Free Press.

Rogers, Everett M. 1986: *Communication technology*. New York: Free Press.

Rogers, Everett M. & Kincaid, D. Lawrence 1981: *Communication networks: Toward a new paradigm for research*. New York: Free Press.

Rogers, Everett M. & Larsen, Judith K. 1984: *Silicon Valley Fever: Growth of High-Technology Culture* New York: Basic Books.

Rosenzweig, Robert M. 1982: *The research universities and their patrons*. Berkeley, CA: University of California Press.

Science 1982: The academic–industry complex. *Science*, *216*, 960–1.

19

Patents and the Measurement of Technological Change

BJØRN L. BASBERG

The measurement of technological change has long concerned economists, economic historians, and historians of technology and research analysis. Although there are many ways of measuring technological change, all with their advantages and disadvantages according to how they are used, none of them is widely accepted. One method of measurement is to use data and statistics on patents. We will call this use of patent statistics a "technology indicator". In this paper we review the literature that utilizes patent information in this way. We emphasize the special problems associated with the method and point out ways of overcoming them.

It is important to emphasize that most available methods of measuring technological change are indirect measures of the process. By this we mean that any indicators we use will shed light only on certain aspects or parts of the process. An "indicator" will be no more than a proxy to help discover trends. General definitions like the following of a technology indicator emphasize its vague character:

> Science and technology indicators are series of data designed to answer a specific question about the existing state of and/or changes in the science and technology endeavour, its internal structure, its relation with the outside world and the degree to which it is meeting the goals set it by those within and without" (OECD/STIU).

The literature on patents may be divided into three main areas. The first deals with the legislation and functioning of the patent system; the second with the rationale of the system; and the third area is the literature that uses patents as technical information. Research in which patent statistics is used as a technology indicator falls within this last category. Some important works deal with all three areas (Penrose 1951; Machlup 1958; Gilfillan 1964; Taylor and Silberston 1973).

The studies in which patent statistics are used as a technology indicator also fall within three broad areas as regards their focus of interest. One direction of research deals mainly with the relationship between technological change, measured by patent statistics, and economic development (Merton 1935; Graue 1943; Beggs 1981). This is probably the main area of research, and Jacob Schmookler's (1966) research probably the most important within this area. His research is described as ". . . so rich and so suggestive that it has to be the starting point for all future attempts to deal with economics of inventive activity and its relationship to economic growth" (Rosenberg 1974, p. 92).

Schmookler's main conclusion that inventive activity is endogenously determined by economic variables has initiated a research debate on "technology push" versus "demand pull" (Utterback 1974; Rosenberg and Mowery 1979). Much of the debate has focused on the extent to which the method used has influenced the conclusions (Duijn 1981).

Schmookler's works have held a central position in the whole methodology debate on patent statistics used as a technology indicator; we will come back to this later. Another use of patent statistics as a technology indicator is in analysis of diffusion of technology from one country to another. This is done by measuring flows of patents between countries and their "balance of technology payment". Data on licenses and royalty payments and receipts are also used in this kind of research (Schiffel and Kitti 1978; Horn 1983).

Finally, a third group of research concerns analysis of the innovation process for assessing and evaluating the output of research activity. This can be done for example by analysis of the relationship between R&D, patents and productivity (Nelson 1981; Scherer 1965).

PATENTS IN THE INNOVATION PROCESS

The Main Questions

The most important reason for using patent statistics is that these data are believed to reflect inventive activity and innovation (Holman 1978; Pavitt 1985). Secondly, the ready availability of the data is of some importance (Sanders 1981). It is possible to put up complete time-series back to the middle of the last century (for some countries even further back) based on the patent registers. Patent statistics may also be used in comparisons across industries and nations.

The main questions in the debate concerning the usefulness of this method have been the following:

- To what extent are patents used commercially? If the patent data are to have any practical value as an indicator of technological mchange, it is important to be able to show that the part of the patent which leads to innovation is high. Related to this is the question of the quality of the patents, which obviously varies and represents a problem when patents are counted. There is also the question of the length of interval from patenting to commercialization. This too may vary to quite a large extent.
- If patent data are to be used in comparisons across firms or industries, it is important to know whether the patent system is used to an equal extent on the participants in the comparison. It is possible to protect an invention in several ways, and attitudes towards the use of patents may vary.
- In comparisons between countries, there is the question whether the patent institutions can be compared. If patent laws and the practices of patent offices vary, the validity and usefulness of patent comparisons will be affected.
- Finally, there are problems in using patent statistics in time-series analysis. A precondition for this is that the institutional framework and attitudes to the use of patenting do not change over time. In the following we will go into these questions in more detail.

Patents in Innovation Models

As a point of departure it might be useful to determine whether patenting plays any role in the innovation process. Since using patents as a technology indicator is obviously a very imprecise method of measurement, it is important to find out what relationship the patenting has with other activities leading to innovation. If patents are to be used as a technology indicator, it has to have a fairly clear and certain relationship to the innovation process.

In stagewise models of innovation, patenting is often seen as an element in certain stages. For example, in a model used by Campbell and Nieves, patenting is placed in a third reduction-to-practice stage before the R&D stages (1979). More common, however, is what Roberts does in letting patent application be the final activity of the development work (his second stage) (1974).

Freeman mentions patents as output indicators of the invent work, which is the second step in his model (1982).

Two additional models where patents are treated very explicitly are: (1) Evenson's categorization of those activities a firm goes through when changing its technology, and how patenting varies within these activities (1984), and (2) Arrow's five-stage model shows activities and details the indicators which are reflecting them (1980).

Most models relate patenting to the development phase, as an output indicator of R&D activity, and a positive relationship between R&D and patenting is, as mentioned earlier, empirically well documented.

The Relationship between Patents, Inventions and Innovation

The question of the relationship between on the one hand patenting and on the other invention and innovation is of crucial importance (Maclaurin 1953). In Figure 1 we show in a general way the share of inventions made for example within a year in a firm, industry or country. Only some of these inventions are patented. Furthermore, a smaller share of the inventions will become innovations. Some of the innovations will be patented. The most interesting part of the figure will often be the innovations. Patent data will obviously contain some innovations, but they will also contain inventions without any commercial value. The sizes of the different shares in the figure are arbitrarily chosen, and may of course vary from one sector to the next and from one point in time to the next.

The figure also gives a picture of the relationship between invention and innovation. We will not go into the many possible explanations as to why inventions are/are not put into practice, but stress instead that there are no simple correlations. One complication is the long and varying time intervals from invention to commercialization (Gilfillan 1935; Nelson 1981).

In Figure 1, a rather small share of the total number of inventions are patented. How large this share is will depend on the sector and the point in time under investigation. Schmookler, for example, has assumed that not more than half the number of important inventions are patented (1966). The share also seems to be decreasing throughout the twentieth century. In a list of important inventions or innovations, it is not difficult to find those which were never patented (Gilfillan 1964).

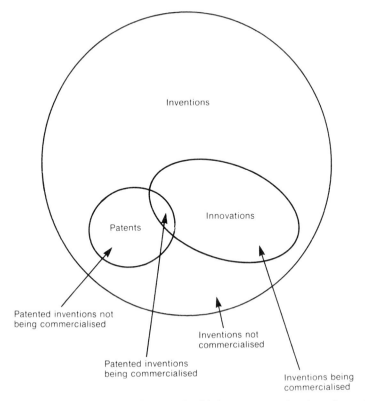

Fig. 1. A generalized picture of the relationship between patenting, invention and innovation.

Patents or Trade Secrets

An invention will usually be protected in one way or another if it has some potential economic value. There are two alternatives: patenting or keeping it a secret. We describe factors that may influence the choice between these two main strategies, and point out several reasons why inventions and innovations are not patented (Noone 1978).

First, an invention might not be patentable because of some explicit exclusion mentioned in the law concerning patents. Legislation may vary from one country to the other. Within new technologies—such as microelectronics or bio-technology—there may also be some uncertainty as to the patentability of new inventions, the inventor therefore choosing to keep the invention a secret. Another reason for secrecy has to do with the economic expectations. If the inventor cannot afford to pay for the patenting, or considers the expected income as uncertain or lower than the costs, he will not patent. A third reason for not patenting is when it is considered easy for a competitor to invent around the patent. Mansfield, for example, has found that patenting costs connected with the process of inventing around are not sufficiently high enough to effectively discourage a competitor.

A patent would often delay the imitation for only a few months, and would therefore be useless (Mansfield et al. 1982). The expected economic life of the invention will also play a role when deciding strategies for protection: If the expected life were much longer than the maximum patent life (usually twenty years), it would be rational to keep the invention a secret. Keeping the invention a secret would also be preferred if on the other hand, the expected life were very short. This is the case in rapid developing areas like microelectronics, where an invention might be outdated before the patent is granted.

Patents and Commercialization

The process of patenting is often long and expensive, and it is reasonable that the inventor or the sponsor expect future profits from the invention (Kuznets 1981). On the other hand, as earlier emphasized, there is of course an element of uncertainty. The prospects may fail; there are examples that "the economic importance of an invention has little relation to its patentability" (Penrose 1951, p. 17).

An American investigation from 1958 found that as many as 75 percent of all patents studied were assumed to have economic importance, but not more than 57 percent were actually in use (Sanders et al. 1958; Schmookler 1966). When looking for reasons why patents are not utilized, one finds that lack of demand and competition are important, but also rapid obsolescence (Sanders et al. 1958). An example of competition is when a firm keeps patents not to use them commercially, but to prevent competitors from using them (National Science Foundation 1979).

There are obviously differences in quality between those patents that are used and those that are not. However, also among patents that are being used commercially there are differences in quality. Some patents have made the basis for new successful products, firms or whole industries. Others, like many improvement patents, may have insignificant economic importance. So, from what has been said so far, it is obvious that patents to some extent are incomparable (Kuznets 1981; Scherer 1965).

Basic Innovations or Improvements?

Aggregated patent statistics do not distinguish between those that lead to basic innovations and those that lead to minor technical improvements (Duijn 1981). When Schmookler deals with his wave-like patent graphs, what is he actually analyzing? What does a peak in patent activity really reflect? Schmookler himself interpreted his data with great care, and assumed that they might indicate inventive activity. But even if innovations are kept out of the interpretation, it is still questionable whether patent statistics indicate basic inventions, improvements, less important inventions or are diffuse. It is possible to argue that a basic invention that leads to a patent will be preceded and succeeded by less important patents within the same technological area. A peak of aggregate patent activity will then indicate the technological breakthrough. This is illustrated in Figure 2(a). If, on the other hand, the purpose is to indicate innovation activity, the use of patent statistics may be insufficient. Schmookler did not separate between patents that lead to innovations and those remaining inventions. It is possible to assume that basic innovation will fluctuate differently from total patent statistics, for example as we have shown in a general way in Figure 2(b). Here, the basic innovations will generate later patent activity, and the picture of technological change would be different if the total patent statistics were interpreted in isolation (Walsh 1982).

Variations across Sectors

If we study the technological development in different firms, industries or nations, there is the problem of considerable variation in patent policy (Marcy 1978). In comparisons across nations, an additional problem is that patent legislation too, may vary. Differences in the number of patents granted therefore do not necessarily reflect differences in the number of inventions alone. Against this background it is maintained that patent statistics should be used as information of technological trends only *within* an industry (Reekie 1973). A precondition is of course that habits are fairly constant within the industry. This, however, is not always the case. Even between firms there can be differences (Scherer 1965). It is, for example, common to assume that patenting varies in accordance with size of firm. Research shows that small firms patent more frequently than larger ones (Schmookler 1975). Furthermore, research shows that small firms will use a higher percentage of their

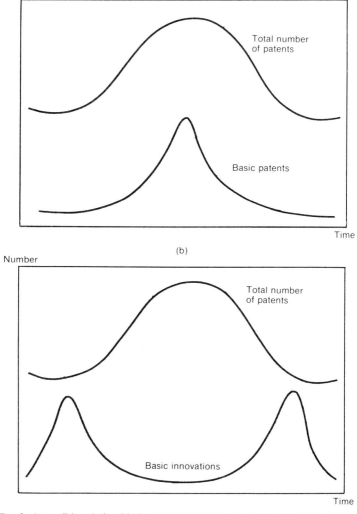

Fig. 2. A possible relationship between aggregated patent statistics, basic patents and basic innovations in the same sector.

patents commercially than large firms (71 percent and 49 percent respectively) (Sanders 1964).

Changes over Time

One advantage of using patent statistics as a technology indicator is the possibility of constructing long and complete time-series. If

such data are going to have any value, however, a major precondition is that the quality of an average patent remains
unchanged. Another precondition is that the relationship between
patents and inventions in an area needs to be constant. Finally,
attitudes to the use of the patent system must remain unchanged
(Nelson 1981). This is not always the case, so time-series of patent
data have to be interpreted with great care (Beggs 1984; Reingold
1960). In several countries, domestic patenting has flattened out
or fallen since World War II. The reason might be more in the
decreasing interest in patenting than in decreasing inventive activity
(Kuznets 1981; Schmookler 1960).

Another reason for the falling trends might be found in the shift
from independent inventors towards increased federal involvement
in research and development. This makes protection through patenting less important (Kuznets 1981).

Patents in the Industry or Firm Life Cycle

A final point concerning times-series studies is how patenting and
patent policy change in the life cycle of the industry or firm. An
example of this problem can be seen in Schmookler's research: It
is maintained that demand conditions do not explain patenting in
new and fast growing industries. This, however, is more the case
in old and mature industries (Freeman 1979). Schmookler's conclusions are based on mature industries like railways, agriculture,
petroleum and paper. In established sectors, it is not surprising
that technological change is induced by demand factors (Duijn
1981).

IMPROVING PATENT INFORMATION

It is possible to maintain that in using aggregated patent statistics,
the large numbers will ensure an average quality of sufficient size
(Beggs 1981). This, however, is too simple a way of getting around
the problems of methodology, and in this section we review several
possible ways of dealing with the quality question. In some ways
the debate concerning patent statistics has not advanced since the
1960s when Schmookler published his important works (Pakes and
Guliches 1984). The same pros and cons are still being used. When
it comes to ways of using and improving the presentation of
patent statistics in a way that the validity problems are overcome,
however, there has been some advance. The following ought to
show this.

"Important" Patents

Perhaps the most obvious way to improve the patent data is to separate the patents in accordance with quality. This is what R. Baker has done (Baker 1976). His criteria is that "an invention is important or significant if the history of some subject could not be written without reference thereto". First, important technological areas are defined, and then important inventions within those areas are picked out. Using Baker's data for English patents back to 1691, J. Clark et al. have made a distinction between "master patents" and "key patents" (Clark et al. 1981).

Master patents are defined as the first patents in an area to become economically successful, while key patents refer to the most important patents overall. Series of both data correlate, but they show trends quite different from time-series for total patenting. This illustrates the important point that conclusions drawn from analyzing patent data depend on the type of data in use. L. Soete et al. (1983) have used Baker's concept and classified innovations according to importance based on patent activity in the following way:

Radical innovation—New Class of patents created
Major innovation—Family of patents
Important innovation—Number of key patents
Minor innovation—Two or more patents
Incremental innovation—Not necessarily patented

A final example is J. Townsend's (1976) work on innovation in the British mining industry. Each patent is given a weight from one to four according to importance. He further shows how his conclusions depend on the way the weights are distributed.

A problem with these methods is their subjectivity in estimating quality. One fairly objective method is to decide quality by the effective life of the patents. In most countries the maximum life of patents is between 15 and 20 years. However, few patents are protected for such a long time (Federico 1958). The annual fee, progressively increasing in most countries, is compared to the expected income from the invention. The more economically successful the invention is, the longer the patent life, will be extended. Patent life will thereby be an indication of importance. Based on such calculations, large differences in quality have been found (Schankermann and Pakes 1983). Changes have also been in effective patent life over time. A question is, however, whether such changes reflect changes in quality or have to do with institutional conditions? (Bosworth 1973; Eisman and Wardell 1981).

Patenting Abroad

There are many reasons for applying for a patent abroad. A patent might protect existing or potential export markets. Licence production will very often have a patent as a precondition (Scherer 1954). Foreign patents have been used as a technology indicator because on average they are of a higher quality than domestic patents. It is reasonable to assume that only inventions with significant profit expectations in a larger market will be patented abroad because of the time and costs involved in such processes (Gilfillan 1964).

International legislation concerning priority for foreign applications also explains the quality of such patents.

Patenting abroad is carried out in two different ways. Taking Norway as an example, we may on the one hand study Norwegian patenting abroad because this is said to indicate the technological level in Norway. On the other hand, we might study foreign patenting in Norway from the viewpoint that this, too, reflects Norwegian technological conditions or trends. In the former case, patenting abroad reflects characteristics of the technology of the country of origin. In the latter case, foreign patenting in Norway reflects conditions in the receiving country. It is possible to argue for both methods.

One reason for using foreign patent data is that comparisons can be made between countries. By comparing the patenting of several countries in a third country, the difficulties associated with different patent legislation are overcome. Which country should be used? One possibility is to study total foreign patenting. In using one country, a natural choice would be to look at foreign patenting in a country with a dominant position both economically and technologically. France, Germany, England and the U.S. are among them. Their positions as important receivers of patent applications from abroad have changed over time, and the U.S. is today the leading country. In a study of the pharmaceutical industries of several nations from 1900 to 1963, however, foreign patenting in the U.K was used since this was the most important patent country at the time (Reekie 1973; Tilton 1971).

Foreign Patenting in the U.S.

K. Pavitt and L. Soete (1980) maintain that "each country has the same propensity to patent in the USA in relation to size of its innovative activities", making such data a useful technology indicator. They argue that these data are of a higher quality than domestic patents. Evidence of this view is found in the higher correlation between domestic R&D and U.S. patenting of a country than between R&D and domestic patenting. Pavitt and Soete use data for foreign patenting in the U.S. to analyze the relationship between technological change and foreign trade, and conclude that "technological performance is the most important trade explanatory variable . . ." (Soete 1980). It is possible, however, to explain patenting abroad, especially in the U.S., in a different way. D. Schiffel and C. Kitti (1978), for example, found that a country's exports to the U.S., together with domestic patenting, explain the U.S. patenting of that country.

Countries that are close trading partners will usually also have strong patent flow between them. Studies concluding that neighbor countries have especially high inter-country patent flows (Slama 1981) are therefore no more surprising than the same countries being close trade partners. In contrast to Pavitt and Soete, therefore, it is maintained that "the propensity to patent in another country is probably related to the perceived potential of the market in that country" (National Science Foundation 1979, p. 17; Marmor 1979).

References and Citations

One way of finding out about the quality of patents is to study what is called patent-to-patent citation networks (Carpenter et al. 1981). Usually an application for patent will refer to other patents of close relationship; like the list of references in a research paper. The more often a patent is referred to, the more important it is likely to be (Carpenter et al. 1981). A very thorough investigation by this approach has been carried out by Campbell and Nieves (1979), who determine the quality of a patent by a combined analysis of own references and citations by other patent applications.

Reclassification

Overcoming the quality problems might be done by studying the patent data at the lowest possible level of aggregation. To study patenting within a firm will give data of less noise than patenting within an industry or a whole nation. Some of the most accessible patent data are data for patent classes or subclasses. However, if the focus of interest is in analysis of the relationship between patenting and economic activity, we are faced with the problem that patent classification and industry classification are not directly

comparable. Patents are classified according to technological systems or principles and therefore have to be reclassified when used
in economic analysis. Schmookler did a pioneer work on this issue
(Griliches and Hurwicz 1972). Later, such reclassifications have
been made permanent in for example the data bases of OTAF,
making the patent data possible to analyze in an economic context
(Grevink and Kronz 1980). The reclassification is not trivial, and
there has been some debate following Schmookler's work. A patent
may be reclassified to the industry where the invention took place,
or to the user-industry. The patent-owner will most likely represent
the industry of origin, and the researcher needs a detailed knowledge of different industries to be able to find the actual user-
industry of the invention. Schmookler himself tried to reclassify
according to the user-industry, and there seems to be general
agreement about that approach (Scherer 1981).

Applications and Granted Patents

Patent legislation may change over time and complicate the interpretation of patent data. For example, the number of years an
application is pending varies. The problem may be reduced,
however, by using applications data instead of patents granted
data. The applications reflect the inventor's interest in obtaining
protection and how he judges the importance of his invention. The
choice is not a simple one however. Using granted patents data
also has advantages. The most obvious one is that the average
quality may be higher because of the standards demanded in the
Patent Office.

   The empirical works in the field have emphasized these arguments differently, and both data for applications and grants are
used. Somewhere in between is Schmookler's (1960) solution,
where patents granted data are used at the point in time when they
were applied for.

Vector Analysis

One way of using the patent data for increasing information and
validity has been developed by the Japanese Patent Office and
adopted by OTAF (Office for Technology Assessment and Forecast
1977). The method combines patent data with assignment data.
While annual changes in the number of patent applications are said
to reflect technological activity, annual changes in assigned patents
from the inventor to a firm are said to reflect the interest for the
technology. A combination of these two measures is assumed to

indicate the degree of maturity of the technology. If both sets of data increase, the technology is in a developing phase. If both show a decrease, the technology is mature and on the way down. Different combinations of the data reflect different stages in the process of maturity between the two extremes.

CONCLUSIONS

In this paper we have variously reviewed the literature on patent data. There is a choice between using information on the single patent, groups and classes of patents and finally data for total patenting activity. Further there is a choice between cross-section data and time-series data. Used in the latter way, patent-data are almost unique, since it is possible to construct continuous series of data back to the eighteenth century for several countries. Data may further be used from different stages in the patenting process. Finally there is a choice between data for domestic and data for foreign patenting. The type of data chosen for purposes of analysis depends of course on the purpose of the analysis itself. If the focus is on how the patent system works, it might be useful to look at the relationship between applications and patents granted, patent policy, etc. An historian of technology might be interested in the patentees, the single patents and their technical specifications. On the other hand, students of long-run development, theories of long waves and the relationship between economic and technological change might find it useful to analyze more aggregated patent statistics in a quantitative way. Finally, cross-section data are useful in comparative studies of industries or countries.

We have pointed out a number of problems connected with the use of patent statistics as a technology indicator. Several of them are difficult to overcome and may be the reason for certain common characteristics of many works in the field: After taking reservations and putting forward a decisive critique of the method in general, the data *are* used as if nothing was mentioned. This is of course a result of some kind of trade-off. On the one hand, the limitations of the data are perceived. On the other hand, however, we know the lack of other adequate technology indicators, especially in the analysis of historical data in a long-run perspective. Schmookler put it as follows: "We have a choice of using patent statistics cautiously and learning what we can from them, or not using them and learning nothing about what they alone can teach us" (Schmookler 1960, p. 56).

REFERENCES

Arrow, K. S. 1980: *A proposed conceptual framework for indicators of R&D inputs, outputs and industrial innovation* (STIU/OECD), Paris.

Baker, R. 1976: *New and improved—inventors and inventions that have changed the modern world* (British Museum, London).

Beggs, J. J. 1984: Long run trends in patenting. In Griliches (ed.).

Bosworth, D. L. 1973: Changes in the quality of inventive output and patent based indices of technological change. *Bulletin of Economic Research, 25* (2).

Campbell, R. S. & Nieves, A. L. 1979: *Technology indicators based on patent data: the case of the catalytic converters* (Battelle Pacific Northwest Laboratories, Richland, Wash.).

Carpenter, M. P., Narin, F. & Woollf, P. 1981: Citation rates to technologically important patents. *World Patent Information, 3* (4).

Clark, J., Freeman, C. & Soete, L. 1981: Long waves, inventions and innovations. *Futures, 13* (4).

Van Duijn, J. J. 1981: Fluctuations in innovation over time. *Futures, 13* (4).

Eisman, M. M. & Wardell, W. M. 1981: The decline in effective patent life of new drugs. *Research Management, 24* (1).

Federico, P. J. 1958: *Renewal fees and other patent fees in foreign countries* (Study of the Subcommittee on Patents, Trademarks and Copyrights of the Committee on the Judiciary, 85th Congr. 2nd sess. Study No. 17), Washington, D.C.

Freeman, C. 1979: The determinants of innovation, market demand, technology and the response to social problems. *Futures, 11* (3).

Freeman, C. 1982: *The economics of industrial innovation*, 2nd edn. (London).

Gilfillan, S. C. 1935: *The sociology of invention* (Chicago).

Gilfillan, S. C. 1960: At attempt to measure the rise of American inventing and decline of patenting. *Technology and Cultures, 1* (3).

Gilfillan, S. C. 1964: *Invention and the patent system* (88th congr. 2nd sess, US Governm. Joint committee print), Washington, D.C.

Graue, E. 1943: Invention and production. *Review of Economics and Statistics, 25.*

Grevink, H. & Kronz, H. 1979: *Evolution of patent filing activities in the EEC. A contribution to the study and assessment of the technological trends developing in the EEC from 1969 to 1975, based on a statistical analysis of patents* (Commission of the European Communities, Brussels).

Griliches, Z. (ed.) 1984: *R&D, patents and productivity* (University of Chicago Press).

Griliches, Z. & Hurwicz, L. (eds.) 1972: *Patents, invention and economic change. Data and selected essays by Jacob Schmookler* (Cambridge, Mass.).

Griliches, Z. & Pakes, A. 1984: Patents and R&D at the firm level: a first look. In Griliches (ed.).

Harris, L. J. et al. 1978: *The meaning of patent statistics* (National Science Foundation, Washington, D.C.).

Holman, M. A. 1978: An analysis of patent statistics as a measure of inventive activity, in L. J. Harris et al.: *The Meaning of Patent Statistics.* Washington D.C.: National Science Foundation.

Horn, E. J. 1983: Technological balance of payments and international competitiveness. The case of the Federal Republic of Germany. *Research Policy, 12* (2).

Kuznets, S. 1962: Inventive activity: problems of definition and measurement. In Nelson (ed.).

Machlup, F. 1958: *An economic review of the patent system* (Study of the subcommittee on patents, trademarks and copyrights of the committee on the judiciary) (Washington, D.C.).

Maclaurin, W. R. 1953: The sequence from invention to innovation and its relation to economic growth. *The Quarterly Journal of Economics, 67* (1).

Mansfield, E. et al. 1982: *Technology transfer, productivity, and economic policy* (New York).

Marcy, W. (ed.) 1978: *Patent policy, government, academic and industry concepts* (ACS symposium series 81) (Washington, D.C.).

Marmor, A. C. et al. 1979: The technology assessment and forecast program of the U.S. patent and trademark office. *World Patent Information, 1* (1).

Merton, R. K. 1935: Fluctuations in the Rate of Industrial Invention. *Quarterly Journal of Economics, 49* (2).

National Science Foundation, *Science Indicators*, Washington, D.C. Biennial.

Nelson, R. R. (ed.) 1962: *The rate and direction of inventive activity* (Princeton).

Nelson, R. R. 1981: Research on productivity growth and productivity differences: dead ends and new departures. *The Journal of Economic Literature, 19* (3).

Noone, T. M. 1978: Trade secret vs. patent protection. *Research Management, 21* (3).

Office for Technology Assessment and Forecasting (OTAF). *Annual reports* (U.S. Department of Commerce, Washington, D.C.).

Pavitt, K. 1985: Patent statistics as indicators of innovative activities: possibilities and problems. *Scientometrics, 7* (1–2).

Penrose, E. 1951: *The economics of the international patent system* (Baltimore).

Plant, A. 1934: The economic theory concerning patents for inventions. *Economica, 1* (1).

Reekie, W. L. 1972: Patent data as a guide to industrial activity. *Research Policy, 2.*

Reingold, N. 1960: U.S. patent office records as source for the history of invention and technological property. *Technology and Culture, 1* (2).

Roberts, R. E. 1974: *Investment in innovation* (National Science Foundation, Washington D.C.).

Rosenberg, N. 1974: Science, invention and economic growth. *The Economic Journal, 84* (1).

Rosenberg, N. & Mowery, D. 1978: The influence of market demand upon innovation: a critical review of some recent empirical studies. *Research Policy, 8* (2).

Sanders, B. 1964: Patterns of commercial exploitation of patented inventions by large and small corporations. *Patent, Trademark and Copyright Journal, 8* (7).

Sanders, B., Rossman, J. & Harris, L. J. 1958: The non-use of patented inventions. *Patents, Trademark and Copyright Journal, 2* (1).

Sanders, R. L. 1971: The commercial value of patented inventions. *Idea, 15* (4).

Schankerman, M. & Pakes, A. 1983: *The rate of obsolescence and the distribution of patent values: some evidence from European data* (NBER/ENSAE, Paris).

Scher, A. V. 1954: *Patents, trademarks and copyrights. Law and practice* (Basel).

Scherer, F. M. 1965: Firm size, market structure, opportunity and the output of patented inventions. *American Economic Review, 55* (5).

Scherer, F. M. 1981: *Demand-pull and technological invention: Schmookler revisited* (Northwestern University, Evanston, Ill.).

Schiffel, D. & Kitti, C. 1978: Rates of invention. International patent comparisons. *Research Policy, 7* (4).

Schmookler, J. 1960: An economist takes issue. *Technology and Culture, 1* (3).

Schmookler, J. 1966: *Invention and economic growth* (Cambridge, Mass.).

Schmookler, J. 1975: Innovation in business. In E. Dale (ed.), *Readings in Management* (New York).

Slama, J. 1981: Analysis by means of a gravitation model of international flows of patent applications in the period 1967–1978. *World Patent Information, 3* (1).

Soete, L. 1980: *The impact of technological innovation on international trade patterns: the evidence reconsidered* (STIU/OECD, Paris).

Soete, L. & Wyatt, S. 1983: The use of foreign patenting as an international comparable science and technology output indicator. *Scientometrics, 5* (1).

Taylor, C. T. & Silberston, Z. A. *The economic impact of the patent system* (London).

Tilton, J. E. 1971: *International diffusion of technology. The case of the semi-conductors* (Washington, D.C.).

Townsend, J. F. 1976: *Innovation in coal mining machinery: "The Anderton Shearer Loader"—the role of the NCB and the supply industry in its development* (SPRU, Sussex).

Utterback, J. 1974: Innovation in industry and the diffusion of technology. *Science, 183*.

Walsh, V. 1982: *The use of patents and other indicators in the study of invention and innovation in the chemical industry* (STIU/OECD, Paris).

20

Market Structure and Innovation

EINAR HOPE

The relationship between market structure and innovation has interested and puzzled economists for a very long time. Despite the very considerable research effort being devoted to this issue, however, it is fair to say that our understanding of it is still rather rudimentary and vague, and few "truths" seem to have been established. There are some good reasons, inherent in the nature of the problem, for such a state of the art to be expected, e.g.:

- Demarcation of the problem. Many investigations purporting to study the market structure–innovation relationship are really focusing on a somewhat different issue. Two main bodies of investigations may be distinguished here. First, there are studies of the relationship between *firm size*, i.e. absolute size, and the rate and direction of innovative activity, and not *relative* size as one aspect of market structure. Secondly, many studies focus on *invention* as the outcome of the process, and not innovation. There are also investigations with the broader perspective of studying the relationship between market structure and *technological change*, expressed e.g. as shifts in production functions. While the subject of this survey is market structure and innovation, we will not draw the line too precisely in relation to firm size and invention.
- The problem of causation. There is a fundamental problem about the causative structure in the relationship between market structure and innovation. Does the causation run from innovation to market structure? Is it the opposite way around or are there more complex causative links? The problem of causation has been neglected in the literature; it is very seldom discussed explicitly and many studies do not even distinguish between short-run and long-run aspects of the market structure–innovation relationship.
- Dynamic aspects. The innovation process itself is dynamic in nature, and the analysis of the economic effects of innovations

should be conducted within a dynamic framework. Still, most of the literature on market structure and innovation is of a static or comparative-static nature.
- Uncertainty. Innovation is an activity characterized by great uncertainty about the final outcome and about the amount and type of resources which are necessary to invest to obtain a commercially interesting result. From a methodological point of view we also find a "discrepancy" here, in the sense that an inherently uncertain (stochastic problem is typically being studied within a deterministic framework.
- Structure–performance relationship. A considerable part of the literature on market structure and innovation postulates a given market structure in the short-run and tries to infer the effects on innovative activity from alternative specifications of the structure. However, a one-to-one correspondence between structure and performance can not be expected in this area, given the nature of the activity in question, as discussed above. A given set of structural market conditions may result in quite different innovation performance, through the intervening influence of strategy and conduct. An oligopolistic market structure is a good example of such a situation.

We will come back to some of these aspects in more detail later.

The studies which have been performed about the relationship between market structure and innovative activity vary greatly in terms of approach, scope and methodology, but they may conveniently be classified into three main groups:

1. Theoretical analyses, relying mainly on a priori reasoning.
2. Econometric studies testing various hypotheses, but usually relating some measure of innovative activity as dependent variable to firm size, industrial concentration, product differentiation, or other structural conditions such as independent variables.
3. Descriptive case studies of particular industries or firms and their innovation performance.

We will structure the survey according to this classification, but will pay little attention to group 3 studies (Kamien and Schwartz 1982; Hope 1973, 1985).

THE THEORY OF MARKET STRUCTURE AND INNOVATION:
THE SCHUMPETERIAN HYPOTHESES

There is a long tradition in the static economic theory of markets
of contrasting perfect competition with monopoly in promoting
innovation and technological progress; the former is invariably
hailed as a stimulating and invigorating force while monopoly is
charged as representing a barrier to, or at any rate, a retarding
influence on such progress. The most distinguished opponent of
this position was Joseph A. Schumpeter, who discussed the issue
in his early writings. The fullest and most eloquent statement of
his position in this matter, however, is to be found in his *Capitalism,
Socialism and Democracy* (Schumpeter 1942). To quote only a few
passages:

> It is not sufficient to argue that because perfect competition is
> impossible under modern industrial conditions—or because it
> always has been impossible—the large-scale establishment or
> unit of control must be accepted as a necessary evil inseparable
> from the economic progress which it is prevented from sab-
> otaging by the forces inherent in its productive apparatus. What
> we have got to accept is that it has come to be the most powerful
> engine of that progress and in particular of the long-run expan-
> sion of total output. . . . In this respect perfect competition is
> not only impossible but inferior, and has no title to being set up
> as a model of ideal efficiency.

Schumpeter stated his theory in rather general terms, and often
came up with vague and unsubstantiated claims. Consequently,
there has been a lot of discussion and disagreement over the precise
content of his theory and the extent of his claims. A major source of
confusion is that he did not pay enough attention to the distinction
between *absolute* size of the firm as opposed to *relative* size, i.e.
in relation to the market; or, to put it differently, that he did not
distinguish clearly enough between structural conditions relating
to the firm as a separate entity, and structural conditions relating
the firm to its economic environment. Schumpeter no doubt differentiated between firm size and market power in his analysis, but
he could have been more explicit in disentangling their separate
effects. Again to phrase his basic position somewhat differently:
even though he sometimes seems to argue on the implicit assumption that monopolists tend to be big, there is, I think, no denying
that he meant that both absolute size and market power would

encourage innovative effort and lead to a higher rate of technological progress than under other structural conditions.

The two basic elements in the Schumpeterian position here may conveniently be termed *demand* and *supply* factors. First, he argues that the monopoly position of a firm will stimulate to a greater demand for innovations, because the firm will be able to use its market power to obtain a higher profit from the innovation than under competitive conditions. It is in a position to *appropriate* the potential returns from the innovation. Secondly, he argues that the monopoly firm will be able to generate a larger supply of innovations, because there are certain advantages open to it which a competitive firm will generally not be able to attain. Presumably, Schumpeter is here thinking primarily of scale economies in the research and development functions and in innovative activity generally, i.e. arguments relating mainly to absolute size.

We have discussed Schumpeter's contribution to the subject under survey in a very summary fashion, but we emphasize it because he has had such a strong influence on thinking concerning the relationship between firm size, market structure and innovation. Another reason is that subsequent work in this area, both theoretical and empirical, has taken a somewhat different direction than was laid down in the schumpeterian system. The main difference is that the emphasis has shifted to some extent from relative size (market power) to absolute size and, consequently, to an analysis of the role of the large firm, taken by itself, in furthering technological progress. Or, to use the dichotomy between demand and supply factors referred to above, the emphasis has shifted from a certain preoccupation with demand factors in Schumpeter's theory over to a stronger influence of supply factors in the theory of his followers.

Galbraith (1952) gave a forceful statement of this new position:

Providence . . . has made the modern industry of a few large firms an almost perfect instrument for inducing technical change. It is admirably equipped for financing technical development. Its organization provides strong incentives for undertaking development and for putting it into use. The competition of the competitive model, by contrast, almost completely precludes technical development.

There is no more pleasant fiction than that technical change is the product of the matchless ingenuity of the small man forced by competition to employ his wits to better his neighbor. Unhappily, it is a fiction. Technical development has long since become the preserve of the scientist and the engineer. Most of the cheap and simple inventions have, to put it bluntly and unpersuasively, been made.

Galbraith's line of argument was supported by several economists at that time and some were also prepared to draw the implicit policy implications, primarily in relation to anti-trust policy, of such a position. It is fair to say, however, that Galbraith and his followers never managed quite to convince the majority of economists of the alleged comparative advantages of monopolies/ oligopolies and large firms in furthering innovation. The disincentives inherent in the monopolistic case, inferred from static economic theory, were thought to be so strong as to outweigh possible scale arguments in the innovation function.

The unsatisfactory nature of the attempts which we have looked at so far to resolve the issue of the relationship between firm size, market structure and innovative activity has stimulated a great deal of research. On the theoretical side researchers have strived for more analytical rigor by stating precisely the assumptions and structure of their models, going from there to deduct the effects on innovative activity of alternative structural forms. We will discuss briefly one such contribution by Arrow (1962), both because it is one of the first formal analyses of the monopoly–competition–innovation issue, because it stirred up some controversies in the literature, and because it illustrates well some of the methodological and analytical problems in comparing structural market forms in relation to innovation performance.

Arrow examined the incentives to innovate for monopolistic and competitive markets, by comparing the potential profits from an innovation with the costs. In the monopolistic situation, he assumed that only the monopoly itself could innovate. He further assumed constant costs both before and after the innovation, the unit costs being c before the innovation and $c' < c$ afterwards. Under these assumptions he comes to the conclusion that the incentive to innovate is less under monopolistic than under competitive conditions.

We notice that Arrow stays completely within a static or comparative-static framework in his analysis; there is no discussion of dynamic effects of possible differences in the magnitude of the incentives to innovate under the two alternatives. It is also important to note how the comparison between monopoly and competition is being made. The two industries or markets do not, in fact, exist simultaneously side by side, it is rather an analysis of how the incentives to innovate in a competitive industry would change if the industry were monopolized.

Arrow's argument lends itself to a simple geometric illustration (see Demsetz [1969]). In Figure 1 let c and c', respectively, be the unit cost of production before and after the innovation. The competitive price before the innovation is c. After the innovation the competitive price is $c' + r = p$, since the innovator selling his innovation to the competitive industry is assumed to set his per unit royalty, r, so as to maximise his remuneration from innovating, that is, the royalty is set so that the quantity demanded is where marginal revenue (MR) is equal to c'. The innovator has an incentive to innovate in this case so long as the cost of the innovation to him is less than the rectangle $\pi' = c'puv$.

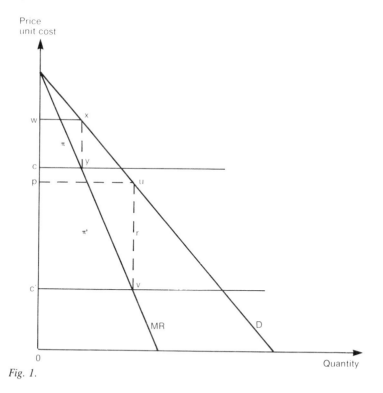

Fig. 1.

In the alternative situation in which the industry is a monopoly owned by the innovator, the monopoly price is set at w before the innovation, where $c = \text{MR}$, yielding a profit of $\pi = cwxy$. After the innovation, the profit maximizing price is p and the new profit rectangle is π'. The increase in profits attributable to the innovation for the monopoly innovator is consequently $\pi' - \pi$. Since this magnitude is always less than π', Arrow arrived at the conclusion that the incentive to innovate is less under monopolistic conditions than under competitive conditions.

In his critique of Arrow's analysis Demsetz maintained that the two situations cannot be compared directly. The innovator, Demsetz argued, not only produces an innovation, he also possesses the monopoly power to discriminate in the royalty charges which he sets for the two industries. Besides, in the monopoly case proper account should be taken of the normal behavior of a monopolist of restricting output to reap monopoly profits, compared with competition.

To adjust for this restrictive effect Demsetz redefined the MR curve in Figure 1 to be the demand curve facing the competitive industry. Now both the monopoly and the competitive industry will produce the same rate of output, for any given unit cost, and both will pay the same royalty to the innovator. Thus, there is no difference in the incentive to innovate under the two regimes. When Demsetz, in addition, introduces rivalry between innovators but keeps adjusting the sizes of the industries to remove the normal restrictive effects of a prior monopoly, he comes to the opposite conclusion of Arrow's, that is, that the rewards from innovation are greater under monopoly than under competition.

This brief analytical exposé thus seems to leave us in an impasse, with contradicting results as to the incentives to innovate under monopoly and competition, depending upon the specific assumptions of the analysis and the methods of comparison being used. One drawback with the Demsetz method of adjusting pre-innovation and post-innovation output in the two situations is e.g. that the demand curve (MR) for the competitive industry is uniformly less elastic than the demand curve (D) facing the monopolist. This may in itself tilt the incentive to innovate in favor of the monopolistic industry. In general, it is not obvious how to eliminate the scale effect in the most appropriate way in a comparison of the two situations.

If we accept Arrow's analysis that the incentive to innovate is greater under competition than under monopoly, does this then refute the Schumpeterian contention that some monopoly power is needed to stimulate innovative activity? This is not necessarily so, however, since Arrow refers to the structure of the industry purchasing the innovation rather than to the structure of the industry producing it. When e.g. rivalry between innovators seeking to develop a similar innovation is introduced, the analysis may lead to different results.

The theoretical and analytical refinements in recent years of the market structure–innovation relationship have substituted the Schumpeterian hypothesis in the singular for a broader set of hypotheses and sub-hypotheses. Some of these refer to the *market structure level*, stating the reasons or conditions for innovation to be expected to be greater in monopolistic industries than in competitive ones (e.g. appropriability, strategies to prevent imitation, profits to finance innovative activities, etc.), while some refer to the *firm level*, giving in a similar way reasons why large firms should be expected to be more innovative than small firms (e.g. economies of scale in R&D, diversification, parallel research efforts, etc.). This research has in itself contributed to better understanding and insights into this rather complex problem and, not the least, has given us a sounder foundation for econometric analysis and testing of the various hypotheses. We now turn to this body of literature.

ECONOMETRIC STUDIES OF THE SCHUMPETERIAN HYPOTHESES

Econometric testing of what we have called the Schumpeterian hypotheses is fraught with difficulties. Some of these relate to problems of *measurement* of the magnitudes involved, while others relate to the *specification* of the econometric model.

In its most general form the econometric model specifies a relationship between a measure of innovative "output", I, as dependent variable and a number of market and/or firm specific factors or determinants as explanatory variables:

$$I = f(m_1, \ldots, m_n, f_1, \ldots, f_k)$$

where m_i is the market structural feature i, and f_j is the firm determinant j.

Some of the problems which are typically encountered with such an approach may be listed as follows:

- Identifying and measuring an innovation. It is not always easy in practice to identify a new product or process, in particular when it constitutes an improvement or a new version of existing products or processes, as is often the case. The most commonly used measure of innovative output is patents, but patent statistics have many drawbacks and weaknesses in this respect. (For a discussion, see the paper by Bjørn Basberg in this volume.) In many econometric studies even a measure of what should be considered an *input* into the innovation process, e.g. R&D expenditures or personnel, is being used as an output measure, for lack of adequate output data.

- Measurement of monopoly power. Two approaches to the empirical measurement of monopoly power are usually tried, one direct and one indirect. The direct approach focuses on constructing indices for the deviation of the actual monopoly price from the price that would prevail under perfect competition; the most commonly used such index being the Lener Index. The indirect approach attempts to find a market structure proxy for monopoly power, e.g. some form of a concentration index. Both types of indices however, are, only imperfect and imprecise empirical substitutes for the tenuous concept of monopoly power under actual market conditions.

- Measurement of firm size. Common measures of firm size are sales, number of employees, and total assets. On neither empirical nor on theoretical grounds is one able, however, to discriminate between them as to being the "best" measure of firm size as a determinant of innovative activity.

- Specification and causality. We refer her to the discussion in the introduction to the paper. Market structure may be the consequence of an innovation rather than a cause. There may thus be a simultaneity problem which ought to be considered in the specification of econometric model.

The above problems are, of course, methodologically not unique to testing the Schumpeterian hypotheses, but they may be more complex and intangible than in many other areas of applied econometric research. Still, a very large number of such studies have been undertaken. It is almost hazardous to try to summarize this literature and to draw some overall conclusions from the reoccurrence of certain patterns in the empirical material at hand. With these reservations in mind, however, the following extract of results may be tentatively stated in a summary fashion:

- The nature of the innovation process seems to reveal a production structure with increasing returns up to a threshold level of resource commitment and non-increasing returns beyond. The threshold level varies across industries. The threshold for efficient operation does appear to constitute an entry barrier in certain industries, although generally not a formidable one.

- The bulk of empirical findings indicate that innovative activity does not typically increase faster than firm size. R&D activity, measured by either input or output intensity, appears to increase with firm size up to a point and then to level off or decline, as is consistent with the evidence on the nature of the innovative process itself.

- There is ample evidence that the resources devoted to innovative activity differ between industries, but the form and the causal links of the relationship between innovative activity and industry structure have not yet been explored in sufficient detail at the empirical level to justify firm and consistent conclusions.

- Little support has been found for the hypothesis that innovative activity increases with monopoly power, commonly measured in terms of concentration ratio. Some evidence is available that a market structure intermediate between monopoly and perfect competition would promote the highest rate of innovative activity in a market structure context, even though the evidence has proved difficult to substantiate empirically.

- Differences in the magnitude of the technological opportunities available seem to exert an influence on innovative activity. Technological opportunity is undoubtedly related to market structure elements, but determinants on the production side may be more important. The concept of technological opportunity is rather vague, however, and needs a sharper definition.

- Different qualities, abilities and commitments are needed on the various stages of the innovation process. In particular, there is ample evidence that small firms generally have greater *inventive* capabilities than large firms.

SOME QUALIFICATIONS, SPECULATIONS AND ALTERNATIVE VIEWS

The relationship between firm size, market structure and innovation is one aspect of the broader question of how structural conditions affect the economic performance of sectors or agents in the economy. The research effort embodied in the literature which we have surveyed here has undoubtedly increased our knowledge of the issues and relationships involved in the structure–innovative activity debate, but at the same time I have the impression that we have now reached a point where diminishing returns quickly set in to further empirical work along the lines of the studies which we have concentrated upon here. More "innovativeness" is evidently needed in the research on innovation. We will therefore conclude by listing very briefly and in arbitrary order some unresolved problems and point to possible ways of attacking some of these to make further progress.

(a) First a question of research methodology and the nature of possible policy implications. As we noted in the preceding section, most of the empirical studies start off from an assessment of Schumpeter's position and then go on to test the relationship between size of firm, market structure and innovation implied by his theory, as the author of the study interprets it. However, there is, in my opinion, a fundamental difference in approach or in philosophy between Schumpeter's position and the one underlying the econometric studies purporting to test his theory. Apart from the essential dynamic character of his system, which the econometric analyses have not been fully able to capture, I will argue that, if we cut through to the heart of the matter of the involved Schumpeterian system, it can best be interpreted as a "threshold" theory: just as there is a threshold size of firm under which innovative activity cannot in general be adequately undertaken, there is also a threshold level of market power necessary to induce firms to undertake research and development and introduce innovations to the market. Consequently, some departure from a state of perfect competition will have to be tolerated in order to give the economic system a vigorous source of technological progress. However, this threshold level was never defined by Schumpeter, and I doubt if he ever thought it possible (or necessary) to define it in quantitative terms. But what he definitely did not mean was that, if we somehow could measure the degree of market power (or the degree of departure from perfect competition), an increase in market power would result in a proportionate increase in the volume of innovations per unit of time. It is, however, such a correspondence in quantitative terms which the econometric studies establish. They tell us how innovative activity can be explained by, or is attributable to, factors such as firm size, concentration, diversification, technological opportunity, etc.; and, in principle, we can also compute from the regression equations how much the volume of innovations would change if the explanatory factors were changed with given amouts. The industrial policy implications of Schumpeter's version and of the "quantitative" version would, of course, also be different.

   The above is not meant as a critique of the econometric approach, but simply to point out that there is a discrepancy in fundamental viewpoint of the innovation process between that the studies intend to do and what they in fact do, when they set out to test "the Schumpeterian hypotheses".

(b) We have seen from the empirical evidence that innovative intensity does not appear to vary systematically with either absolute or relative firm sizes, and the suspicion creeps up that we might well be focusing on the wrong, or more peripheral, issues in being so preoccupied with the role of size and market structure as determinants of innovative activity and technological change. Perhaps we should concentrate more on clarifying the *incentives* to, rather than the *capacity* of, firms to engage in innovative activity and be technologically progressive. The incentives will naturally be related to structural elements, but broader and more complex forces may well be at work here. If the hypothesis therefore is that the factors determining firm size might be more important than is firm size itself in explaining the progressiveness of a firm, the obvious implication is that we should look to the theory of the size and the growth of firms for further guidance.

(c) An argument in the same vein as above is that we may have been too preoccupied with the extremes in considering the relationship between size of firm, market structure and innovation. This is expressed in two ways. Firstly, there has been a tendency among many writers to let perfect competition be faced squarely with monopoly, or small firm size with big firm size, and nothing in between. On the other hand, many researchers have aimed at determining *the* optimum firm size in relation to innovative activity. Both approaches are, in my opinion, misdirected. What we have to consider together, given the overall objective of having an efficient innovative system for creating and introducing new products and processes, is (a) a *size distribution* of firms above a minimum threshold size, and (b) an appropriate *decomposition* of the innovation process into identifiable stages; and then inquire whether, despite decomposability of the process, a single size or form of organization has optimum properties with respect to all stages. If we come up with a negative answer, so that firms of different sizes have a comparative advantage at different stages of the process, a next step would be to investigate to what extent firms specialize according to their comparative advantage and to consider the mechanisms by which knowledge and "semi-products" are transferred among the agents in the innovation process. Such an approach would open up for new avenues of research and interpretation.

(d) The literature about size of firm and innovative activity has been concerned with this relationship in terms of *overall* firm size, e.g. production, sales, total employment, etc. In other words, arguments of scale economies in this activity have been related to total firm size. However, we would, in my opinion, obtain greater insights into the nature of scale effects if we distinguished between scale economies (or diseconomies) in the innovative function (e.g. the research and development unit) itself (internal or functional scale economies), and scale economies in the rest of the organization (or even outside the firm) having an effect upon the performance of the innovative function. The latter type of economies (external or overall economies) would be related to total firm size, e.g. economies in finance, marketing, production, etc., making it possible to perform innovative work more efficiently with increasing firm size. Surprisingly little research has been done about the nature of scale effects in innovative activity, especially in consideration of the great effort invested in the literature which we have surveyed here. I feel, however, that the potential pay-off might be significant to an investigation into these effects starting off from the simple distinction between internal and external economies described above.

(e) Items c) and d) above taken together imply that the organizational form of both the innovative function itself and the overall organization of the firm is an important determinant of the performance of firms in innovation. Much has been made of the problems of the large and complex organization in dealing with innovative activity, but most of this discussion has implicitly taken the organizational form as given. A more fruitful approach would be to find out what *specific* advantages or disadvantages in particular types of innovation firms of different sizes and organizational forms would have, and then to see, if we found certain advantages attached to small size, what organizational changes would have to be effectuated in large firms to make it possible to capture these advantages and still preserve their specific advantages, thereby increasing the technologically progressive performance of the firm. For instance, the multi-divisionalization hypothesis, much discussed in the organization literature, might be relevant here.

(f) A considerable part of the literature on innovative activity has been concerned with invention rather than innovation, and in addition the focus has often been on radical, breakthrough inventions. For many purposes, however, for example in relation to designing an industrial policy to foster overall technological progressiveness, a shift of emphasis is needed. Firstly, invention is only a part of the process and we must recognize the importance of the final stages too, i.e. we must pay more attention to innovation.

Secondly, radical inventions are the exception rather than the rule, most inventions being of the secondary, improvement type. Consequently, more attention should be concentrated on the analysis of the determinants and character of these kinds of inventions. Such a shift of emphasis could also have implications for perception one might have of the role of small and large firms in the process. The delivered literature on invention in general, and on radical inventions in particular, attributes a significant role to the individual inventor and the small firm, while at the same time there is considerable evidence, as discussed above, (a) that a great many inventions have been developed and taken to the final point of successful, commercial innovation by larger firms, and (b) that the research and development activity of large firms, for several reasons, may be biased towards secondary, improvement inventions. If the indicated shift of emphasis is accepted as valid, this would consequently assign a relatively greater role to play for the large firm in the innovation process.

(g) There is ample evidence that the role and character of innovative activity differ among industries. Two issues may be singled out here in relation to firm size, market structure and innovation. First, there is the concept of technological opportunity, mentioned earlier. It is important to recognize that this is essentially a *dynamic* phenomenon, and that we therefore, in addition to analyzing variations across industries, also must consider the effects on the industrial and technological development over time of the individual industries of variations in their technological opportunity foundation. It could well be that some form of a life-cycle model would capture these effects. The dynamic aspects of the technological opportunity concept have not yet, to my knowledge, been analyzed formally or empirically (Pavitt 1984).

The second issue is to a certain extent related to the first; it concerns changes over time in the basic requirements necessary to do innovative work. The evidence is not conclusive, but there are tendencies discernible in most industries which point to the fact that there has been a considerable increase in the commitment of resources required to do successful innovative work. This is not the place to go into the reasons for this development, but some contributing factors might be enumerated very briefly: increased requirements to engage in basic research to couple up with basic research done external to the firm and to keep up with developments in science and technology; increased complexity of the scientific and technological foundation of innovations and consequently the need to understand and to integrate the knowledge of a broader range of specialist disciplines; increased complexity of instrumentation and test gear and increased costs of testing operations (pilot plants, etc., becoming more prevalent); increased requirements of design, tooling, production set-up, etc. These developments may well lead to a different division of innovative effort (or ability) between small and large firms than has been the coventional view up to now.

(h) The effects of market structure on innovation are very imperfectly known, but there is a great need to include other structural dimensions in the analysis, in addition to the elements which attention has been focused upon till now. Two issues are of particular interest, in my opinion. Firstly, innovative activity as a *competitive strategy*, and patterns of reactions among rivals in the market, should be considered. It is important to realize that a significant portion of the research and development effort in firms is imitative and competitive rather than innovative in the Schumpeterian sense, and we ought to know more about the factors underlying the decisions of firms to innovate or to imitate. Secondly, the effects of *entry conditions* should be considered. There are two forces at work here. On the one hand, barriers to entry may have an effect upon the rate and direction of the innovative effort of the established firms in the market, while on the other hand the cost of innovation may in itself form an effective barrier to the entry of new firms; or the established firms may use innovative activity as a strategy to deter entry.

(i) In this survey we have concentrated our attention on firm size and market structure as determinants of the innovative activity. It is evident, however, that much more complex forces are at work behind the innovation process, and there is now a tendency to play down structural conditions of this type as independent and sole factors in explaining the rate and direction of innovation. A major difficulty with many of the other explanatory factors which have been drawn in, e.g. communication and information, research conception of the management of the firm, experience and understanding of user requirements, market research, etc., is that they are ill-defined and hard to quantify. This makes it still more difficult to generalize about the effects of the various factors.

(j) A final point is again primarily directed to the methodology of studies of innovation. Innovation is a dynamic process characterized by great uncertainty, and the tools of analysis should, of course, be tailored to capture these characteristics. We have too long been attacking economic issues about innovation within the framework of static economic theory under certainty. Formidable analytical problems are encountered in extending the analysis to take uncertainty and intertemporal relationships into account, but some interesting theoretical work has already been done along these lines (Kamien and Schwartz 1982).

REFERENCES

Arrow, Kenneth 1962: Economic welfare and the allocation of resources for inventions. In R. R. Nelson (Ed.), *The rate and direction of inventive activity.* Princeton: Princeton University Press.
Basberg, Bjørn 1987: This volume.
Demsetz, Harold 1969: Information and efficiency: Another viewpoint. *Journal of Law and Economics.*
Galbraith, John Kenneth 1952: *American capitalism.* Boston: Houghton Mifflin.
Hope, Einar 1973: The effects of firm size and market structure on innovation: a survey. *Discussion Paper, 8.* Norwegian School of Economics and Business Administration.
Hope, Einar 1985: Innovation in high-technology industries. *Working Paper, 9.* Center for Applied Research, Norwegian School of Economics and Business Administration.
Kamien, Morton I. & Schwartz, Nancy L. 1982: *Market structure and innovation.* Cambridge: Cambridge University Press.
Pavitt, Keith 1984: Sectoral patterns of technological change: towards a taxonomy and a theory. *Research Policy.*
Schumpeter, Joseph A. 1942: *Capitalism, socialism and democracy.* New York: Harper & Row.

21

Continuity in the Potential for Innovation

REIDAR GRØNHAUG

THE IMPORTANCE OF EVOLVING CONTEXTS

In the 1950s and 1960s the social sciences at large started their attack on the problem of change and innovation and a large inventory of ideas and insights about process and dynamics was developed. Looking back on the achievements, it may strike us today that we are better equipped for explaining change than for explaining continuity, and we have focused more on questions of production than on questions of destruction. We have been insufficiently explicit about the interconnections of micro-events and the dynamics of the larger systems providing the preconditions for repetitive or innovative events at the micro level. I would say we need a better conceptualization of what the critical entities are in socio-cultural evolution with its dialectics of internal elements, internal–external, and micro- and macro-relations. When we study "change" and "innovation", I believe we have much to gain by more clearly identifying life world systems as the interesting entities which move in time and display dynamical properties like transformation, development, and evolution.

Most researchers today would agree that the important thing to understand is not the individual innovative act taken in isolation, but rather the processes of its creation and repercussions, and that ". . . this general viewpoint shifts our attention from *innovation* to *institutionalization* as the critical phase of change" (Barth 1967, reprinted 1981, p. 117). This taken for granted, the question comes up of how to delineate the systemic interconnections of the processes that produce innovative acts and make up their repercussions as they are being institutionalized. In each case, we want to discover the socio-cultural net effects of innovative events and to map the constructive and destructive consequences of innovations for actors as well as for their life world as a totality. For each phase of change in a life world, we want to evaluate the marginal effects upon its potential to provide inhabitants with the tools for further (mal-)adaptive initiatives—with the consequences this may have for the viability changes of the life world at large.

As I understand it, such a study should still be actor- and process-oriented and deal with feedback connections between the determinants of individual events and the larger social forms as aggregates of such events (Barth 1967). Our attention, however, should more be turned at the proper dynamics of the super-individual activity fields that make up the situated life worlds moving in time (cf. Grønhaug 1974, 1978). More explicitly we should utilize the insight that what evolves in evolution is not species of individuals, but feedback systems of interconnections between such individuals and the ecological niches to which they adapt for upkeeping viability.

Niche properties make individual adaptation possible, but the form of individual adaptation in its turn gives shape to niche properties: ". . . Turf was the evolving response of the vegetation to the evolution of the horse. It is the *context* which evolves" (cf. Bateson 1973, p. 128). Hence, it seems wise to let the study of the "intensity of interaction" be informed by the study of "the qualitative structure of contexts". Contexts have properties of not only energy householding and social organization, but also of communication, and they can "themselves be messages" (Bateson 1973).

This last point of Bateson should be related to the important analytical problem about connecting values and motivation at the level of actors not only with actional constraints given with organized context, but also with the symbolic and normative rules that are implied in such contexts as institutionalized practices. This is what the job is all about if we are to discover how innovative innovative acts are—on the background of institutionalized values and action, and how widely and deeply a case of innovation brings about changes in the situated ecological–social–cultural personality context that has given rise to it.

With this question in mind we shall turn to the East Antalya scene.

DIFFUSION AND INNOVATION BY LOCAL TRANSFORMATION

Arguably, what we term innovation is the instance of some cultural element being imported from outside into a localized social system, where it is worked upon, integrated with pre-existing inventory, and transformed into a new organizational solution to old problems in some sector of production. The local innovation, a form of syncretism, may be given a form which restricts its use to the locality where this new syncretism was made, or it may result in a cultural element of such an objectified nature that it can be transported across local and national situations and be put to use elsewhere, in other localized situations, and enter into processes of innovative syntheses there.

The critical issue is the way and the consequences of a new element's being integrated into the total inventory of a local nature, i.e. how it affects the viability chances of the actors inhabiting the system and thereby the overall system's capacity to provide the preconditions for viability to the people encompassed by the system.

We shall define 'innovation' as the construction of locally new organizational elements providing a new kind of value output. Such an output may concern only a few of the members in the system, and we shall deem an instance of innovation more interesting, i.e. more innovative, if it results in a raised output and resource control for the encompassed population as a whole. Even more critically, however, are the effects of relevance for the system's potential further on to renew the adaptational capacity of actors and to strengthen their viability chances. Especially, we should try to identify those interconnections of events that have destructive versus productive effects upon the social system's potential to guarantee sufficient adaptability and viability chances to its member populations pursuing their strategies under shifting ecological circumstances.

"Innovation" is seen here as inventiveness by synthesis and syncretism occurring in a localized social setting, in a relation of import and export of cultural elements vis-à-vis the outside world. In a sense, innovation seems to be mere borrowing, but it cannot just be flat copying, because it has to be integrated into the local system on that system's own premises and put to use by way of a new organizational solution. Most generally, our scenario is one of widespread, largely global communicational flow between multiple localized points, functioning as transformation scenes where foreign elements are adopted, considered, and in a more or less modified form put to homely use in ways that in turn may serve as models for others.

The theoretically interesting point will then be to try to conceptualize these transformation scenes, the "hotbeds", "kettles", "vessels", or whatever metaphor we want to apply to designate

those delineated life world systems which enable us to be members
of global society and still have a home, to take part in world culture
and still produce the standards we feel are ours and to operate
as worthwhile partners for others in relations of symbiosis and
competition.

CASE: SHIFT IN NICHE OCCUPANCY

An important case of historical change in Southern Turkey was the
substitution of Muslim Turks for Christians of Greek and other
backgrounds in trade and crafts. Historically, Greeks, Armenians,
and others had specialized in such activities, and together with
representatives of European firms under the "Capitulation"
system, holding a monopoly of long-distance trade between the
Ottoman area and Europe. Thus, in the late Ottoman period,
Muslims were typically found in agro-pastoral production and in
military and civilian administration, and only marginally in trade
and crafts. Such was the situation in Antalya, too, where Christians
held important positions in trade and were also dominating locally
important crafts like masonry and carpentry.

This was the local variant of the Ottoman "Millet" system, i.e.
the practice of differentiation between people according to religion,
mainly Muslim, Christian, and Jews (cfr. Berkes). In practice,
religious identity was combined not only with a specific legal status,
but also a special occupational adaptation and thereby a typical
position of power and rank in society. In situated contexts, given
one's religious identity, one's life course was channeled into one
of these few types of life situations with the cluster of competences
and statuses implied. The actors themselves emerged as belonging
to one of a few types of social persons, locally labeled as e.g.
"Muslim" versus "Christian" referring to different "packages" of
religous, legal, occupational, economic, power, rank, kinship, residential and other status aspects.

This process of agglomerating a large number of statuses into a
few types of social persons has aptly been termed "status summation" (Nadel, Barth, Grønhaug). The connection of occupation
with religious identity in pre-Republican Turkey had the net effect
of recruiting mostly non-Turks to important positions in trade.
Consequently, skills, expertise, and forms of middleman- and
entrepreneurship associated with certain forms of trade and crafts
were more frequently found among local Christians than among
Turks.

After World War I and the establishment of the Turkish Republic, Turkey and Greece agreed to undertake a population exchange of Muslims in Greece for Christians in Turkey. The local Greek Christians left Antalya, and gradually local Turks moved into the niches in trade and crafts opened up. Initially, they lacked much of the technical expertise to run such businesses. They did not, however, lack the more elementary capacity to readapt by learning the new occupations and gradually performing quite well in export (of e.g. timber and agricultural product) and import (consumer goods), and in the crafts demanded in house construction.

The critical factor was hardly the (non)-existence of certain trade-related psychological traits among local Turks, as nineteenth-century European observers used to maintain. The decisive matter was rather one of informational distribution combined with opportunity costs, of the attractiveness of new niches as compared to old ones combined with relative access to the resources needed to start up a certain activity. Once Turks of local background started to consider long-distance trade as more profitable than mere land holding, or they could economically supplement or replace a family farm with carpentry work, Turks moved into trade or crafts.

In one sense, these changes meant no more than one set of personnel taking over a niche after another, even needing some time before they could perform as well as their predecessors. Significant novelty was involved, however. The dissociation of religious identity from positions in trade and crafts meant a weakening of status summation as a factor in the social construction of social persons and groups. The new situation opened up for a higher mobility of personnel across economic sectors and more flexible role permutations in the making of individual careers. Considering the preconditions for the readaptation, we can point out the following structural factors.

Firstly, people treated the economic scene as a structure of niches with cultural and organizational barriers determining the distribution of the opportunities and constraints which actors must consider in their adaptive choices and value conversions. Secondly, the recruitment to certain occupational niches was determined by a larger set of life circumstances, some of which were more imperative and critical than others, and which elements were more or less imperative changed with time. Thirdly, a seemingly simple step of readaptation, moving from agriculture into crafts, implied—as my short sketch indicated—a considerable change in the whole mode of constructing social persons and organizing their interrelations. But fourthly, although patterns of careers and life courses changed as people shifted from old to new circumstances, they followed the same basic values of person and identity, of personal honor and dignity.

We are thus faced with the important question in studies of innovation—about "how wide" in terms of organizational implications, and of "how deep" in terms of values defining human identity and the translation of these values into concrete life courses. Disregarding whether the Turks taking over after the Greeks in Antalya should be called innovation or not, the instance brings up questions about transformation in culture and organization that are always associated with innovation and change and which I shall pursue more systematically below.

THE IMPRECISENESS OF ENTREPRENEURSHIP

Barth (1963) has made the general statement that in an economic structure where values circulate partly in separate spheres, and where a value discrepancy exists by specific values not being directly convertible into each other, entrepreneurship is likely to be expected to emerge at such points of value discrepancy. At these points values can be converted into each other in new ways with new and so far unutilized ways of making profit. Entrepreneurship will emerge as personnel are mobilized in new ways, values converted in new ways, with new forms of making profit and of investing it to perpetuate the enterprise. The entrepreneur will be followed by others, and the new niche filled up to the point where further careers of that type are impossible. The important point is then to look for the constraints upon innovation and its institutionalization.

For the ecological setting of East Antalya, I shall show how entrepreneurship emerges as actors in control of resources in one field combine them in new ways with resources they gain access to in other fields, and how organization of the activity fields themselves brings about boundaries and differences between resource spheres that allow for innovative forms of resource management at various levels of scale.

THE INSTITUTIONALIZATION OF INNOVATION

In the following I present a picture of rather self-sustained economic growth as a series of innovative events and transformations of socio-cultural arrangements in the region of East Antalya, Southern Turkey since the 1920s, but with my primary focus on the years approximately between 1940 and 1980. The more general topic I shall discuss is the institutionalization of innovation. With this I mean the social implementation of individual innovative acts, the social system's absorbing and adjusting itself to such changes, as well as its production of further innovation with the organizational readjustments following from them.

Although I shall point at individual actors with remarkable achievements as entrepreneurs and modernizing models for their fellow citizens, I shall be more concerned with the social pre-conditions for such innovative persons and acts, i.e. how they emerge and how the effects and repercussions they produce are being fed into the socio-cultural infrastructure that made them possible. I am especially concerned with the conditions for the maintenance of that infrastructure as a source of further adaptive change for the system and the population as a whole.

This is a typical unit of analysis problem of a kind I believe is not sufficiently clarified in the social scientific discourse about innovation and change. Granted that social roles and acts, "repetitive" as well as "innovative", are socially constructed, we must describe innovative persons and acts as embedded in and an integral part of the larger socio-cultural systems that bring them about. The focal issue is then how given innovative events affect the preconditions for the continued maintenance of the larger socio-cultural entity as a reproductive system.

At the focus of my analysis shall be questions about viability and adaptability pertaining to innovation and change in people's traffic in business, politics, and social activities in a wider sense, as these express an underlying set of cultural values and rules which themselves may change along the way. My concern is to point out productive versus destructive effects of innovative events at the level of those social fields that appear to be imperative in maintaining viability for people in East Antalya under the shifting circumstances of their adaptations. The case of East Antalya over time displays a characteristic balance between institutional reproduction and transformation. This experience may be generalized to some comments upon current theories about innovation and social change, as discussed towards the end of my paper.

LANDSCAPING IN THE EVOLVING CONTEXT

The area is situated on the Mediterranean coast, and can roughly be divided into three zones, viz. lowland plains, foothills, and higher valleys hills and mountains with peaks up to 2,700 meters above sea level. On the inland side the mountain ranges slope

down towards the Anatolian plateau at roughly 1,000–1,200 meters'
altitude. The name of one of the adjacent districts on the Anatolian
side of Bozkir, literally "grey steppe", while Antalya is frequently
celebrated as "green Antalya". The contrast of green versus grey,
of wetter versus drier, aptly sums up the contrast between the
natural resource abundance in Antalya as against the restricted
ecological regime in Anatolia. Western winds bring rain falling
mainly in the slopes turned to the sea. On the coast itself there
are local enclaves, like Alanya, with an average precipitation of
. . . . . mm annually, the figure for Bozkir being merely . . . . a
year. For millennia, organized irrigation has been an important
part of Antalya's cultural landscape, and agriculturalists at any
altitude operate canal systems of varying size and complexity.
Fertile ground is found at all altitudes, but differences in topogra-
phy, temperature and first of all water provisions create considerable variation in the conditions for economic utilization of the
land.

Our story is very much one of how economic changes have
brought about a step-wise re-landscaping of the area into successive
patterns of differentiated opportunity zones, that in turn have
provided the cards for the economic game in the next round.
Especially, it is a story of how this young moving differentiation
into separate zones of ecological niches and economic opportunities, from phase to phase, has been a wellspring of individual
entrepreneurship and innovation over and over again.

Through time, ecological and economic changes have been
accompanied by readjustments in social organization having made
possible economic development along with demographic growth,
rather than runaway changes serving a few in limited sectors
while undermining the existence of the many. The core of these
readjustments in social organization pertains to control over
resources from below in the local community as well as in institutions of larger scale in the area.

The economic and political culture in the area includes blatant
individualism, but also collectivistic rules for the safeguarding
of shared resources. The dialectics of the Antalya variants of
individualism and collectivism has provided much of the cultural
source for the kind of evolution that Antalya has experienced.

TRANSFORMED IN AGRICULTURE

Up to 1940 most land was owned by big landlords. Lowland
landlords also held land in the higher zones, highland villagers

renting land from them on sharecropping contracts, implying an annual net flow of agro-pastoral produce from the higher areas to the owners in the lowland. Some of these owners were also traders, exporting agro-produce to cities in Turkey and importing sugar, tea, soap, luxury goods, and in turn selling these goods to villagers. Characteristic for the pattern was a set of multi-purpose bonds between proprietors–traders–patrons in the lowland and tenants–customers–clients in the highland. Personal ties of patron–client relations went across class lines and integrated lowland and highland into one economic system of production and distribution.

In the 1940s the state began building roads tying Antalya to the national centers, and there was an increase in the demand from cities and abroad for agricultural produce from Antalya. Antalya lowlands were especially well suited for producing fruits and vegetables, and big landowners now intensified their market-oriented production of these goods. They sold off land in the mountains, investing the money in land at the coast, turning it into orchards and fruit groves. Capital was put into clearing land, making gardens, fertilizing, storing, transport.

The old types of sharecropping contracts were abolished on the initiative of owners. In fruit production, there was no need for a permanent large labor stock. Besides a small group of core personnel, only seasonal labor,especially for fruit-picking, was necessary. This labor was found in the mountain villages. Mountain farmers were now becoming freeholding peasant households and had their most intensive work in grain production in summer and autumn. Winter time meant leisure on the family farm and, traditionally, long distant seasonal labor migration, e.g. to Western Turkey. Now there was a large demand for labor in the winter fruit-picking season in the Alanya lowland. Labor was recruited among men in the villages, working up to a month, living cheaply, returning with their cash to the household at home. Mountain villagers needed money for debts, more land, brideprices and other ritual expenses. As a net effect, mountain villagers more and more changed into free, independent self-owning peasants, and the lowland area became more and more capitalized, dominated by capitalist farms. A new type of relation between highlanders and lowlanders emerged. The old multipurpose patron–client bond disappeared and was replaced by market relations between capital owner and wage labor, working mostly on daily or other short time bases. The area as a whole was gradually being restructured into a modernizing lowland sector and an apparently traditional highland peasant sector.

ECOZONES, ENTREPRENEURSHIP, AND BOUNDARY
CREATION

In the 1920s a form of entrepreneurship was the following. Animals
were a source of money, and pastoralist owners could make good
profits by selling animals and animal produce. Already at this time
pastures were being encroached upon, and life was becoming
difficult for pure pastoralists. One strategy was to settle down and
start a combined agro-pastoral form of adaptation, but it was
difficult to do this so as to profit from both one's agricultural and
pastoral sectors of activities. This was especially difficult in the
lowlands, where pastures were being eaten up by fields. In some
mountain villages, it was possible to rent and buy agricultural land,
plus make use of the local village's home and mountain pastures.
In one village, there is the case of a rich nomad settling down in
the 1920–30s, operating as a large herdowner, selling off the flock,
investing in agricultural land, renting surplus land to local share-
croppers, and still in the 1960s and 1970s being one of the largest
local landowners plus the decidedly most significant flockowner.
In the 1940–50s this type of career had become difficult to copy,
especially because land prices even in the mountain villages were
rising much faster than the price of animals.

   In the 1940–50s, however, illustrious entrepreneurial careers
were made in the opposite geographical direction. Mountain agro-
pastoral families, which by then were in the possession of some
wealth, could sell off their mountain resources and buy still available
land in the lowlands. Several families managed this conversion,
selling what they had of value in the mountain area and investing
it in as productive and cheap as possible land in the lowlands area.
The typical investment object would be fields or uncleared land
which only by some further investment and input could be turned
into fruit and vegetable gardens for market production.

   Such entrepreneurs contributed to the sharpened price difference
between landed property in the lowlands versus highland villages.
Rather quickly lowland land had become so precious and expensive
and highland land by comparison so cheap that further entrepreneurial careers from highland agro-pastoral adaptation to lowland fruit and vegetable production became virtually impossible.

FURTHER GROWTH IN THE LOWLANDS

Families making a success as fruit growers in the lowlands could
invest their profit in further (1) more intensive agricultural production, e.g. glass houses, for select high profit production, (2) means

of transport, (3) processing and the production e.g. of conserves, (4) shops and trade at various scales, (5) more land for agricultural purposes, but especially (6) sites for high value urban purposes. There is the case of a successful family of the type mentioned in the paragraph above, which quite early bought some land within a small town nearby. The town grew rapidly to become the commercial and administrative center of the area, and the value of centrally situated sites grew especially fast. The former mountain peasant turned fruit grower crowned his enterprise by erecting on one of his urban sites the so far highest building in town, moving into one of the flats himself, renting the rest of the house out for apartments and offices, and running the main office for his various activities in the front room at street level.

The increase in agricultural and commercial activities in the area also brought on an increase in public investment and several public activities, offices, banks, etc., were located in Alanya supplementing the growing sectors of agriculture, processing, transport, and all kinds of trade, plus a growth in all kinds of services—like lawyers, doctors, engineers, architects, and so forth. In turn, this created an increase in the demand for both skilled and unskilled labor, e.g. in construction work.

INNOVATIVE READAPTATIONS IN THE MOUNTAINS

In the 1960s mountain villagers could say: "*Aga devri gecti*" (the era of the old patron landowners has passed). The net effect of the development had in fact been to bring about a veritable land reform from below. On the other hand, mountaineers were aware of the changing difference between themselves and the rich people in the lowlands. Formerly it has been the close patron-client relation of owners and tenants, but ". . . today, they (the wealthy lowlanders) do not say hello to people like us any longer". The class difference changed in form and content, and it did not become any less marked through the 1950s and 1960s, but simultaneously the mountaineers became better, not worse, off.

In 1960 labour migration from Turkey to Europe started. At the end of the decade very few people from the mountain area had left for Europe, and still in 1980 this area displayed a remarkably low percentage of migrants to Europe. The overwhelming majority stayed on and strengthened their economic viability by utilizing the resources to which they had access in their home village and the surrounding district.

Gradually, cultivable land in the mountain villages became scarce in a new critical sense. The population continued to grow, but the available land was insufficient for providing all new families with a farm of their own. Nevertheless, young people and newly wed couples continued to stay and establish themselves in the village. The village council made the decision that new families wanting to remain in their home village should be granted a house site at a symbolic sum. In the 1970s many families took advantage of this and the number of houses grew rapidly. But how did these people manage to make a living?

Agricultural productivity was increased somewhat by the introduction of motor pumps making local irrigation practices more efficient. Motorable roads eventually linked the mountain villages to urban markets, and it became possible to market surplus grain, fruit and vegetables, which up until then had been almost impossible. But most of the additional income came from sources outside agro-pastoralism. The village itself saw the establishment of a few saw-mills, carpenter shops and a few new smitheries and grocery stores. But the real cash was earned in the lowland. The growth in construction activities had been important for a long time, and in the 1970s came a new boom, that of tourism in the form of a considerable number of hotels and pensions at the sea side. This created a large demand for building materials, and highlanders became active in establishing shops and small factories for the processing of various types of construction elements, while others became construction workers or were employed in the hotel and restaurant sector.

This meant a whole new labor market for men. Their old seasonal cash source, viz. fruit-picking, was still significant, but it was considerably less well paid, besides there were few vacant men in the winter fruit season. Hence the village women moved into and took over this seasonal labor market.

The individual mountain village in the 1980s obtains its income from many more sources than twenty years earlier, but most of the old elements are still there. The agro-pastoral production is still carried on in more or less the same way, and the old ideas of cooperation and contracts are still in operation, but the division of labor has changed. Formerly, several brothers inheriting their parents' land would split and establish their own holdings. This split is still done as a specification of individual rights, but today one brother may stay in the village with his own and his brother's families as full-time farmer, while the others are employed as craftsmen or laborers in the lowland and commuting between home and work area. Each individual remains a member in a peasant farm run with traditional technology and for the main purpose of making its owners self-sufficient in food. In addition, various of its members, alone or together with others, may run a small firm producing construction material, in transport or in retail trade, or are employed as craftsmen or unskilled workers. Female members, besides carrying a heavy load in the farm and the family, earn cash in winter season fruit-picking, in addition to throughout the year sewing, weaving and knitting for sale.

I shall indicate the range of the elements in these families' occupational combinations. The last time I visited the village, we were about to leave as passengers in a pick-up run by one of the neighbors trafficking between the village and the lowland. The vehicle was parked at one of the new saw-mills, but few people were around, as many were in the mountain pastures looking after the animals, while others were busy in the lowlands. A big heavy sack was brought to the car to be carried to a lowland bazaar. It contained small pieces of flint, sharp as knives. They were fixed to a type of wooden sledge which, drawn by oxen, is used to cut up the straw after the harvest and the grain has been collected. It is an age-old tool, and, as for the oxen plow, it is difficult to find something more economical to replace it in the family farm in Antalya.

CONDITIONS FOR DEVELOPMENT IN ANTALYA

The story so far has been one of growth in the sense of an increasing regional product feeding an increasing regional population at a better level than earlier. There are losers in the game, but in this short presentation we can stress that for East Antalya it is remarkable that over the last decades development has been both quantitative and structural. There are more people than earlier and most of them have a higher disposable income. At least development in East Antalya has not been accompanied by gross proletarization or impoverization of whole sectors of the population. Development has been structural in the sense of not only bringing about a more diverse division of labor in the area, but also ways for plowing back into the area and its localities comparatively much of the economic value actually produced by those living there.

THE VILLAGE: HOTBED OF BACKWARDNESS OR PROGRESS?

The Turkish village in the literature on modernization from the 1950s was branded as the source of backwardness itself. "Village" and "Villager" were almost synonymous with "fanatic", "traditionalism" and the like. Today, we can see that this was very much an ideological stance, i.e. that of people of urban background wanting development but not knowing how to bring it about, and blaming the villagers for it, whom they did not sufficiently know.

One lesson from the story of East Antalya is that (1) the village as a type of local organization is an important tool of adaptation, and given that (2) the village is given the conditions for evolving its organizational potential, the village itself works as a source for innovative adaptation and as protection for the maintenance of achieved levels of resource control. I shall take a case from the mountain village of Beden, a name meaning "body", but also—incidentally—"castle" and "bastion". We shall note especially the character of the village as an arena where individualist and collectivist values can be kept in balance.

Agricultural land for grain fields, orchards, wine groves, and vegetable gardens is privately owned by individual households; it circulates by inheritance, purchase, and share-cropping and hire contracts. Most irrigation sources are run by many households together, so most irrigated land, the gardens, is concentrated in a specific area, close to the nucleus of houses in the village. The hills with rain-fed grain fields circumscribe the village. The total area of these fields is divided in two halves for a biannual cyclus of cultivation and grazing cum fertilization. This arrangement is vital for maintaining the high yields of these fields (up to 1:12 of high quality wheat and barley); it was introduced in the last century and is still practiced by common decision among the villagers.

Each household also owns a flock of animals, while other animal owners, too, pass through the village on their way to mountain pastures. An armed guardsman, appointed and paid by the village, takes care that animals do not enter cultivated fields, that they pasture only on waste land, and that foreigners pass through the village rapidly enough. He also fines trespassers.

Yayla is the term for the pastures proper, especially the area of mountain pasture which belongs exclusively to each village community, and where each family has a hut where they stay in the summer season with their animals. The yayla is public land. Yayla borders between the various communities are ultimately sanctioned by the state. The use of the yayla is controlled collectively by the village and is a theme of debate in the two decision-making fora of the village meeting with representatives from all households. The council of elders with the village headman.

The village yayla is critical in maintaining the pastoral part of the family household's occupational combination, i.e. the production of a most valuable part of the family's food plus items like wool, hides, etc., and also a very important part of the cash income. Individual flock management therefore goes together with maintaining the collective village frame for local pastoralism. There are three typical threats against this collective arrangement, viz. overgrazing, local theft of pasture land by local villagers trying to clear new fields, and legal claims upon public village land documented as old private land. Let us see how the village can handle these threats.

Overgrazing is prevented by the village council deciding to close a specific part area with its vegetation for pastoral use for a period. I remember also the case of a big flock-owner who had already invested much labor in herding, and who therefore wanted to import animals from neighboring villages on a product-sharing contract between himself as a herdsman and the owners. This was immediately taken up at the village meeting where the council of elders decided against it and forced the man in question to send the outsiders' animals back.

Individual theft of public land is practically impossible in this village, as the paid guardsman and everyone else would discover it immediately and report to the village headman.

There was the case of a rich man from the lowlands who, formerly having sold off most of the family's land in the mountain village, returned with a team of several men and pairs of oxen and started to plow up a considerable part of the yayla. When the villagers turned up to find out what was happening, he showed them a legal document indicating that the land belonged to his family. The document was by no means clear, and the villagers banded together to put direct pressure on the man and to hire a lawyer to handle the case for them. The outsider lost, and the land is still pasture.

The village as a corporation could decide to turn part or whole of its yayla into agricultural land, but could only do so by a clear majority based decision and with the consent of the state authorities. So far, the village has not been interested in doing so, since the pastoral component is a continuous vital part of the individual households' arrangement for adaptation.

Foodstuffs like grain, vegetables, fruit, meat, cheese, yoghurt, etc., become more and more expensive as they are demanded by a growing population of people in the non-agricultural sectors. The individual family in the village can therefore keep up an economic freedom of action the more it can keep itself self-sufficient with these items. Hence we see the arrangement of mountain villagers of family farm background, each keeping their ownership and rights to part of the produce in a farm, run by one of their relatives as a full-time farmer, while the others are economically productive in sectors and places outside the village.

THE FLEXIBILITY OF THE VILLAGER HOUSEHOLD

The flexibility of household organization is in itself an important part of the general adaptational flexibility. It consists in the capacity to distinguish between various functions in the household, to organize these functions into separate teams, and to combine and recombine these teams according to the needs of the individual members as they negotiate their interrelationship. In the setting of the traditional family farm, working full-time in agro-pastoral production primarily for its own consumption needs, the patriarchal family of a man, his wife, their unmarried children, and their married sons plus daughters-in-law plus grandchildren, made up a multipurpose unit of common property, management, production, income division, and consumption, all arranged within one and the same residence group sharing one and the same or a set of adjacent dwellings. In the situation of village-based households with a much wider range of occupational combination, the pattern is one of separate residential units, each running its own non-agricultural activities in or outside the village, but nevertheless remaining corporate around a piece of land and a flock of animals which it owns together as a property unit, but which one of the member families runs as a part-owner on a share-producing partner basis, distributing the agro-pastoral produce of the year in portions to the various members. This arrangement allows for a maximum of mobility for everyone. One keeps a home, a small income, and a protective environment in the village, and is free to experiment with enterprises in trade, crafts, transport or even to go as labor migrant to other places in Turkey, to Europe, or the Arab countries.

The general set up is age-old and arch-traditional in the sense that the story of each village and family through time is nothing more than ever new combinations of this sort and new transformations. The story is one of a small family-based corporate group, where the core values pertain to the patriarchic ideas which regulate authority and cooperation between men as husbands, fathers, sons, brothers, and women as wives, mothers, daughters and sisters, rules of inheritance, of individual ownership combined with common management, and a differentiation between separate tasks and teams that allows for a great number of organizational solutions varying from the highly compact, densely cooperating unit of the patriarchal multi-generational extended household to a set of small nuclear member families together forming a corporation e.g. specifically designed to take care of shared property or to appropriate more of a critical resource.

THE HIGHLAND–LOWLAND CONNECTION

An important form of organizational arrangement that has been preserved through all transformations is the critical linkage between lowlands and highlands, between the expansive, capitalizing lowlands and the highlands evolving on their own premises. Throughout the whole period from about 1920 to 1980 the organization of division of labor in the area has had this "vertical" link where, undershifting circumstances, the highlands have delivered labor to the lowlands, and that part of the value created in the lowlands has been ploughed into the highlands to allow for a rising population in the villages there.

Throughout the period, East Antalya has been an exporter of agro-pastoral products; to a growing extent it has produced more and more goods and services consumed on a rising scale within the area. The lowlands have also received a considerable immigrant population of all kinds of skilled and professional people. But the area is remarkable in the sense that emigration has been insignificant. East Antalya after 1960 never became one of the areas in Turkey which exported large percentages of people to Germany and Europe.

These forms of highland–lowland connections have been built on very simple ties. In a context of public investment, especially the building of roads within the area, and between it and the outside world, it is individual actors, employers and employees, producers and consumers, buyers and sellers who—in various forms of market relations, from the old share-cropping contracts to modern entrepreneurial deals—have maintained the ties that in their aggregate effects amount to the vertical connection. This linkage has served over time as a moving pattern of differentiated niches and opportunities, the basis both for complementarity in the division of labor, a productive distribution of value in the area, as well as for ever new possibilities of small and large innovative steps in economic adaptation.

EAST ANTALYA AND THE WORLD SYSTEM

I have sketched three dynamic patterns and their articulation into a whole that together show us something of the capacity of the East Antalyan people to adapt to shifting circumstances and stay viable under new conditions. These activity fields were (1) the highlands–lowlands connexion of labor division, (2) the local community, and (3) the household. The more complete story would make it necessary to include a number of other fields, e.g. those of locality—state political relations, of religion, and others, which we cannot go into here (cf. Grønhaug 1971, 1974). Suffice it here to say a few words about our small cluster of fields as related to the large-scale fields of international relations, i.e. the world system of economic and political relations between centers and peripheries in Europe and the Middle East and adjacent areas over time.

European imperialism gained momentum in the Mediterranean area with the Napoleonic era. The various European powers— Britain, France, Austria, Russia, etc.—established themselves in the region and penetrated in various ways the territory of the Ottoman empire ruled from Istanbul. Important institutions were the so-called capitulations, which gave European mercantile representatives protection to do business in the area. Local Christians worked closely with them, and this became a basis for the success of various indigenous Christian populations in various kinds of trade. Throughout the last century these people were dominant in the trade that tied Antalya to the outside world. And for that period we can imagine how the highlands–lowlands connection articulating with local Christian traders controlled exports and imports.

With the population exchange between Turkey and Greece in the 1920s, the foreigners and the local Christians disappeared and Turks moved into the vacant niches. Gradually, local landowners turned the surplus from their land into trade enterprises, a development which was strongly stimulated by World War II and a demand for agricultural produce from Turkey in Europe. From then, internal population growth and increasing urbanization in Turkey meant an increasing demand for produce from Antalya, and in the 1950s economic organization in East Antalya had been transformed to the new form of linkage between an intensively capitalizing lowland and a highland turning itself into an area of freeholding peasants. In the 1960s, as mentioned, few people in the area took part in the emigration to Europe, but the increasing amount of money in Turkey, partly deriving from remittances from Turks in Europe, made its effect felt in East Antalya, resulting in many new forms of investment, and especially, in the 1970s growth in local tourism. But agricultural exports from East Antalya continued to grow, as part of the remarkable growth in Turkish agricultural production and export in general in that period. Turkey has not so far become a member of the Common Market, but has successfully made a re-entry into large markets in the opposite direction. When the USA terminated its grain exports to the USSR at the end of the 1970s, Turkey stepped in to cover part of this market. Food provision in Iran was a disaster several years before the Shah fell, and some Turkish agricultural exports, went to both Imperial and Revolutionary Iran. But most important have been the Arab countries, rich in oil, but poor in foods. Along with this there has been a not unremarkable development of Turkish mechanical and other industry, covering more and more of the home market and competing e.g. with East European produce in the Middle East. East Antalya proper has seen little or nothing of this industrial development, but its career of development in agriculture is part of the overall economic change in the Turkish Republic arriving at new and probably more adaptive relations to the economies first of all in the adjacent countries in the Mediterranean and Middle Eastern area.

MAINTENANCE OF THE INNOVATIVE SCENE

Throughout these changes over the last two hundred years or so, East Antalya seems to have managed to maintain a form of economic organization which literally has made up a kind of bounded, localized enclave controlled from below. Many elements together explain the strength of this enclave and its capacity to readapt to shifting circumstances both from within and from the macro environment, and much more remains to be said about the actual pattern of regional export and import, the role of the state in local development, and other things.

OUTLINE OF THE PROCEDURE

To identify the relevant processes of evolution in the area we have used an approach integrating the elements of (1) cultural ideas and rules, (2) population and ecology, (3) groups and networks in social organization, and (4) identities and statuses as combined in individual careers and social persons. From there we have moved on to delineate (5) the activity fields of importance for adaptation, and (6) the way several such fields of varying content and scale articulate with each other in the construction of persons and society in Antalya. The analytical unit as a whole appears as a system of systems, a linkage of linkages, displaying both the structures of individual action and the aggregate features of society in the area. The most critical question has concerned the interconnection of moves and strategies into individual action courses and careers, and the interplay between them and the aggregate level of organizational circumstances. That is, the preeminent locus for the discovery of innovative as well as repetitive action, of modernism or traditionalism, the values and goals as well as the consequences and repercussions of the initiatives taken by specific actors (cf. Grønhaug 1974, 1978.)

CONTINUITY, INNOVATION, AND COMPLEX SOCIAL STRUCTURE

Southern Turkey is one of those places in Eurasia that for millennia has been densely inhabited by populations highly skilled in multiple productive activities, settled in regions with cities and rural communities, with complex modes of labor division integrating local and regional economies with international networks of trade. It is a place with civilizations tying local rites and beliefs to the great traditions of world religions, and where local politics through the ages have articulated with states and empires, composed by culturally diverse and stratified populations.

As in the distant past, local life today is embedded in various macro structures. People's lives are constituted by participation in a number of small- and large-scale contexts and by integrating values from them into individual life courses and local traditions. With the tempests of history, changes have raged over the area, but besides drastic shifts and alterations (rise and fall of empires, invasions, resettlements), there has also been a remarkable continuity through the centuries and millennia. Cities and villages are situated today where they were millennia ago, and in local production we still find elements dating back to neolithic times alongside machinery of the latest technological fashion. In this context, "adapting" means first of all handling one's context of societal structure regulating the use of the natural environment properly. Part of the individual's acquired culture, therefore, is the capacity to adapt to changing local circumstances as well as to changes in the various macro-systems affecting one's life. Long ago, the basic lesson for people in this area became one of how to combine scarcities from diverse sources and learn how to be a social person with relations to partners socially and culturally different from oneself and living in other places.

A social structure of this complex, civilizational kind allows for a certain freedom in the individual's combination of roles to shape one's life career. It makes possible sharing a view of basic values and identities with others but still to differ from them in the way one's life career evolves. The significance of this feature has varied with time and circumstances, but in a sense it is correct to say that change and innovation have been and remain part of tradition in Antalya. Basically, this springs from the diverse and composite character of social structure itself. The structure has changed in drastic ways over time, and above we have given a glimpse of part of the story. In a sense, however, changing Antalya is very much an instance of remaining the same the more it has been changing.

This may seem strange from viewpoints of theories e.g. about modernization (see e.g. Lerner) or unequal exchange (cf. e.g. Amin or Gunder Frank). I will argue that this depends more on the faults of such theories, which have not been very useful for identifying the proper units of analysis in the study of change and innovation, i.e. the analytical units of those organizational entities that integrate micro- and macro-relations in the constitution of people's life worlds. Part of the argument I have tried to make concerns the question of a procedure for defining the units of analysis in the study of long- and short-term innovation, change, and continuity.

INNOVATION OR TRADITIONALISM?

Ultimately, local development is being made locally through the way local actors arrange their interrelations and evolve their careers, creating thereby both the incentives to produce and the organizational solusions for realizing their aims. I have shown a case of localized evolution taking place as a transformation in a socio-cultural infrastructure of some basic ideas of identities and roles and organizational ways of translating them into life courses, networks and between groups. I have pointed especially to the shifting organizational pattern of the highlands–lowlands linkage and its connection with fields of varying scales, from local community and household to macro-fields like the state and the European world system.

The reader may charge me for having written much about adaptation by putting old wine in new bottles, and little about innovation proper. This would be true in so far as e.g. very few of the technological ideas involved in our story can be said to have originated in East Antalya. My cases of innovative action are more examples of putting to local use tools and ideas imported from the outside, or even to make new use or repeat the old use of old, in many cases ancient elements. Neither the use of modern hydro-science nor of the flint stones in East Antalya are in themselves testimonies of an especially innovative mind.

The kind of innovation we have described consists rather in borrowing, of importing something discovered in one field of action and putting it to use in another. What is new is the localized adaptive organization itself, with a series of new solutions to adaptational problems. This form of transformation displays an underlying talent for guarding one's socio-cultural viability, for being aware of new circumstances—threats as possibilities, and for shifting to new organizational solutions more satisfying to the individual as well as allowing more people to live better. This is the kind of growth and development which planners and govern-ments in so many parts of the world try to design, but so rarely manage to move from the stage of aid to that of "self-sustaining growth".

The innovation I have been speaking of is first of all organ-izational innovation, the art of designing and institutionalizing new organizational solutions to the economic and general adaptational tasks that spring from the basic values of the people in question. The story of East Antalya would not be complete without portraits of some of the individuals who in various positions and points of time have appeared as the primary movers and models for others. On the other hand, even big men are socially constructed, and in my short sketch I have given priority to pointing out the super-individual and collective circumstances that give rise to novelty and greatness in individual action.

The material I have presented, however, should be sufficient to substantiate the argument that innovation in the form of devel-opment and growth takes the form of novelties in organizational solutions that in many ways *are* old wine in new bottles. The question is basically one of utilizing a pre-existing infrastructure of socio-cultural competences, which we normally call tradition, and transforming them under new circumstances into new ways of realizing basic values known to people beforehand in the contexts of family, local, occupational, and general social life. East Antalyan people have been so innovative because they have been able to mobilize and evolve their basic and highly traditional ideas about what an honorable person is. This is in contrast to the frustrating stalemate of so many formally designed development projects where the plan from outside only poorly articulates with the way local people arrange their life context.

MODERNIZATION?

By way of analogy we could term development in East Antalya as a case of "Japanization"—developing by transforming traditional infrastructure into a new organizational form allowing for economic growth as well as preservation of the traditional discourse of values. Different from Japan, the course of action has not so much been planned by the state from above, but rather achieved in local and regional contexts as actors, firms, households and individuals strive to realize their individual goals in a framework of shared values.

Traditionalization: the revival of Islamic teaching, Koran courses, and "fundamentalist" Islam, financed by the profit of economic development. Religion, family life, etc., not less Turkish or more Western. Modernity of tradition. Basically: the integration of developed economy, vital Islam, and Republican politics into a new cluster, a form of society on its own terms, not weakened in its basic identity by the partners with which it does business. Long-term innovation: re-establishing terms for interrelations between the Northern societies which took the lead in development and the various states of the Middle East, Asia, etc., which now may have some advantages because they did not get the disadvantages of taking the lead, to a level of equality à la about 1500, which is to say that making history repeat itself may be said to be one kind of innovation.

Unequal exchange: Center eats periphery. I have not discussed the general trend in the terms of trade between Turkey and USA, Common Market. But my point has been to show that due to local conditions, especially the organizational potential in local circumstances, a small district may adapt and fare quite well in the main period of the European world system, disregarding whether the battleships in the ocean nearby have flown British, Italian, American or Soviet flags to make clear where the center of the world system has been.

REFERENCES

Amin, S. 1970. *L'accumulation à l'échelle mondiale*. Paris: Anthropos.
Barth, F. 1963. *The Role of the Entrepreneur in Social Change in Northern Norway*. Oslo: Universitetsforlaget.

Barth, F. 1967. 'On the study of social change'. In: *American Anthropologist* 69, pp. 661–669.

Barth, F. 1972. 'Analytical dimensions in the comparison of social organizations'. In: *American Anthropologist* 74, 207–220.

Bateson, G. 1973. *Steps to an Ecology of Mind. Collected Essays in Anthropology, Psychiatry, Evolution and Epistemology.* Frogmore: Paladin.

Bendix, R. 1966–67. 'Tradition and Modernity Reconsidered'. In: *Comparative Studies in Society and History* IX, pp. 292–346.

Berkes, N. 1964. *The Development of Secularism in Turkey.* Montreal: McGill University Press.

Eisenstadt, S. N. (ed.) 1970. *Readings in Social Evolution and Development.* Oxford: Pergamon Press.

Frank, A. G. 1969. *Capitalism and Underdevelopment in Latin America.* New York: Modern Readers Paperbacks.

Grønhaug, R. 1974. *Micro-macro Relations. Social Organization in Antalya, Southern Turkey.* Bergen Studies in Social Anthropology, No. 7. Bergen: Dept. of Social Anthropology.

Grønhaug, R. 1978. 'Scale as a variable in analysis: Fields of social organization in Herat, Northwest Afghanistan'. In: F. Barth (ed.) *Scale and Social Organization.* Oslo: Universitetsforlaget.

Lerner, D. 1958. *The Passing of Traditional Society. Modernizing the Middle East.* The Free Press of Glencoe.

Nadel, S. 1965. *The Theory of Social Structure.* London: Cohen and West.

Nisbet, R. A. 1969. *Social Change and History. Aspects of the Western Theory of Development.* New York: Oxford University Press.

Name Index

Subject Index